Adaptive Methods – Algorithms, Theory and Applications

Edited by
W. Hackbusch
G. Wittum

Notes on Numerical Fluid Mechanics (NNFM)   Volume 46

Series Editors: Ernst Heinrich Hirschel, München (General Editor)
Kozo Fujii, Tokyo
Bram van Leer, Ann Arbor
Keith William Morton, Oxford
Maurizio Pandolfi, Torino
Arthur Rizzi, Stockholm
Bernard Roux, Marseille

Volumes 1 to 25 are out of print.
The addresses of the Editors are listed at the end of the book.

Adaptive Methods – Algorithms, Theory and Applications

Proceedings of the Ninth GAMM-Seminar
Kiel, January 22–24, 1993

Edited by
Wolfgang Hackbusch and
Gabriel Wittum

Produced by W. Langelüddecke, Braunschweig

ISSN 0179-9614

ISBN 978-3-528-07646-7   ISBN 978-3-663-14246-1 (eBook)

DOI 10.1007/978-3-663-14246-1

Preface

The GAMM Committee for "Efficient Numerical Methods for Partial Differential Equations" organizes workshops on subjects concerning the algorithmical treatment of partial differential equations. The topics are discretization methods like the finite element and finite volume method for various types of applications in structural and fluid mechanics. Particular attention is devoted to advanced solution techniques.

The series of such workshops was continued in 1993, January 22-24, with the 9th Kiel-Seminar on the special topic

"Adaptive Methods
Algorithms, Theory and Applications"

at the Christian-Albrechts-University of Kiel. The seminar was attended by 76 scientists from 7 countries and 23 lectures were given.

The list of topics contained general lectures on adaptivity, special discretization schemes, error estimators, space-time adaptivity, adaptive solvers, multi-grid methods, wavelets, and parallelization.

Special thanks are due to Michael Heisig, who carefully compiled the contributions to this volume.

November 1993
Wolfgang Hackbusch
Gabriel Wittum

Contents   Page

Contents (continued)

GALERKIN/LEAST-SQUARES-FEM AND ANISOTROPIC MESH REFINEMENT

A. Auge

Dresden Univ. of Technology, Fluid Mech.
Departm., Mommsenstr.13, D-0-8027 Dresden

G. Lube

Magdeburg Univ. of Technology, Math.Departm.,
Postfach 4120, D-0-3010 Magdeburg

D. Weiß

Magdeburg Univ. of Technology, GKMBI,
Postfach 4120, D-0-3010 Magdeburg

SUMMARY

The Galerkin /least-squares finite element method is considered as a tool for solving singularly perturbed partial differential equations of elliptic type on adaptively refined grids. Local error estimates in subdomains away from boundary and interior layers are uniformly valid with respect to the small parameter. Boundary and interior layers can be resolved using locally anisotropic mesh refinement. Numerical examples are given for scalar convection-diffusion equations and for incompressible Navier – Stokes flow problems.

INTRODUCTION

In this note, we consider the numerical solution of singularly perturbed elliptic boundary value problems on adaptively refined grids. In particular, we analyze as a model problem the scalar convection-diffusion-reaction problem

$$L_\varepsilon u \equiv -\varepsilon \Delta u + \mathbf{b}\cdot\nabla u + cu = f \quad \text{in } \Omega \subseteq \mathbf{R}^d$$
$$u = g \quad \text{on } \partial\Omega . \tag{L_ε}$$

(L_ε) is the FOURIER-KIRCHHOFF equation modelling the energy balance in the welding process [We] or in Boussinesq's model of non-isothermal, incompressible flow problems [AL]. For both problems, the resolution of thermal boundary layers is of interest.

The Galerkin/least-squares (GLSFEM) and streamline diffusion

1

(SDFEM) finite element methods are now widespread used numerical methods for solving singularly perturbed problems. The goal of the present paper is to study the robustness of GLSFEM with respect to adaptive mesh refinement.

GALERKIN / LEAST-SQUARES FINITE ELEMENT METHOD

Let $W^{k,p}(G)$, $G \subseteq \Omega$ denote the usual Sobolev spaces with $k \in N_0$, $1 \leq p \leq \infty$. $(\cdot, \cdot)_G$ is the inner product in $L^2(G)$, $\|\cdot\|_{k,G}$ and $|\cdot|_{k,G}$ denote the norm and seminorm in $H^k(G) \equiv W^{k,2}(G)$, respectively. $\|\cdot\|_{\infty,G}$ is the norm on $L^\infty(G)$.

Let $\bar{\Omega} = \underset{i}{\cup} \bar{K}_i$ with finite elements $K_i \in T_h$. V_h denotes a finite element space consisting of piecewise polynomial basis functions of degree $k \in N$

$$V_h = \{ w_h \in W_0^{1,2}(\Omega): \quad w_h|_{K_i} \in P_k(K_i) \quad \forall \quad K_i \in T_h \} .$$

Then the Galerkin/least-squares method reads [FFH]:

Find u_h s.t. $u_h - g_h \in V_h$ and $\forall \quad v_h \in V_h$:
$$a_\delta(u_h, v_h) = l_\delta(v_h) \tag{GLS}$$

where

$$a_\delta(u, v) \equiv \varepsilon(\nabla u, \nabla v)_\Omega + (b \cdot \nabla u + cu, \ v)_\Omega + \sum_i \delta_i \ (L_\varepsilon u, L_\varepsilon v)_{K_i} \tag{1}$$

$$l_\delta(v) \equiv (f, v)_\Omega \qquad + \sum_i \delta_i \ (\ f, \ L_\varepsilon v)_{K_i} . \tag{2}$$

g_h is a suitable interpolation of an arbitrary $\tilde{g} \in W^{1,2}(\Omega)$ such that g is the trace of $\tilde{g}$.

The classical Galerkin method ($\delta_i = 0$) suffers from nonphysical oscillations unless the mesh is sufficiently refined. In case

of c=0 and piecewise linear elements on triangles, (GLS) reduces to the well-known streamline diffusion method.
Henceforth we assume (for simplicity)

(H.1) $\mathbf{b}$, div $\mathbf{b}$, $c \in L^\infty(\Omega)$; $f \in L^2(\Omega)$; $g_h \in W^{1,2}(\Omega)$;

$c(x) \geq \alpha^2 \geq 0$, div $\mathbf{b} = 0$ a.e. in Ω.

STABILITY OF (GLS)

Then holds a stability estimate w.r.t. a "stabilized" norm

$$\|\!\|u_h\|\!\|_\delta^2 \equiv \varepsilon\|\nabla u_h\|_{0,\Omega}^2 + \alpha^2\|u_h\|_{0,\Omega}^2 + \sum_i \delta_i \{ \|\mathbf{b}\cdot\nabla u_h\|_{0,K_i}^2 +$$
$$+ \|-\varepsilon\Delta u_h + cu_h\|_{0,K_i}^2 \} \leq C \{ \|f\|_{0,\Omega}^2 + \|g_h\|_{1,\Omega}^2 \} \tag{3}$$

without restrictions on the mesh and the parameter set $\{\delta_i\}$.
Hence there exists a unique solution of (GLS).

A discrete maximum principle is not valid for (GLS). Indeed
small oscillations of the discrete solution may occur in the
neighborhood of boundary and/or interior layers. For a modified streamline diffusion method there holds in 2D and for
piecewise linear elements (k=1) on triangles [JSW]

$$\|u_h\|_{\infty,\Omega} \leq C h^{-1/4} \ln^{3/2}(h^{-1}) \|f\|_{\infty,\Omega} . \tag{4}$$

As a remedy, Johnson et al. propose the "shock-capturing" variant of the streamline diffusion method with a modified dif-
fusion coefficient $\tilde{\varepsilon}|_{K_i} = \max \{\varepsilon; C_1 h_i^{3/2} ; C_2 h_i^2 |L_\varepsilon u_h - f| \}$
resulting in a nonlinear scheme [EJ].

3

GLOBAL A-PRIORI ESTIMATES AND PARAMETER DESIGN

Let (H.1) and the following assumptions be valid:

(H.2) $\qquad$ $u \in W^{l+1,2}(\Omega)$, $1 \leq l \leq k$

(H.3) T_h is shape-regular, i.e. the ratio of the circum-
scribed ball for $K \in T_h$ to that of the inscribed ball
for K is bounded independently of T_h and h.

In particular, (H.3) allows locally refined "isotropic"
meshes. Then there holds w.r.t. the "stabilized" norm (cf.(3))

$$\interleave u-u_h \interleave_\delta^2 \leq C \sum_i h_i^{2l} B_i(\varepsilon,h_i,\delta_i) \ |u|_{l+1,K_i}^2 \qquad (5)$$

$$B_i \equiv \delta_i \|b\|_{\infty,K_i}^2 + \min \{\delta_i^{-1}; \varepsilon^{-1}\|b\|_{\infty,K_i}^2\} \ h_i^2 + \varepsilon + \|c\|_{\infty,K_i}^2 \ h_i^2 \ .$$

Minimization of the coefficient B_i w.r.t. δ_i results with the
local Peclet number $P_i \equiv \varepsilon^{-1}h_i\|b\|_{\infty,K_i}$ and $R_i \equiv \varepsilon^{-1}h_i\|c\|_{\infty,K_i}$ in

$$\delta_i = \frac{h_i^2}{\varepsilon \sqrt{1 + P_i^2 + R_i^2}}, \qquad (6)$$

and $\qquad$ $B_i \leq \varepsilon + \|b\|_{\infty,K_i} \ h_i + \|c\|_{\infty,K_i} \ h_i^2 \quad [Lu]$. $\qquad$ (7)

The parameter design condition **(6)** with $R_i = 0$ is now frequently used in Computational Fluid Dynamics [Sh],[Te],[Be].

A boundary and/or interior layer of the solution of (L_ε) is
located at a (at most) (d-1)-dimensional manifold. So it would
be desirable to allow the use of anisotropic mesh refinement.
In case of d=2, k=1 (piecewise linear elements on triangles),
assumption (H.3) which is essentially a "minimum angle condition" can be replaced by a "maximum angle condition" (denoting

by $\xi(K)$ the maximum angle of triangle K):

(H.3)[*] $\quad \xi(K) \le \xi_o < \pi \qquad \forall \ K \in T_h$.

Then there holds (denoting by $C(\xi_i)$ an increasing function of the maximum angle ξ_i of triangle K_i) [La]

$$\| u - u_h \|_\delta^2 \le \sum_i C(\xi_i) \ \{ \varepsilon + \| b \|_{\infty, K_i} h_i + \| c \|_{\infty, K_i} h_i^2 \} \ \| u \|_{2, K_i}^2 . \quad (8)$$

LOCAL A-PRIORI ESTIMATES

The singularly perturbed problem (L_ε) with $0 < \varepsilon \ll 1$ can be characterized by the presence of narrow boundary and/or interior layers where the solution of the first order limit problem for $\varepsilon = 0$ does not satisfy the boundary conditions of (L_ε) or is not smooth. Let (for simplicity) be $b(x) \ne 0$ in Ω. Then typically one has "ordinary" boundary layers at outflow parts of $\partial\Omega$ of width $O(\varepsilon \ \ln\frac{1}{\varepsilon})$ and "parabolic" boundary and /or interior layers of width $O(\sqrt{\varepsilon} \ \ln\frac{1}{\varepsilon})$ (cf. also Fig. 1).

It is well-known from numerical experiments that boundary and interior layers are detected by the GLSFEM even on a "rough" mesh. Furthermore one has high accuracy of the numerical solution outside the layers. This is reflected in local error estimates using some technical assumptions.

For a Lipschitzian subdomain $G \subseteq \Omega$ we denote by $(\partial G)_+$, $(\partial G)_-$ and $(\partial G)_o$, respectively, the subsets of ∂G where $b \cdot \nu_G$ is positive, negative or vanishing. ν_G is the outward pointing unit normal vector on ∂G.

(H.4) Suppose that $b(x) \neq 0$ in Ω and no closed characteristic curves of the first order operator L_o exist in Ω.
Let $\Omega' \subseteq \Omega''$ be subdomains of Ω as given in Fig. 1.

The following result is useful in case of red and green ("isotropic") refinement of the mesh.

Theorem: Suppose that (H.1), (H.3), (H,4) and

    (H.5)   Let $u \in W^{l+1,2}(\Omega'')$, $1 \leq l \leq k$.

are valid. Then there holds with $\boldsymbol{\varpi}_i = \max\{\varepsilon, h_i\} \leq \boldsymbol{\varpi} = \max\limits_i \boldsymbol{\varpi}_i$:

$$\||(u-u_h)|_{\Omega',\delta}^2 \leq C \sum_i h_i^{2l} \boldsymbol{\varpi}_i \{ \|f\|_{0,K_i}^2 + |u|_{l+1,K_i \cap \Omega''}^2 \} . \tag{9}$$

Proof: It follows the lines of the proof of Th. 4.9 in [Nä]
using a cut-off function technique. One has to take care of
the Galerkin/least-squares modification of the original
streamline diffusion method and of the locally variable
δ-set.   **#**

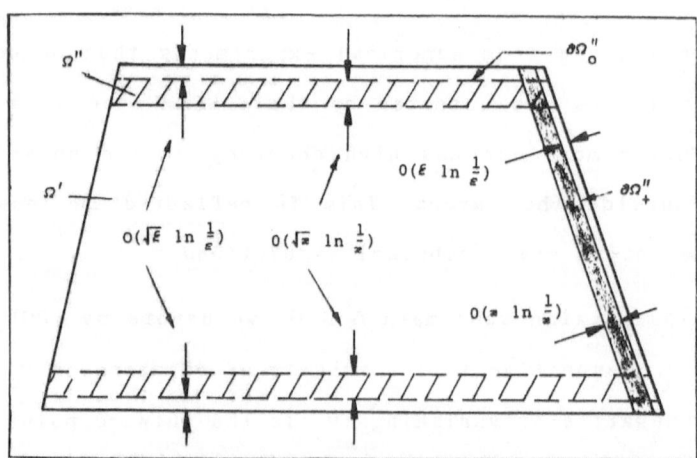

Fig. 1   Typical situation for local a-priori estimates

Remark. Local L^∞-estimates were given for the modified SDFEM (mentioned above) in [JSW], [ZH].   #

RESOLUTION OF BOUNDARY AND INTERIOR LAYERS

This foregoing result seems to be not sufficient if one looks for resolution of boundary and interior layers. Let Σ be a (d-1)-dimensional manifold in $\bar{\Omega}$ where a layer of width $O(\varepsilon^\sigma \ln \frac{1}{\varepsilon})$ is located. For resolution of the layer with an isotropically refined mesh we need $O(\varepsilon^{-\sigma(d-1)} \ln \frac{1}{\varepsilon})$ elements.

For piecewise linear elements on triangles in a plane domain Ω, we try to resolve the layers using anisotropic elements (typically arising in the blue refinement method [KR]) which are adapted to the flow direction **b**. α denotes the aspect ratio of the element (cf. Fig. 2).

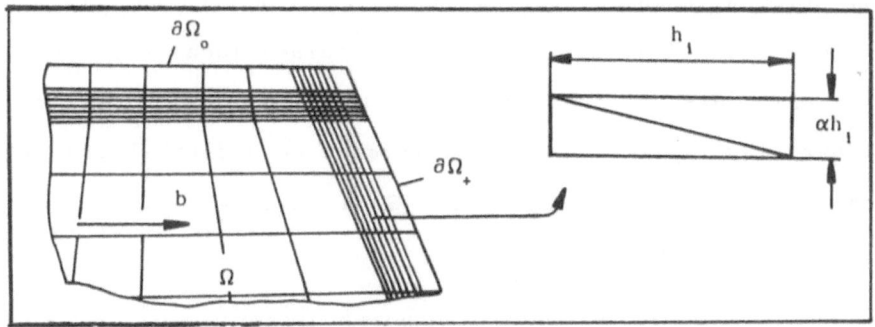

Fig. 2   Anisotropic finite elements in
              boundary or interior layers

Resolution of a layer of width $O(\varepsilon^\sigma \ln\frac{1}{\varepsilon})$ requires now only

$O([\alpha\varepsilon^{-\sigma}]^{d-1} \ln \frac{1}{\varepsilon})$ anisotropic elements.

Now we modify assumption (H.4) as follows:

(H.4)[*]   Let (H.4) be valid with the exception of

$$\text{dist}(x, (\partial\Omega'')_+) \geq C(1) \, \alpha h \, \ln \frac{1}{\alpha h} , \quad \forall \ x \in \Omega'$$

$$\text{dist}(x, (\partial\Omega'')_0) \geq C(1) \, \sqrt{\alpha h} \, \ln \frac{1}{\alpha h} , \quad \forall \ x \in \Omega'$$

and with $h = \max_i h_i$, $0 < \alpha \leq 1$, $1 \leq l \leq k$.

Local error estimates as stated in the Theorem above remain valid for this situation under the assumptions (H.3)[*], (H.4)[*] and (H.5). Furthermore it is possible to use an anisotropic mesh in the layers which is graduated in the gradient direction of the discrete solution.

Remarks.   (i) It is an open question whether this result holds in 3D on a tetrahedral mesh with piecewise linear elements and for some higher order elements too.   #
(ii) Grids with anisotropic boundary layer refinement are also considered in [Shi] and [GS]. They obtained local error estimates which are uniformly valid w.r.t. ε in case of parabolic boundary layers for a finite difference method [Shi] and outflow layers for the SDFEM [GS]. Such an result seems to be open for the GLSFEM.   #

NUMERICAL RESULTS FOR CONVECTION - DIFFUSION PROBLEMS

Numerical experiments were performed with a 3D code GLSFEM We]. Let for simplicity be $\Omega = (0, 1.2) \times (0, 1)^{d-1}$ with d=2 or d=3.

Example 1: (Interior and boundary layers)

Let be $\varepsilon=10^{-6}$, $b=(1.2, 0.5)$ and $c=f=0$. Data are specified at the inflow part of the boundary with $x_1+1.2 \ x_2 \leq 1.2$ as follows:

$u = 1$ if $x_2 < 0.25$, $u = 0$ if $x_2 \geq 0.25$.

The inflow data jump generates an interior layer. At the outflow part of the boundary with $x_1+1.2x_2 > 1.2$ we set $u = 0$ which results in an ordinary boundary layer for $x_2 \geq 0.75$.

On a mesh with 193 nodes and only 40 shape-regular elements outside the layers we found an almost exact solution (cf. Fig. 3). It turns out that the anisotropic "layer elements" (with aspect ratio $\alpha = 0.00833$ in the interior layer) have to be carefully adapted to the flow direction b only in the interior layer. Note that the solution of GLSFEM (identically with streamline diffusion method) has still small oscillations even on a 100x100 **regular** mesh. On the other hand, the shock-capturing streamline diffusion method tends to "smear out" the interior layer.

Example 2: (Interior layer)

Let be $\varepsilon=10^{-6}$, $b=(1.2 \ x_2, 1-x_1)$ and $c=f=0$. The inflow data are at $x_2= 0$: $u = 0$ if $0 \leq x_1 < 0.6$, $u = 1$ if $0.6 \leq x_1 \leq 1.2$ and at $x_1= 0$: $\frac{\partial u}{\partial \nu} = 0$. The initial data jump again generates an interior (but curved) layer. Outflow data for $x_1+ 1.2x_2 > 1.2$ are $\frac{\partial u}{\partial \nu} = 0$.

On a mesh with 247 nodes (118 nodes outside the layer) we obtain an almost exact discrete solution.

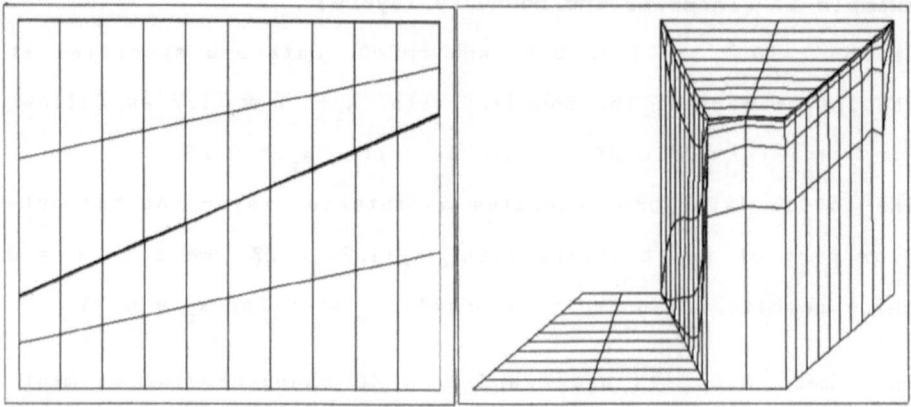

Fig. 3   Mesh and discrete solution of Example 1

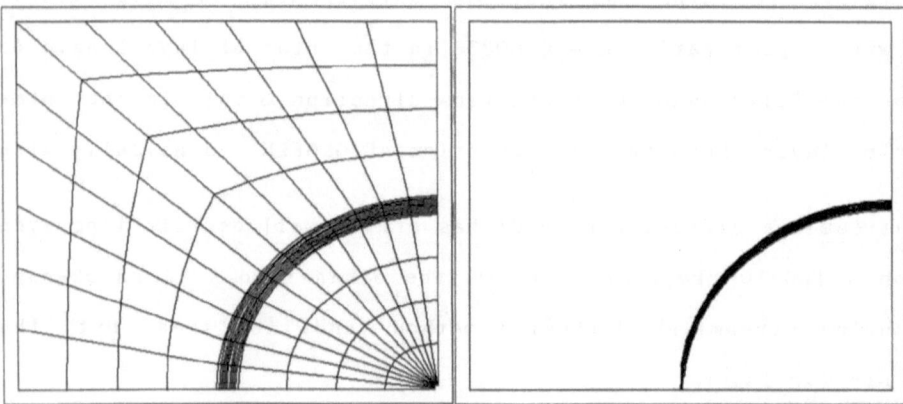

Fig. 4   Mesh and isolines of Example 2

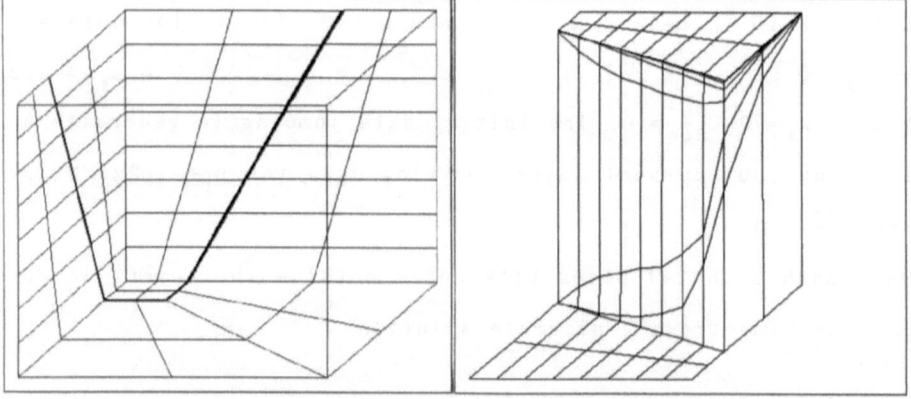

Fig. 5 Mesh and solution plot (at $x_1 = 1,2\, x_2$) for Example 3

Example 3: **(Interior layer in 3D)**

Let $\varepsilon=10^{-6}$, $\mathbf{b} =(0.6, 0.5, 1.0)$ and $c=f=0$. Boundary conditions are specified as follows: For $x_3=1$ let $u = 0$. For $x_3 < 1$ set

$$u = 1 \quad \text{if} \quad x_1, \ x_2 \leq 0.2 + \frac{1}{2} x_3, \qquad u = 0 \quad \text{else}.$$

An interior layer is generated along the characteristic curves (straight lines) starting from the inflow data jump at $x_3 = 0$. On a tetrahedral mesh with 1440 elements and 440 nodes (440 elements outside the interior layer) we obtain again an almost exact discrete solution (cf. Fig. 5).

EXTENSION TO NONLINEAR PROBLEMS

We extend the approach to nonlinear elliptic problems

$$\text{Find} \quad u \in V : \qquad N_\varepsilon(u) u = f \in V^* \tag{10}$$

or in a weak formulation with a bilinearform $a_o(w; \cdot, \cdot)$

$$\text{Find} \quad u \in V : \qquad a_o(u; u, v) = <f, v> \quad \forall \ v \in V. \tag{11}$$

The Galerkin/least-squares formulation of (11) reads:

$$\text{Find} \quad u_h \in V_h : \ a_\delta(u_h; u_h, v) = l_\delta(u_h, v) \quad \forall \ v \in V_h \tag{12}$$

with

$$a_\delta(w; u, v) \equiv a_o(w; u, v) + \sum_i \delta_i \ (N_\varepsilon(w) u, \ N_\varepsilon(w) v)_{K_i} \tag{13}$$

$$l_\delta(w; v) \equiv <f, v> + \sum_i \delta_i \ (\ f \quad , \ N_\varepsilon(w) v)_{K_i}. \tag{14}$$

A linearization procedure for (12)-(14) is given by the following defect correction method:

o Let $\quad u_h^{(m)} \in V_h \quad$ be given $\quad (m \in \mathbf{N}_o)$.

o Find $\quad z^{(m)} \in V_h \quad$ s.t. $\quad \forall \ v \in V_h$:

$$a_{\tilde{\delta}}(u^{(m)}; z^{(m)}, v) \;=\; l_{\delta}(u^{(m)}; v) \;-\; a_{\delta}(u^{(m)}; u^{(m)}, v) \tag{15}$$

o Update : $\qquad u^{(m+1)} \;=\; u^{(m)} \;+\; \omega_m z^{(m)}$.

The forms $a_{\tilde{\delta}}(\cdot; \cdot, \cdot)$ and $a_{\delta}(\cdot; \cdot, \cdot)$ may differ. The $\tilde{\delta}$-set is choosen such that $a_{\tilde{\delta}}$ generates a "robust" iteration matrix.

The incompressible Navier-Stokes problem for velocity u and pressure p (denoting by $\hat{u} = (u, p)$)

$$N_{\varepsilon}^1(u)\,\hat{u} \;\equiv\; -\,\varepsilon\,\Delta u \;+\; u\cdot\nabla u \;+\; \nabla p \qquad =\; f \quad \text{in} \quad \Omega$$

$$N_{\varepsilon}^2 u \;\equiv\; \qquad\qquad\qquad\qquad \text{div } u \;=\; 0 \quad \text{in} \quad \Omega \tag{16}$$

$$u \;=\; g \quad \text{on} \quad \partial\Omega.$$

can be written in the form (10) or (11). With $V = X \times Q$, $X = W_o^{1,2}(\Omega)^d$ and $Q = L^d(\Omega)/R$, we introduce

$$a_o(w; \hat{u}, \hat{v}) \;\equiv\; \varepsilon(\nabla u, \nabla v)_{\Omega} \;+\; (w\cdot\nabla u, v)_{\Omega} \;-$$

$$-\; (p, \text{div } v)_{\Omega} \;+\; (q, \text{div } u)_{\Omega} \tag{17}$$

$$a_{\delta}(w; \hat{u}, \hat{v}) \;\equiv\; a_o(w; \hat{u}, \hat{v}) \;+\; \sum_i \delta_i^1 \, (N_{\varepsilon}^1(w)\,\hat{u}, \; N_{\varepsilon}^1(w)\,\hat{v})_{K_i}$$

$$+\; \sum_i \delta_i^2 \, (N_{\varepsilon}^2 u, \; N_{\varepsilon}^2 v)_{K_i} \,. \tag{18}$$

$$l_{\delta}(w; v) \;\equiv\; (f, v)_{\Omega} \;+\; \sum_i \delta_i^1 \, (f, \; N_{\varepsilon}^1(w)\,\hat{v})_{K_i}. \tag{19}$$

The defect correction method (15) requires the solution of linearized Navier-Stokes problems of Oseen type. Stability and error estimates for such problems are given in [AL] w.r.t. the following stabilized norm with $w = u^{(n)}$

$$\|\hat{u}\|_{\delta}^2 \;\equiv\; \varepsilon\|\nabla u\|_{o,\Omega}^2 \;+\; \frac{\varepsilon}{\varepsilon + \|w\|_{\infty,\Omega}} \, \|p\|_{o,\Omega}^2 \;+ \tag{20}$$

$$+\; \sum_i \{\, \delta_i^1 \, \|N_{\varepsilon}^1(w)\,\hat{u}\|_{o,K_i}^2 \;+\; \delta_i^2 \, \|\text{div } u\|_{o,K_i}^2 \,\}.$$

The parameter sets are choosen as follows [AL]

$$\delta_1^1 = \frac{h_1^2}{\varepsilon \sqrt{1 + Re_1^2}} \quad , \quad \delta_1^2 = \varepsilon \sqrt{1 + Re_1^2} \qquad \text{(21)}$$

using the local Reynolds number $Re_1 \equiv \varepsilon^{-1} h_1 \|w\|_{\infty, K_1}$.

Note that for the linearized discrete problem holds a modified Babuska-Brezzi condition thus allowing arbitrary combinations of velocity-pressure interpolation functions (which possibly do not pass the standard inf-sup condition).

In case of 2D problems with piecewise linear finite elements on triangles it is again possible to use anisotropic elements. As an numerical example we considered the standard "driven cavity" problem in $\Omega = (0,1)^2$ with $\varepsilon^{-1} = Re = 3200$ on different meshes with 32 x 32, 48 x 48 and 64 x 64 nodes: A) equidistant mesh, B) equidistant anisotropic mesh refinement in the boundary layer, C) non-uniform anisotropic refinement in the boundary layer.

The secondary recirculation flow domains in the left and right lower and left upper corners are well resolved on all meshes. For mesh C it is even possible to identify "tertiary" eddies in the lower corners on the 32 x 32 mesh. In Fig. 6 we present   a) mesh C,   b) the vector plot for the 64 x 64 mesh C,

c) a zoom of the "tertiary" eddy in the lower right corner on the 64 x 64 mesh C   and   a comparison of   d) the vertical and e) horizontal velocity components at $x_2 = 0.5$ and $x_1 = 0.5$, respectively.

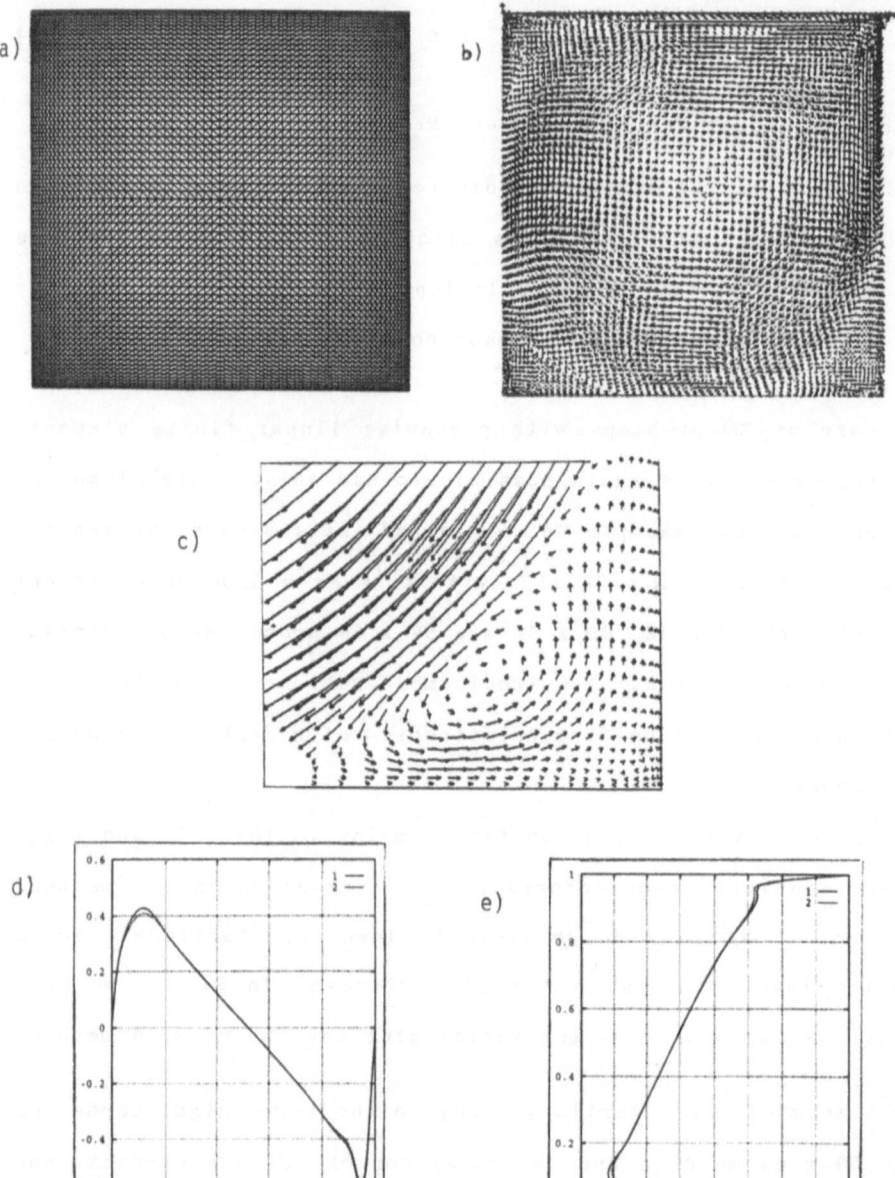

Fig. 6 Driven cavity problem (Re = 3200, 64 x 64 mesh)
a) Mesh C, b) Vector plot for mesh C,   c) Zoom for mesh C
Comparison of the   d) vertical velocity component at x_2 = 0.5
e) horizontal velocity component at x_1 = 0.5   (mesh A, mesh C)

14

REFERENCES

[AL]   Auge, A. , Lube, G. : Regularized Mixed Finite Element Approximations of Incompressible Flow Problems. Preprint TU Magdeburg 1993.

[Be]   Behr, M. : Stabilized Finite Element Methods for Incompressible Flows with Emphasis on Moving Boundaries and Interfaces. Thesis, Univ. of Minnesota 1992.

[GS]   Guo, W. , Stynes, M. : Pointwise Error Estimates for a Streamline Diffusion Scheme on a Shishkin Mesh for a Convection-Diffusion Problem. Preprint 1993 - 2, Dept. of Math. , Univ. College. Cork, Ireland 1993.

[EJ]   Eriksson, K. , Johnson, C. : Adaptive Streamline Diffusion Finite Element Methods for Convection-Diffusion Problems. Math. Comp.. (to appear).

[FFH]   Franca, L. P. , Frey, S. L. , Hughes, T. J. R. : Stabilized finite element methods. Part I. Comp. Meths. Appl. Mech. Engrg. 95 (1992), 253 - 276.

[JSW]   Johnson, C. , Schatz, A. H. , Wahlbin, L. B. : Crosswind Smear and Pointwise Errrors in Streamline Diffusion Finite Element Methods. Math. Comp. 49 (1987) 25-38.

[KR]   Kornhuber, R. , Roitzsch, R. : On Adaptive Grid Refinement in the Presence of Internal or Boundary Layers. Preprint SC 89-5, K. Zuse-Zentrum Berlin (1989).

[La]   Lang, J. : An Adaptive Finite Element Method for Convection-Diffusion Problems by Interpolation Techniques. Technical Report TR 91-4, K. Zuse-Zentrum Berlin 1991.

[Lu]   Lube, G. : Stabilized Galerkin Finite Element Methods for Convection Dominated and Incompressible Flow Problems. Proc. 37. Sem. Numer. Analysis and Math. Modell. 1991. Banach Center Public. Warsaw 1993.

[Nä]   Nävert, U. : A Finite Element Method for Convection-Diffusion Problems. Thesis, Chalmers Univ. Göteborg 1982

[Sh]   Shakib, F. : Finite Element Analysis of the Compressible Euler and Navier-Stokes Equations. Ph. D. Thesis Stanford Univ. 1989.

[Shi]   Shishkin, G. I.:   Approximation of the solution of sin-
        gularly perturbed boundary value problems with a parabo-
        lic boundary layer (Russ.).   Sh. Vytsch. Matem. Mat. Fis.
        29 (1989) 7, 963 - 977.

[We]   Weiß, D.: Feldberechnungen mit der FEM zur Simulation
        von Temperatur- und Gefügeverteilungen bei Schweißpro-
        zessen. Forschungsbericht TU Magdeburg, GKMBI 1992

[Zh]   Zhou, G.: An Adaptive Streamline Diffusion Finite Ele-
        ment Method for Hyperbolic Systems in Gas Dynamics.
        Thesis, Univ. Heidelberg 1992.

Adaptive Multigrid Methods: The UG Concept

P. Bastian   G. Wittum

IWR, Universität Heidelberg, Im Neuenheimer Feld 368, 69120 Heidelberg, Federal Republic of Germany, email: wittum@iwr.uni-heidelberg.de

Abstract

In the present paper we discuss the development and practical application of a flexible software toolbox for multigrid methods on unstructured and locally refined grids. Our first aim is to combine modern optimal multigrid methods with the robustness strategies developed for structured grids and in the second part we discuss a parallel implementation of the programming environment on multiprocessors with distributed memory. The various techniques are illustrated with many practical experiments.

1 Introduction

In the first part of this paper we discuss the development and practical application of robust multi-grid methods to solve partial differential equations on adaptively refined grids. Since a couple of years multi-grid methods are well established as fast solvers for large systems of equations arising from the discretization of differential equations. However, it is still a substantial unresolved question to find robust methods, working efficiently for large ranges of parameters e.g. in singularly perturbed problems. This applies to diffusion-convection-reaction equations, arising e.g. from modelling of flow through porous media, the basic equations of fluid mechanics and plate and shell problems from structural mechanics.

Multi-grid methods are known to be of optimal efficiency, i.e. the convergence rate κ does not depend on the dimension of the system, characterized by a stepsize h. Following [28] we call a multi-grid method robust for a singularly perturbed problem, if

$$\kappa(h,\varepsilon) \leq \kappa_0 < 1, \ \forall \varepsilon > 0, \ h > 0, \tag{1}$$

ε denoting the singular perturbation parameter. Up to now multi-grid methods satisfying (1) have been studied in the literature only for special model cases using structured grids, see [24], [25], [17], [27], [28], [29].

Problems of the type mentioned, typically show degenerations in hyperplanes. To resolve these zones special dynamic grid adaptation techniques are necessary. Here it is necessary to rethink standard multi-grid techniques. In §2 we classify several multi-grid approaches for adaptively refined grids. On the one hand adaptively refined grids can substantially weaken the robustness requirement (1) as outlined in §3. On the other hand the unstructured grids generated by adaptive refinement require special numbering techniques so that the smoother does a good job on the problem. It is one of the main objectives of the present paper to present a strategy to combine the techniques of robust multi-grid and adaptivity.

The practical use of the techniques mentioned so far requires a substantial programming effort. Compared with traditional approaches on structured uniformly refined grids in logically rectangular domains, code complexity is at least an order of magnitude higher according to our experience. It is the aim of the ug code described in §4 to find proper programming abstractions that allow reuse of the code for many different applications. This is especially important in the parallel version (§5), where the dynamic management of the distributed data structure complicates the code substantially. It should be noted here that most of the robustness techniques described in the first part of the paper are very hard to parallelize efficiently, so the parallel version uses only simpler parallelizable smoothers at the moment.

Several practical examples illustrating the robustness techniques in the serial version and speedup results for the parallel version are presented in §6.

2 Multi-Grid Strategies

2.1 Basic Multi-Grid Techniques

Let the linear boundary-value problem

$$Ku = f \text{ in } \Omega \tag{2}$$
$$u = u_R \text{ on } \partial\Omega$$

with a differential operator $K : U \to F$ between some function spaces be given on a domain $\Omega \subseteq \mathbf{R}^d$. Let (2) be discretized by some local discretization scheme on a hierarchy of admissible grids (cf. [14])

$$\Omega_l \quad , \quad l = 0, \ldots, l_{max} \tag{3}$$
$$\Omega_l \subseteq \Omega_{l+1} \subseteq \Omega \quad .$$

We use nested grids only for ease of presentation. Most of the methods discussed below can readily be applied to general loosely coupled grids violating (3). The discretized equations on Ω_l are denoted by

$$K_l u_l = f_l \text{ in } \Omega_l, \text{ for } l = 1, \ldots, l_{max} \quad , \tag{4}$$
$$u_l = u_{R,l} \text{ on } \partial\Omega_l$$

with

$$K_l : U_l \to F_l \quad , \tag{5}$$

U_l, F_l denoting the discrete analoga of U and F with finite dimension n. We assume that the discretized equations are sparse. Further let some "smoother"

$$S_l : U_l \to U_l \text{ for } l = 0, \ldots, l_{max} \quad , \tag{6}$$

and "grid transfer operators"

$$p_{l-1} : U_{l-1} \to U_l, \quad r_{l-1} : F_l \to F_{l-1}, \text{ for } l = 1, \ldots, l_{max} \quad , \tag{7}$$

be given.

Multi-grid methods are fast solvers for problem (4). We basically distinguish between additive and multiplicative multi-grid methods. The multiplicative method is the well-known classical multi-grid (cf. [13]) as given in algorithm 2.1:

Algorithm 2.1 *Multiplicative multi-grid method.*

```
mmgm(l, u, f)
integer l; grid function u, f;
{ grid function v, d; integer j;
    if (l = 0) u := K_l^{-1} f;
    else {
        u := S_l^{ν_1}(u, f);
        d := r_{l-1}(K_l u - f);
        v := 0;
        for j:=1 step 1 to γ do mmgm(l - 1, v, d);
        u := u - p_{l-1}v;
        u := S_l^{ν_2}(u, f);
    }
} .
```

The additive multi-grid method is given by the following algorithm.

Algorithm 2.2 *Additive multi-grid method.*

```
amgm(l, u, f)
integer l; grid function v[l], d[l];
{ integer j;
    d[l] := K_l u - f; v[l] := 0;
    for j:=l step -1 to 1 do { d[j − 1] := r_{j−1}d[j]; v[j − 1] := 0;}
    for j:=1 step 1 to l do v[j] := S_j^ν(v[j], d[j]);
    v[0] := K_0^{-1}d[0];
    for j:=1 step 1 to l do v[j] :=v[j] + p_{j−1}v[j − 1];
    u := u - v[l];
} .
```

The structure of both algorithms can be seen from Figs. 1(a) and 1(b). The main difference between these two variants is that in the multiplicative method smoothing and restriction of the defect to the next coarser level are performed on one level after the other sequentially, while in the additve method smoothing on the different levels

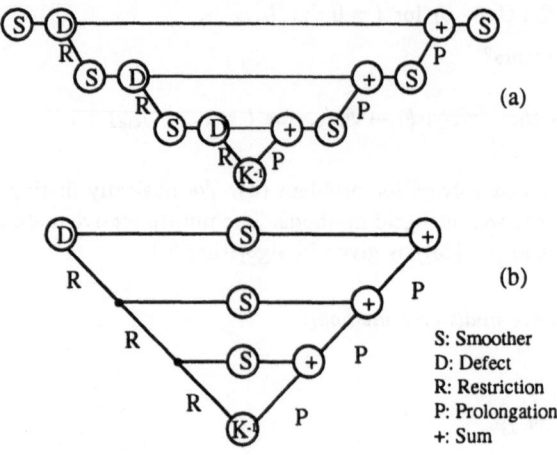

Figure 1: *Outline of the V-cycle multiplicative multigrid algorithm mmgm (a) and of the additive multigrid algorithm amgm (b).*

can be performed in parallel. Restriction and prolongation, however, are sequentially in the additive method too. Usually, the additive methods are applied as preconditioners, since acceleration methods like cg directly pick an optimal damping parameter, the multiplicative methods are used as solvers and as preconditioners. According to [31], these methods can be formulated as additive Schwarz methods.

Applying multi-grid methods to problems on locally refined grids one has to think about the basic question, how to associate grid-points with levels in the multi-grid hierarchy. Consider the hierarchy of grids $\{\Omega_l, l = 0, \ldots, l_{max}\}$ from (3). Early multi-grid approaches smooth all points in Ω_l. This may cause a non-optimal amount of work and memory of $O(n \log n)$ per multi-grid step. This problem was the starting point for Yserentant , [32], and Bank-Dupont-Yserentant, [1], to develop the method of hierarchical bases (HB) and the hierarchical basis multi-grid method (HB/MG). These were the first multi-grid methods with optimal amount of work per step for locally refined grids. This is due to the fact that on level l only the unknowns belonging to points in $\Omega_l \setminus \Omega_{l-1}$ are treated by the smoother. However, the convergence rate deteriorates with $\log n$. For the first time this problem was solved by the introduction of the additive method by Bramble, Pasciak and Xu, [9], (BPX). There on level l the smoother treats all the points in $\Omega_l \setminus \Omega_{l-1}$ and their direct neighbours, i.e. all points within the refined region.

Table 1 gives an overview of the multi-grid methods used for the treatment of locally refined grids and classifies the variant we call "local multi-grid". The methods mentioned above differ in the smoothing pattern, i.e. the choice of grid points treated by the smoother. The methods in the first two lines are of optimal complexity for such problems. The amount of work for one step is proportional to the number of unknowns on the finest grid. However, only the methods in the second line, BPX and local multi-grid converge independently of h for scalar elliptic problems. The basic advantage of the multiplicative methods is that they do not need cg-acceleration and thus can be directly

Table 1: *Multi-grid methods for locally refined grids.*

smoothing pattern	basic structure	
	additive	multiplicative
(1) new points only	*HB* Yserentant, 1984, *[32]*	*HBMG* Bank, Dupont, Yserentant, 1987, *[1]*
(2) refined region only	*BPX* Bramble, Pasciak, Xu, 1989, *[9]*	*local multi-grid,* *[21], [10], [6]*
(3) all points	*parallel multigrid* Greenbaum, 1986, *[12]*	*classical multi-grid, [11]*

applied to unsymmetric problems, further they show a better convergence rate and on a serial computer the additive process does not have any advantage. The local multi-grid scheme is the natural generalization of the classical multi-grid method to locally refined grids, since in case of global refinement, it is identical with the standard classical multi-grid method.

The local multi-grid has first been introduced by Rivara in [21] and first been analyzed in 1991 by Bramble, Pasciak, Wang and Xu, [10]. They considered it as a multiplicative variant of their so-called BPX-method, [9]. However, they did not consider robustness. Without knowledge of this, the authors developed this method as a variant of standard multi-grid based on the idea of robustness (cf. [6]). The main advantage of this approach is that the application to unsymmetric and non-linear problems is straightforward (cf. [6]). Robustness for singularly perturbed problems is achieved by combining local multi-grid with robust smoothers (cf. [6]), as explained in the next section.

3   Robustness Strategies

3.1   Robust Smoothing

Already in 1981, Wesseling suggested the first robust multi-grid method for singularly perturbed problems discretized on structured grids [24], [25]. The main idea is to apply a smoother which solves the limit case exactly. This is possible e.g. for a convection-diffusion equation using a Gauß-Seidel smoother and numbering the unknowns in convection direction. Wesseling however, suggests to use an incomplete LU-smoother, since this handles the convection dominated case as well as the anisotropic diffusion (cf. [17], [28]). Main ingredients, however, are the use of structured grids and a lexicographic numbering.

A simple analysis of the hierarchical basis methods (HB, HB/MG) shows that the smoothing pattern is too poor to allow robust smoothing.

Remark 3.1 *The hierarchical basis method and the hierarchical basis multigrid method* *do not allow robust smoothing for a convection-diffusion equation. The smoothing pattern used in these methods does not allow the smoother to be an exact solver for the limit case. This holds for uniformly as well as for locally refined grids.*

Based on this observation, we extended the smoothing pattern, adding all neighbours of points in $\Omega_l \setminus \Omega_{l-1}$. This allows the smoother to solve the limit case exactly, provided the grid refinement is appropriate. This is confirmed by numerical evidence given in Chapter 5.

Up to now some theory is contained in [28],[29] and the new papers by Stevenson [22], [23] for uniformly refined grids. This theory shows that the basic requirement that the smoother is an exact solver in the limit case is not sufficient to obtain robustness. Additionally it must be guaranteed that the spectrum of the smoother is contained in $[-\vartheta, 1]$ for $0 \leq \vartheta < 1$. This can be achieved by modification (cf. [28], [23]).

3.2 A Robust Smoother for Convection-Diffusion Problems

The construction of a robust smoother, which is exact or very fast in the limit, is the kernel of a robust multigrid method and makes up the main problem when applying this concept to unstructured grids. Here we need special numbering strategies.

In the following we present a strategy for the convection-diffusion equation

$$- \varepsilon \Delta u + c \cdot \nabla u = f \ , \tag{8}$$

with the convection vector c, and $\varepsilon > 0$. Discretizing the convection term by means of an upwind method, we can assign a direction to each link in the graph of the stiffness matrix. If the directed graph generated by this process is cycle-free, it defines a partial ordering of the unknowns. This partial ordering can be used to construct an algorithm for numbering of the unknowns, which brings the convective part of the stiffness matrix to a triangular form. The following numbering algorithm performs such an ordering on general unstructured grids, provided the convection graph is cycle-free.

Algorithm 3.1 *downwind_numbering.*

1. Assign the downwind direction from the discretization of the convective term to each link in the stiffness martix graph. Indifferent links are marked by 0.

2. Put $n =$ number of unknowns.

3. Find all vertices with minimal number of incoming links and put them in a fifo F.

4. Derive a total order from the directed acyclic graph

    For all vertices L initialize Index$(L) = 0$;
    While (F not empty) do
        get E from F;
    (4a)   Put $Index(E) := 1$; Put E in fifo FP; $i := 1$;
    (4b)   While (FP not empty) and ($i < n$) do
            Get K from FP;
            For all neighbors L of K do

If (L downwind from K) and ($\text{Index}(L) \leq \text{Index}(K)$)
$i := \text{Index}(L);$
$\text{Index}(L) := \text{Index}(K)+1;$
Put L in $FP;$

5. Call quicksort with the vertex list and the criterion $Index(L) < Index(K) \Rightarrow L < K$. Output: Ordered vertex list.

Remark 3.2 *If the edge graph is cycle-free, loop (4b) terminates in $O(n)$-steps with $FP = \emptyset$. Loop (4) has complexity $O(q \cdot n)$ where q is the number of minimal elements in the edge graph, which is small. Because of calling quicksort in (5) the complexity of the whole algorithm equals $O(q \cdot n \ln n)$.*
If loop (4b) terminates with $FP \neq \emptyset$ and $i \geq n$, the edge graph contains a cycle.

This method has been used for the computations described in Section 5 . Meanwhile it has been improved by Bey (cf. [7]). Cycles in the matrix graph may occur, if there are vortices in the convection c. If c is vortex-free, cycles can occur if several triangles with sharp angles are neighbouring each other and are almost perpendicular to the flow direction (cf. [7]). These numerically caused cycles, however, can be simply eliminated by finding and cutting elementwise cycles. This is possible with $O(n)$ work count.

3.3 Semi-coarsening

Another strategy to obtain a robust multi-grid method is the so-called semi-coarsening approach (cf. [8]). The basic idea is to improve the coarse grid correction instead of the smoother. Starting with a fine and structured grid, coarsening is performed only in those co-ordinate directions, in which the scale of the equation is already resolved. E.g. for the anisotropic model problem

$$- (\varepsilon \partial_{xx} + \partial_{yy})u = f , \quad \text{in} \Omega = (0,1) \times (0,1) \tag{9}$$

with corresponding boundary conditions one would coarsen an equidistant cartesian grid in case of small ε as shown in Figure 2(a).

Remark 3.3 *Such a sequence of coarse grids yields a robust multi-grid method for the anisotropic model problem (9) without using a special smoother, since the coarse grid resolves the scale in the direction where the smoother does not work.*
This semi-coarsening approach, however, is based on the use of fine grids which do not resolve the differential scale, otherwise there would be no semi-coarsening. Consequently this approach is not applicable as soon as the finest grid resolves the problem scale, which is crucial when solving differential equations.

This does not apply to so-called multiple semi-coarsening approaches [20], since these methods are able to construct sequences of coarse grids from any structured fine one, no matter if the scale is resolved. Solving practical problems we mainly have to look for an approach which allows to adapt the grid to the differential scale by adaptive refinement and to solve efficiently on the hierarchy of grids generated this way.

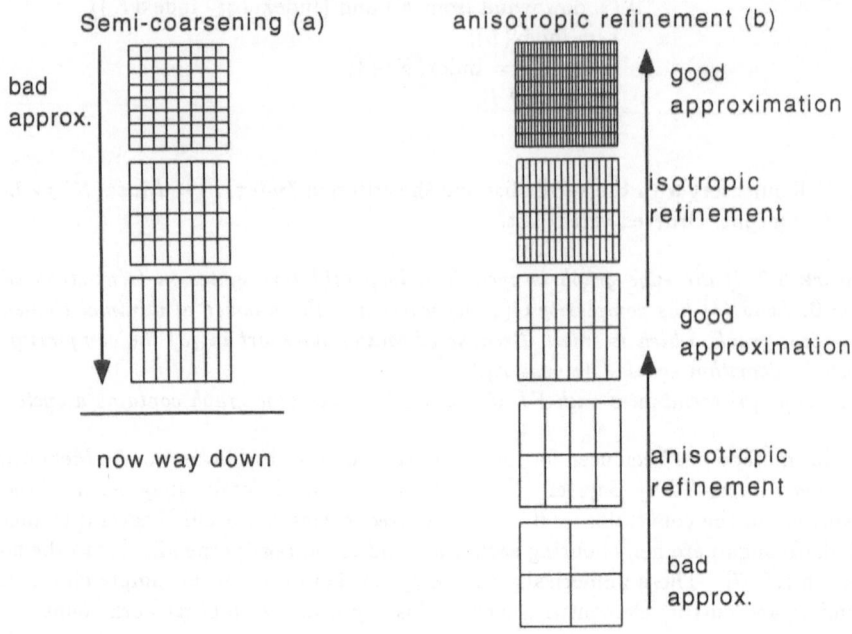

Figure 2: *Illustration of semi-coarsening (a) and anisotropic refinement (b).*

3.4 Anisotropic Refinement

Instead of starting with a fine grid and constructing the grid hierarchy by coarsening we start with a coarse grid and refine that anisotropically in order to resolve the scale successively. Such a refinement process is given e.g. by the "blue refinement strategy" due to Kornhuber, [18]. The basic idea is just to refine quadrilaterals with a "bad aspect ratio" by halving the longer edge. Bad aspect ratios can be introduced either by element geometry or by anisotropic coefficients in the equation. This is shown for the anisotropic model problem (9) in Fig. 2(b). Note that the discretization error is balanced on the *coarsest* grid for semi-coarsening, while it is balanced on the *finest* grid for the anisotropic refinement approach. Kornhuber described how to generalize this approach to triangular unstructured grids. Following this process we finally obtain a grid Ω_l which resolves the scale of the problem.

From this grid on we refine regularly and so the multi-grid process will obviously work without problems.

Remark 3.4 *A proof of robust multi-grid convergence is straightforward since the asymptotic behaviour is determined by the isotropic problem. So we need a robust method only for a finite sequence of grids up to a fixed $h > 0$, weakening the robustness requirement (1) to the relative robustness:*

$$\kappa(h,\varepsilon) \leq \kappa_0 < 1, \quad \forall \bar{\varepsilon} \geq \varepsilon \geq \underline{\varepsilon} > 0, \ \forall h \geq \underline{h} > 0 \,, \tag{10}$$

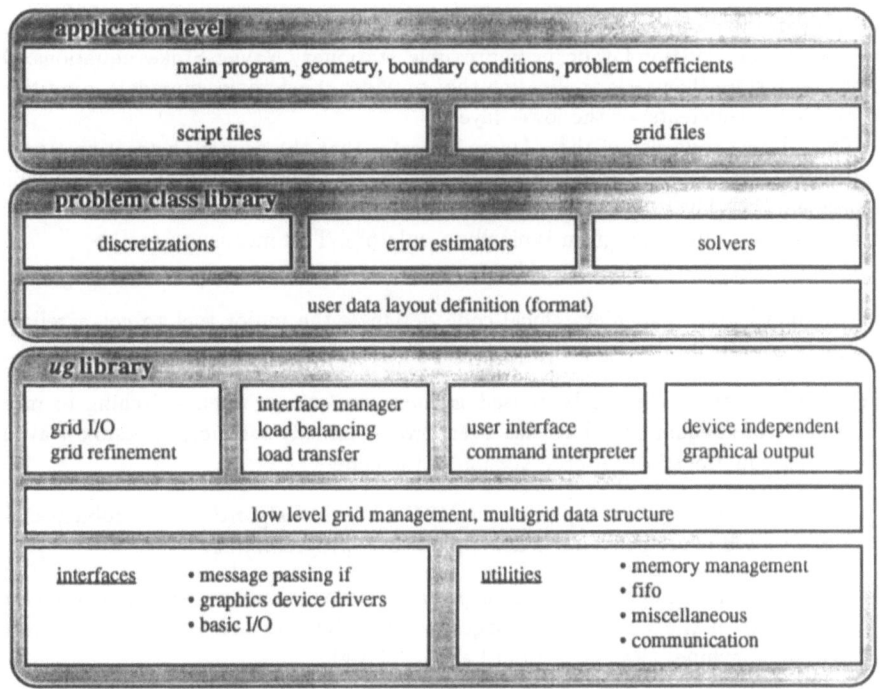

Figure 3: *Overview of the internal structure of the ug code.*

which makes the job much easier. Thus it is sufficient in many cases to use just a lexicographically numbered ILU_β, since we do not need the property that the smoother is exact in the limit case. It is sufficient that it reasonably accounts for the "main connections" up to a fixed range of $\varepsilon > 0$ and for finite h.

Since this process improves the approximation of the differential problem at the same time, this will be the appropriate approach to follow.

An example of that type is the skin problem described in §5.

4   The Software Toolbox ug

A big problem in practice with the modern adaptive multigrid methods introduced above is code complexity when implementing these methods in a computer program. Especially on the parallel computer this is an important point since the dynamic parallel management of the data structure is very complicated. Therefore we developed the software environment *ug* ("unstructured grids") as a problem independent "toolbox" for (parallel) unstructured adaptive multigrid applications. It is a layered construction of several libraries, see Fig. 3 for an overview. The bottom layer contains all components that are totally independent of the PDE to be solved, e. g. grid I/O, grid refinement, device independent graphical output, user interface and dynamic grid management with load balancing in the parallel version. The next layer is the so-called problem class library that implements discretization, error estimators and solvers for a whole class of PDEs, e. g. a scalar conservation law or incompressible, stationary Navier-Stokes equations. On top of that resides the user's application that provides the domain, boundary conditions and problem coefficients for the lower layers.

The relative code size of these layers indicates that the proper abstractions (interfaces) have been chosen: The *ug* layer typically makes about 75-80% of the executable, the problem class layer takes 15-20% in the convection-diffusion case (with many different solvers) and a main program typically is only 5%. This means in practice:

- Modularization and hierarchical code design is the major tool to get a reliable piece of software.

- 75-80% of the code can be reused *without any change* when switching to more complicated equations. This has been proved already for incompressible Navier-Stokes equations in the serial version of the code.

- The user interested in implementing new numerical algorithms (a problem class library) will never be concerned with low level programming.

- As a consequence of that his code is portable since machine dependencies typically arise only in the *ug* layer. Message passing interfaces are available for several parallel machines (Parsytec, Intel Paragon, PVM).

- The final aim is to have a consistent software environment in 2 and 3 space dimensions on serial and parallel machines.

5 Parallelization

Here we will only cover briefly some basic aspects of the parallel implementation, for a more detailed discussion we refer to [5]. The primary parallelization approach is data partitioning, i.e. the parallelism inherent to multigrid methods is exploited. This poses a restriction on the smoother that can be used in the parallel code. Robustness strategies such as ILU or the special numbering strategies presented in paragraph 3.2 can not be parallelized efficiently, at least on unstructured grids (for structured rectangular grids see [3]). Therefore the parallel implementation currently uses either point Jacobi or inexact Block-Jacobi smoothers (with one or several steps GS or ILU as inner solver).

The data partitioning uses a unique mapping of the elements (of all levels) to the set of processors. For each element t assigned to a processor p, this processor holds also a copy of the nodes of t and of the father element of t and its nodes. This leads to an overlapping storage scheme, where e.g. several copies of one node are stored in different processors. The parallel data management module provides high level routines for exchange of data between these copies.

The stiffness matrix is only assembled per processor and is fully parallel and does not need any communication. Restriction and prolongation proceed without communication as long as each son of an element t is mapped to the same processor as t. The nested grid refinement algorithm can also be parallelized efficiently if there is a refinement rule for each edge refinement pattern possible. In this case the iteration in the green closure of the triangulation can be avoided.

Theoretical investigations and practical experience shows that multiplicative multigrid methods are twice as efficient as additive multigrid methods in terms of computer time on a single computer, since the number of iterations to reach a convergence criterion is doubled, but one iteration of amgm costs not much less than one of mmgm. On a parallel machine the picture may be different for two reasons: First the granularity of amgm is coarser. In the smoother amgm can send all updates for levels $1, \ldots, j$ in one large message, while mmgm must send the updates on all levels seperately, i.e. j smaller messages. The second and more important point is, that different load balancing methods can be used for both methods. In mmgm each grid level must be distributed equally onto all processors while in amgm it is sufficient that the number of elements on all levels is about the same for each processor (if there is no communication in restriction and prolongation).

The most complicated part of the parallelization is the load balancing module which is divided into two parts: The load balancer determines only the new assignment of elements to processors while the load transfer part actually moves the parts of the data structure to their new position. The basic idea for the load balancing part is to first assign the elements to clusters (subsets of elements) and then to assign the clusters to the processors. The clustering process reduces the complexity of the load balancer dramatically, the number of clusters is only proportional to the number of processors. Since the clusters are built via the element hierarchy, they are the key to find a compromise between inter- and intragrid communication. The central element of the cluster assignment algorithm is currently a recursive coordinate bisection strategy. More elaborate techniques will be tested in the future when nonlinear systems are to be solved. Due to the high efficiency of the multigrid method for the simple scalar elliptic problems, care must be taken that the load balancer does not dominate the computation time.

6   Numerical Results

The first four problems show applications of the serial ug version, the last two examples give some insight into the performance of the parallel version (more examples can be found in [5]).

6.1   The Skin Problem

As a first test problem we take the following one which is used to model the penetration of drugs through the uppermost layer of the skin (stratum corneum). The stratum corneum is made up of corneocytes which are embedded in a lipid layer. The diffusion is described by the diffusion equation

$$
\begin{aligned}
-\nabla(D(x,y)\nabla u) + \frac{\partial u}{\partial t} &= 0 & \text{in}\Omega & \qquad (11)\\
u &= 1 & \text{on}\Gamma_u &\\
u &= 0 & \text{on}\Gamma_o &\\
\frac{\partial u}{\partial n} &= 0 & \text{on}\Gamma_r \cup \Gamma_l &
\end{aligned}
$$

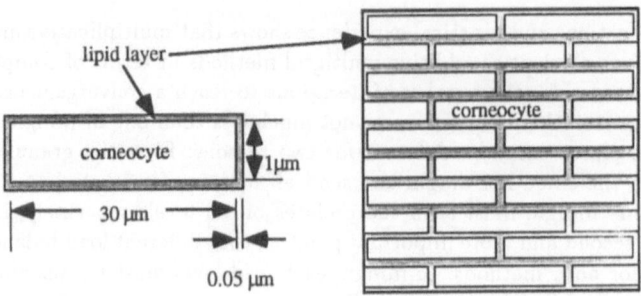

Figure 4: *Right hand side: Structure of skin made up from corneocytes (white) and lipid layers (gray/black). The considered block of stratum corneum is 11μm by 60.2μm. Left hand side: Elementary cell consisting of a corneocyte surrounded by one half of the lipid layer.*

Table 2: *Convergence rate of a (1,1,V)-mmgm applied to the stationary skin problem for various values of D_2 ($D_1 = 1$). The number of unknowns was 54385 on level 5 (6 grid levels).*

D_2	1	10^{-1}	10^{-2}	10^{-3}	10^{-4}	10^{-5}	10^{-6}
ρ	0.08	0.22	0.39	0.41	0.45	0.45	0.43

where Ω is the unit square and the diffusion coefficient $D(x,y)$ is given by

$$D(x,y) = \begin{cases} D_1 & \text{if } (x,y) \in \text{lipid} \\ D_2 & \text{if } (x,y) \in \text{corneocyte} \end{cases},$$

i.e. it may jump by some orders of magnitude across the corneocyte edges. The corneocytes are very flat and wide cells which in a two-dimensional cross-section are approximated by thin rectangles as shown in Fig. 4.

From Fig. 4 we see that the lipid layer is $0.1\mu m$ thick while the corneocytes are 1 by $30\mu m$ of size. Since the permeability may jump by some orders of magnitude between lipid and corneocyte, we must align the coarse-grid lines with the interfaces. So we just take the corners of the corneocytes as points for the coarse grid connecting them to form a tensor product grid. Thus we get rid of the problems induced by jumping coefficients. However, we obtain highly anisotropic grid cells in the lipid layer with an aspect ratio of approx. 1:150. Since such an aspect ratio makes the approximation strongly deteriorate and the multi-grid method as well, we use the anisotropic ("blue") refinement strategy to derive a robust multi-grid method and to create a grid which after 5 levels of blue refinement has elements not exceeding an aspect ratio of 1:5. Above that level we refine uniformly. To obtain a robust method on the coarser grids we use an ILU$_\beta$-smoother, cf. [28]. Average convergence factors for a (1,1,V)-cycle are given in Table 2. For more details on this problem see [19].

Table 3: *Robustness of a (1,1,V)-mmgm with ILU-smoother and downwind numbering. The method used 8 locally refined grids to discretize problem the convection-diffusion problem with over 10.000 unknowns on level 8. The convergence rate $\kappa(10)$ is averaged over 10 steps and refers to the finest grid.*

ε	1	10^{-1}	10^{-2}	10^{-3}	10^{-4}	10^{-5}	10^{-6}	10^{-7}
$\kappa(10)$	0.068	0.067	0.075	0.102	0.092	0.068	0.033	0.018

6.2 Convection-Diffusion Equation

As a second example we show results for the convection-diffusion equation

$$- \varepsilon \Delta u + c \cdot \nabla u = f \tag{12}$$

in the unit square with Dirichlet boundary conditions. We choose c as follows

$$c = \left(1 - \sin(\alpha)\left[2\left(x + \frac{1}{4}\right) - 1\right] + 2\cos(\alpha)\left[y - \frac{1}{4}\right]\right)^4 (\cos(\alpha), \sin(\alpha))^T \tag{13}$$

where α is the angle of attack. The boundary conditions are: $u = 0$ on $\{(x,y) : x = 0, 0 \le y \le 1\} \cup \{(x,y) : 0 \le x \le 1, y = 1\} \cup \{(x,y) : x = 1, 0 \le y \le 1\} \cup \{(x,y) : 0 \le x < 0.5, y = 0\}$ and $u = 1$ on $\{(x,y) : 0.5 \le x \le 1, y = 0\}$. The jump in the boundary condition is propagated in direction α. We have divc $= 0$ and c varies strongly on Ω such that the problem is convection dominated in one part of the region and diffusion dominated in another part. As discretization we use a finite volume scheme with first order upwinding for the convective terms on a triangular grid. The grid is refined adaptively using a gradient refinement criterion. As smoother we took a Gauß-Seidel scheme with downwind numbering using algorithm 3.1 in a (1,1,V)-cycle mmgm. It is important to note that the smoother itself is not an exact solver. Thus we should see the benefit of multi-grid in the diffusion dominated part and of the robust smoother in the convection dominated one. This is confirmed by the results given in Table 3. There we show the residual convergence rate averaged over 10 steps for problem (12) on adaptively refined unstructured grids versus ε.

For the same problem with $\varepsilon = 10^{-7}$ the same mmgm but without downwind numbering shows a convergence rate of 0.95 averaged over 40 steps and taking the smoother with downwind numbering but without coarse grid correction as a solver, we end up with a convergence rate of 0.949 as well. This confirms the outlined concept of robust multi-grid. Results of 3d computations can be found in [7].

6.3 Drift Chamber

This problem solves the Laplacian $-\Delta u = 0$ in the domain given by Fig. 5. The boundary conditions are of Dirichlet and Neumann type as indicated in the figure. The feature of this problem are the small wires with Dirichlet boundary conditions that must be resolved on the coarse grid. The smallest wire has a radius of 0.005 mm, while the whole chamber is 4 mm wide and 1 mm thick. So one has to trade off between a coarse grid with

Table 4: *Results for different solver/smoother combinations for the drift chamber problem. Multigrid data: (2,2,V) cycle for Jacobi smoother, $\nu = 1$ for amgm, (2,2,V) cycle for all other smoothers, initial solution $u = 0$, numbers are iterations for a reduction of the residual by 10^{-6} in the euclidean norm. The grid nodes have been ordered lexicographically, iteration numbers exceeding 100 are marked with an asterisk, diverging iterations are marked with ↑.*

highest level		3	4	5	6
grid nodes		3809	14785	58241	231169
mmgm	djac	*	*	*	*
	gs	79	99	*	*
	sgs	48	59	66	70
	ILU	33	↑	↑	↑
	ILU$_\beta$	9	9	9	9
mmgm+cg	djac	31	38	43	43
	sgs	13	16	17	18
	ILU	10	↑	↑	↑
	ILU$_\beta$	6	6	6	6
amgm+cg	djac	74	99	*	*
	sgs	36	46	53	57
	ILU	62	↑	↑	↑
	ILU$_\beta$	20	24	25	26

6.4   A Shape Design Problem

A nonlinear example computed with ug is the following shape design problem. An infinitely long bar consisting of two different materials (hard and soft) and square cross section is to be designed. The ratio of the two materials in a cross section is prescribed and the torsional rigidy is to be maximized. The materials may be mixed at the interface, for details see [16]. The differential equation modelling this situation is given by

$$-\mathrm{div}(\varphi(\|\nabla u\|^2)\nabla u) = 1 \ \text{in} (0,1)^2 \tag{14}$$
$$u = 0 \ \text{on} \ \partial\Omega$$

where φ is given by

$$\varphi(x) = \begin{cases} \mu_1 & \text{if } x \geq \frac{2\lambda\mu_2}{\mu_1} \\ \sqrt{\frac{2\lambda\mu_1\mu_2}{x}} & \text{if } \frac{2\lambda\mu_1}{\mu_2} < x < \frac{2\lambda\mu_2}{\mu_1} \\ \mu_2 & \text{if } x \leq \frac{2\lambda\mu_1}{\mu_2} \end{cases}$$

and the constants are: $\lambda = 0.008, \mu_1 = 1, \mu_2 = 2$.

The problem has been solved with global and local grid refinement using either a damped fixed point iteration or a nonlinear multigrid method (with a fixed point iteration as smoother). Table 5 shows the average reduction rates per iteration for the four different methods and different refinement depths.

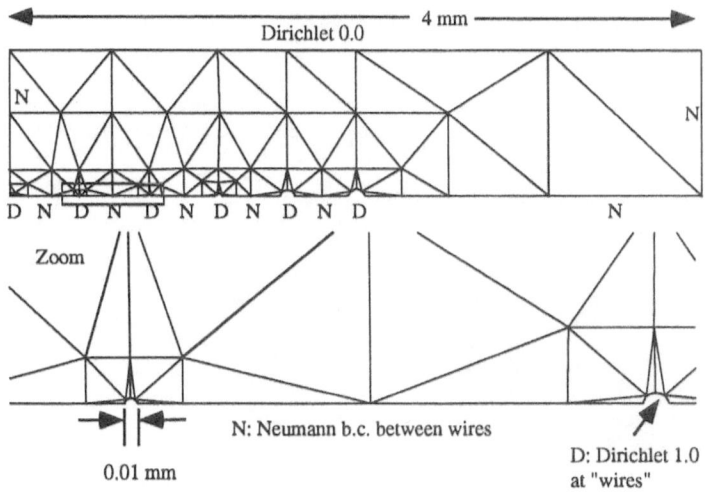

Figure 5: *Problem definition, coarse grid and zoom for the drift chamber problem.*

few unknowns but a large aspect ratio in grid cells and a coarse grid with equal sized triangles but a large number of unknowns. The grid in Fig. 5 is a reasonable compromise with 85 nodes and 112 triangles but still aspect ratios are large and a robust smoother is required.

Table 4 shows the results of multiplicative and additive multigrid with several different smoothers applied after 3, 4, 5 and 6 levels of uniform refinement. Specifically the smoothers were damped Jacobi with $\omega = 2/3$ (djac), (symmetric) Gauß-Seidel (gs, sgs) and ILU without modification and with $\beta = 0.35$ (ILU, ILU$_\beta$). We make the following remarks:

1. h independent convergence is only achieved with the ILU$_\beta$ smoother. The optimal value was $\beta = 0.35$ but the choice is not very sensitive and good results are achieved with values between 0.2 and 0.5. This corresponds nicely with the theory in [28].

2. The additive method shows qualitatively the same behaviour as the multiplicative multi-grid method but has worse numerical efficiency.

3. Multiplicative multi-grid with a symmetric Gauß-Seidel smoother used as preconditioner in a conjugate gradient method is the only combination giving also relatively satisfactory results, being only a factor 3 slower in computation time than the ILU$_\beta$ smoother.

4. The diverging iteration for ILU without modification can be explained by accumulating roundoff errors. Since the global stiffness matrix is symmetric positive definite but *not* an M-matrix due to obtuse angles the diagonal elements in the ILU decomposition can become very small which leads to instabilities. The modification helps in this case too, since it enlarges the diagonal.

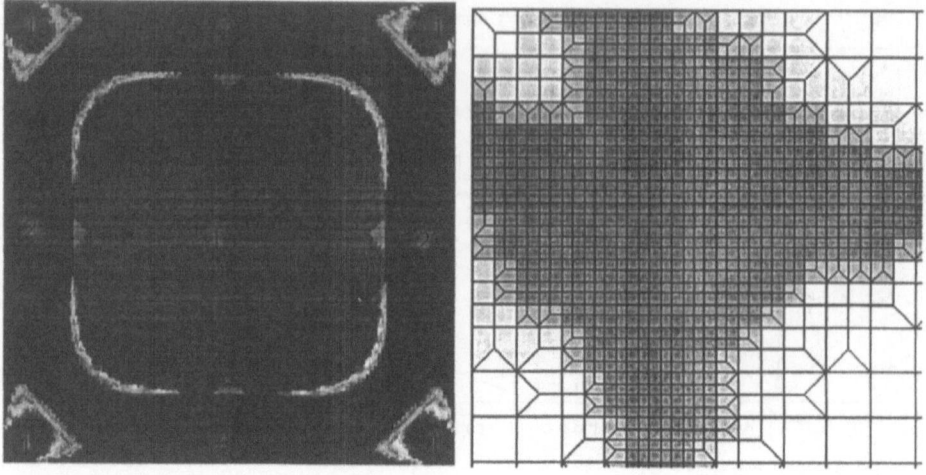

Figure 6: *Left picture shows the function $\varphi(\|\nabla u\|)$ for the shape design problem. Regions labeled with (1) consist of the soft material, regions labeled with (2) consist of the hard material. The right picture shows a zoom into the local refinement in the corner.*

Table 5: *Average convergence rates over 10 iterations for the nonlinear shape design problem using multiplicative multigrid with Gauß-Seidel smoother, and a (2,2,V) cycle.*

j	global refinement		local refinement	
j	nonlinear MG	fixed point	nonlinear MG	fixed point
1	0.05	0.36	0.01	0.1
2	0.40	0.49	0.49	0.1
3	0.73	0.87	0.52	0.57
4	0.62	0.85	0.72	0.76
5	0.78	0.92	0.59	0.62
6			0.65	0.72
7			0.71	0.65

6.5 German Bight

This example is intended to show that multigrid methods on unstructured grids can be parallelized as efficiently as on structured grids. The scalar, time-dependent convection-diffusion equation is solved in the domain given in Fig. 7. The flow field is fixed but nonuniform (it is taken from measurements by the BSH, Hamburg). The coarse grid is already very fine (about 1200 triangles) and is mapped to the processors via orthogonal recursive coordinate bisection [26] (Fig. 7 uses a 3 by 3 processor array).

All parallel results reported below have been computed on a Parsytec transputer system with PARIX 1.2 and 4 MBytes RAM per processor.

Table 6 shows iteration times for a multiplicative multigrid method for up to three levels of refinement (4 grids) and various processor numbers. Execution time increases

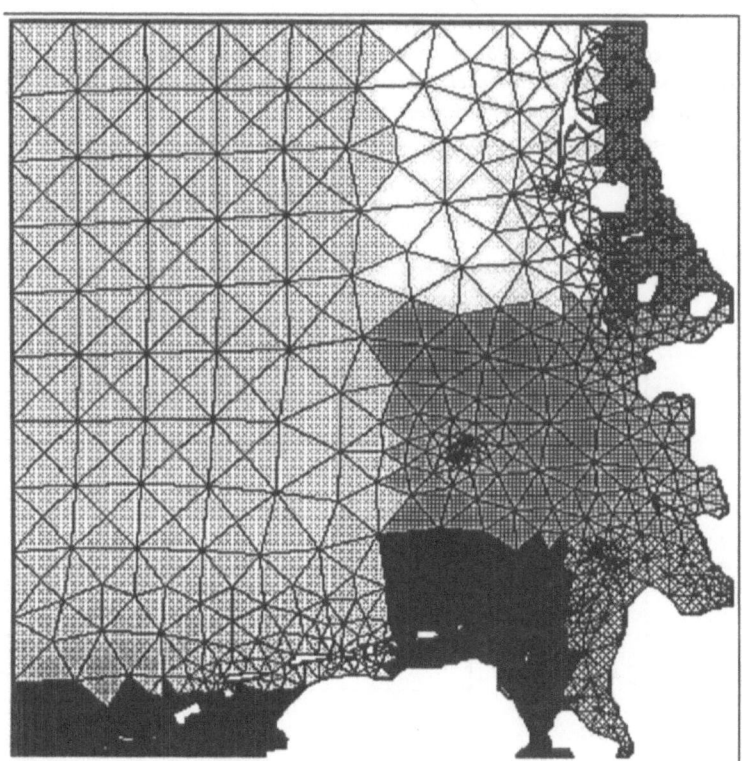

Figure 7: *Triangulation of the German Bight and mapping onto 3 by 3 processors.*

Table 6: *Iteration times in seconds for a (2,2,V) multiplicative multigrid cycle with Block-Jacobi smoother with 2 steps symmetric Gauß-Seidel as inner solver. The residual in the level 0 equation has been reduced by 10^{-4}.*

		Processors						
j	Elements	1	2	4	5	10	16	20
1	5956	7.3	3.8	2.5	1.9	1.3	0.9	0.8
2	23824		16.4	8.8	7.0	3.9	2.6	2.2
3	95296					13.8	8.9	7.3

from 7.3 seconds to 8.9 seconds (22% increase) when going from one to 16 processors with problem size also increased by a factor of 16. Table 7 shows a computation of 50 timesteps on level 2 with various processor numbers from 2 to 20. The computation on 20 processors is 7.1 times faster than on two processors for a fixed problem size.

6.6 A Simple Locally Refined Example

This example illustrates the case of local grid refinement and gives a comparison of additive and multiplicative multigrid in terms of numerical efficiency. The Laplace equation

Table 7: *Total solution time in seconds for 50 steps with implicit euler time integration. Three grid levels have been used with 23824 triangles on the finest grid (fixed problem size for all processor numbers). Within each timestep the residual has been reduced by 10^{-5} with the multigrid method from the previous table.*

	Processors					
	2	4	5	10	20	SGI Crimson
Time	3610	1913	1668	964	510	1310
Speedup	1	1.9	2.2	3.7	7.1	

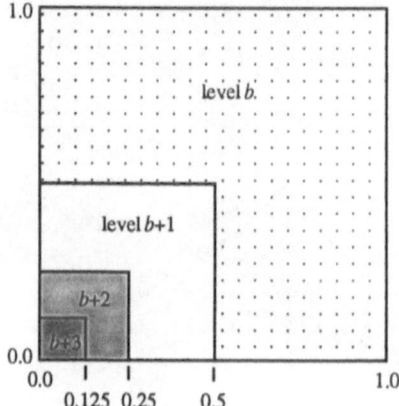

Figure 8: *Grid refinement for the locally refined model problem with $w = 1$.*

with Dirichlet boundary conditions is to be solved in the unit square:

$$
\begin{aligned}
-\Delta u &= 0, & \text{in } \Omega = (0,1)^2 \\
u(x,0) &= \tfrac{1}{4}x, & x \in [0,1[\\
u(1,y) &= \tfrac{1}{4} + \tfrac{1}{4}y, & y \in [0,1[\\
u(x,1) &= \tfrac{1}{2} + \tfrac{1}{4}(1-x), & x \in [0,1[\\
u(0,y) &= \tfrac{3}{4} + \tfrac{1}{4}(1-y), & y \in]0,1[
\end{aligned}
\tag{15}
$$

The initial triangulation T_0 consists of 4 quadrilaterals with one unknown. Refinement is uniform up to some prescribed level b and beginning with level b refinement is such that the grid on level k is restricted to the rectangle $[0, s_k]^2$ with $s_k = (\sqrt{w}/2)^{k-b}, (k > b)$ (see Fig. 8). The factor w is called the "growth factor", since the number of elements on level $k+1$ is defined recursively by $\#T_{k+1} = w\#T_k$. With $w = 4$ one gets uniform refinement and $w = 1$ indicates a case where there is no geometric growth in the number of unknowns. Table 8 shows the results for $w = 1, 2, 3, 4$ and additive and multiplicative multigrid. Note that the problem size is increased with the number of processors. However it is not possible to get always the same number of unknowns per processor for different values of w.

The conclusions drawn from this test are:

Table 8: *Results for different locality of refinement (w, see text) and a varying number of processors. T_{SOL} (total solution time) is for a 10^{-6} reduction in residual norm on level j after a nested iteration. Multigrid data: $\nu_1 = \nu_2 = 1, \gamma = 1$ for mmgm, $\nu = 1$ for amgm with a point Jacobi smoother. Refinement was uniform up to level 4 ($h = 1/32$) except for cases $w = 1$ and $P > 1$ where refinement was uniform up to level 5. N is the number of unknowns after j adaption steps, E_{IT} is the parallel efficiency in one multigrid iteration.*

w		mmgm+jac					cg+amgm+jac				
	P	1	4	16	32	64	1	4	16	32	64
4	j	4	5	6	7	7	4	5	6	7	7
	N	1089	4225	16641	66049	66049	1089	4225	16641	66049	66049
	T_{SOL}	5.95	5.87	6.55	12.49	6.83	9.33	10.17	11.04	20.98	11.43
	E_{IT}		85	75	77	71		88	79	82	76
3	j	5	6	7	7	8	5	6	7	7	8
	N	3553	10657	31656	31656	94440	3553	10657	31656	31656	94440
	T_{SOL}	18.14	15.59	13.09	7.99	15.46	31.38	25.81	21.44	11.97	17.85
	E_{IT}		86	76	71	47		90	80	71	71
2	j	5	7	8	9	10	5	7	8	9	10
	N	2768	12223	24767	49974	99638	2768	12223	24767	49974	99638
	T_{SOL}	15.59	19.05	11.77	12.60	13.44	26.00	28.43	18.68	19.77	20.92
	E_{IT}		78	64	59	55		86	71	74	63
1	j	6	6	10	13	15	6	6	10	13	15
	N	2753	7425	20225	29825	36225	2753	7425	20225	29825	36225
	T_{SOL}	15.39	11.13	10.17	10.69	11.14	24.37	18.49	15.68	13.93	9.96
	E_{IT}		78	60	41	24		85	64	53	45

- The additive method has always better or equal (only one case) efficiencies than the multiplicative method. This is due to coarser grained parallelism and the possibility of using decompositions with smaller interfaces. The latter is the more important point which can be seen by considering the results for $P = 64$ and comparing efficiencies for different w. For $w = 4$ both load balancing methods yield the same decomposition and efficiencies differ not much. For w getting smaller the differences become greater.

- The solution time is always *smaller* for the multiplicative method than for the additive method, except the case of $w = 1, P = 64$. This is not a result of the small number of unknowns per processor. In contrary to the case $w > 1$, one can observe for $w = 1$ that efficiencies do not increase with problem size above level 10 for a fixed number of processors. This is due to the fact that the unknowns do not grow geometrically with the number of levels (no decrease of the surface to volume ratio).

- In the case where additive multigrid is equal or better than multiplicative multigrid in terms of total computation time both methods have to be considered as inefficient. Since additive multigrid is a factor of two (roughly) more expensive on a serial computer, parallel efficiency for multiplicative multigrid must be below 50% in order to allow additive multigrid to be better. Since there are also losses in the latter method, the break even point happens to be at 25% efficiency for multiplicative and 50% efficieny for additive multigrid in this example.

References

[1] R. E. Bank, T. F. Dupont, H. Yserentant: *The Hierarchical Basis Multigrid Method* , Numer. Math., **52**, 427-458 (1988).

[2] R. E. Bank: *PLTMG: A software package for solving elliptic partial differential equations. Users Guide 6.0.* SIAM, Philadelphia, 1990.

[3] P. Bastian, G. Horton: *Parallelization of Robust Multigrid Methods: ILU Factorization and Frequency Decomposition Method.* SIAM J. Sci. Stat. Comput., **12**, No. 6, pp. 1457-1470, 1991.

[4] P. Bastian, G. Wittum: *On Robust and Adaptive Multigrid Methods.* In: Proceedings of the 4$^t h$ European Multigrid Conference, Amsterdam, July 1993, to appear.

[5] P. Bastian: *Parallel Adaptive Multigrid Methods.* IWR Report 93-60, Interdisziplinäres Zentrum für Wissenschaftliches Rechnen, Universität Heidelberg, 1993.

[6] —: *Locally Refined Solution of Unsymmetric and Nonlinear Problems.* In: Hackbusch, W., Wittum, G. (eds.): Incomplete Decompositons - Theory, Algorithms, and Applications, NNFM, vol. 41, Vieweg, Braunschweig, 1993.

[7] J. Bey, G. Wittum: *A Robust Multigrid Method for the Convection-Diffusion Equation on locally refined grids.* In: Adaptive Methods, Proceedings of the Ninth GAMM Seminar, Notes on Numerical Fluid Mechanics, Vieweg Verlag, Braunschweig, 1993, to appear.

[8] A. Brandt: *Guide to Multigrid Development.* in Hackbusch W., Trottenberg U. (eds.): Multigrid Methods. Proceedings Köln-Porz, 1981. Lecture Notes in Mathematics, Bd. 960, Springer, Heidelberg, 1982.

[9] J. H. Bramble, J. E. Pasciak, J. Xu: *Parallel Multilevel Preconditioners*, Math. Comput., **55**, 1-22 (1990).

[10] J. H. Bramble, J. E. Pasciak, J. Wang, and J. Xu, *Convergence estimates for multigrid algorithms without regularity assumptions*, Math. Comp., **57**, (1991), pp. 23–45.

[11] R. P. Fedorenko: *Ein Relaxationsverfahren zur Lösung elliptischer Differentialgleichungen.* (russ.) UdSSR Comput Math Math Phys 1,5 1092-1096 (1961).

[12] A. Greenbaum: *A Multigrid Method for Multiprocessors.* Appl. Math. Comp., **19**, 75-88 (1986).

[13] W. Hackbusch: *Multi-grid methods and applications.* Springer, Berlin, Heidelberg (1985).

[14] —: *Theorie und Numerik elliptischer Differentialgleichungen.* Teubner, Stuttgart, 1986.

[15] —: *The Frequency Decomposition Multi-grid Method.* Part I: Application to Anisotropic Equations. Numer. Math., 1989.

[16] B. Kawohl, J. Stara, G. Wittum: *Analysis and Numerical Studies of a Shape Design Problem.*, Archive for Rational Mechanics, **114**, 349-363, 1991.

[17] R. KETTLER: *Analysis and comparison of relaxation schemes in robust multi-grid and preconditioned conjugate gradient methods*. In: Hackbusch,W., Trottenberg,U. (eds.): Multi-Grid Methods, Lecture Notes in Mathematics, Vol. 960, Springer, Heidelberg, 1982.

[18] R. KORNHUBER, R. ROITZSCH: *On Adaptive Grid Refinement in the Presence of Boundary Layers*. Preprint SC 89-5, ZIB, Berlin, 1989.

[19] R. LIECKFELDT, G. W. J. LEE, G. WITTUM, M. HEISIG: *Diffusant concentration profiles within corneocytes and lipid phase of stratum corneum*. Proceed. Intern. Symp. Rel. Bioact. Mater., 10 (1993) Controlled Release Society, Inc.

[20] W. A. MULDER: *A New Multigrid Approach to Convection Problems*. J. Comp. Phys., **83**, 303-323 (1989).

[21] M. C. RIVARA, *Design and data structure of a fully adaptive multigrid finite element software*, ACM Trans. on Math. Software, 10 (1984), pp. 242–264.

[22] R. STEVENSON: *On the robustness of multi-grid applied to anisotropic equations: Smoothing- and Approximation-Properties*. Preprint Rijksuniversiteit Utrecht, Wiskunde, 1992.

[23] —: *New estimates of the contraction number of V-cycle multi-grid with applications to anisotropic equations*. In: Hackbusch, W., Wittum, G. (eds.) : Incomplete Decompositions, Algorithms, theory, and applications. NNFM, vol 41, Vieweg, Braunschweig, 1993.

[24] P. WESSELING: *A robust and efficient multigrid method*. In: Hackbusch, W., Trottenberg, U. (eds.): Multi-grid methods. Proceedings, Lecture Notes in Math. 960, Springer, Berlin (1982).

[25] —: *Theoretical and practical aspects of a multigrid method*. SIAM J. Sci. Statist. Comp. **3**, (1982), 387-407.

[26] R. D. WILLIAMS: *Performance of Dynamic Load Balancing Algorithms for Unstructured Mesh Calculations*, Report C3P 913, California Institute of Technology, Pasadena, CA., (1990).

[27] G. WITTUM: *Filternde Zerlegungen - Schnelle Löser für große Gleichungssysteme*. Teubner Skripten zur Numerik Band 1, Teubner, Stuttgart, 1992.

[28] —: *On the robustness of ILU-smoothing*. SIAM J. Sci. Stat. Comput., **10**, 699-717 (1989).

[29] —: *Linear iterations as smoothers in multi-grid methods*. Impact of Computing in Science and Engineering, **1**, 180-215 (1989).

[30] J. XU: *Multilevel theory for finite elements*. Thesis, Cornell Univ., 1988.

[31] —: *Iterative Methods by Space Decomposition and Subspace Correction: A Unifying Approach*, SIAM Review, **34**(4), 581-613, (1992).

[32] H. YSERENTANT: *Über die Aufspaltung von Finite-Element-Räumen in Teilräume verschiedener Verfeinerungsstufen*. Habilitationsschrift, RWTH Aachen, 1984.

Finite volume methods with local mesh alignment in 2-D

Rüdiger Beinert
Institut für Angewandte Mathematik
Universität Bonn, Wegelerstr. 6, D-5300 Bonn, West-Germany

Dietmar Kröner
Institut für Angewandte Mathematik
Universität Freiburg, Hermann-Herder-Str.10, D-7800 Freiburg, West-Germany

Summary

In this paper, we shall discuss recent results on finite volume methods on unstructured grids in 2-D for solving systems of conservation laws. Although finite difference methods on structured grids are used most frequently for this kind of problems, for local mesh refinement and coarsening it is much more convenient to work on unstructured meshes. We shall present some numerical experiments on aligned grids for solving the Euler equation of gas dynamics. This means, the shape and the alignment of the triangles is automatically controlled by a mesh indicator. The indicator, e.g. the gradient of the Mach number, is still a heuristical tool but it turns out, that the resolution of shocks are much better on locally aligned grids than on locally refined grids without alignment.

Introduction

Since finite volume methods on unstructured meshes are still not used very frequently we shall first describe two different schemes and shall report on some recent theoretical results concerning convergence.

First we shall describe first and higher order schemes on unstructured grids. Then the local mesh alignment method in the sense of Kornhuber and Roitzsch [5] applied to nonlinear systems of conservation laws will be described and at the end we shall present some experimental numerical results, where we apply the local mesh alignment to the system of the Euler equations of gas dynamics

$$\partial_t U + A(U)\partial_x U + B(U)\partial_y U = 0, \tag{1}$$

where

38

$$U = \begin{pmatrix} \rho \\ u \\ v \\ p \end{pmatrix}, \quad A(U) = \begin{pmatrix} u & \rho & 0 & 0 \\ 0 & u & 0 & \frac{1}{\rho} \\ 0 & 0 & u & 0 \\ 0 & \gamma p & 0 & u \end{pmatrix}, \quad B(U) = \begin{pmatrix} v & 0 & \rho & 0 \\ 0 & v & 0 & 0 \\ 0 & 0 & v & \frac{1}{\rho} \\ 0 & 0 & \gamma p & v \end{pmatrix}. \tag{2}$$

We consider the flow for an ideal gas and therefore we have to take into account the equation of state:

$$p = e\,(\gamma - 1)\left(e - \frac{\rho}{2}(u^2 + v^2)\right).$$

Additionally we have to impose initial values and boundary conditions.

First order methods on unstructured grids

Let us consider a regular mesh of triangles T_j, $j = 1, ..., N$ of a polygonal domain Ω in $\mathbf{R}^2$. We are going to describe the methods for scalar conservation laws

$$\partial_t u + \partial_x f_1(u) + \partial_y f_2(u) = 0 \quad \text{in } \mathbf{R}^2 \times \mathbf{R}^+ \tag{3}$$

where f_1 and f_2 are arbitrary functions in $C^1(\mathbf{R}, \mathbf{R})$ satisfying $f_1(0) = 0$ $f_2(0) = 0$. We shall use the following notation (see Figure 1).

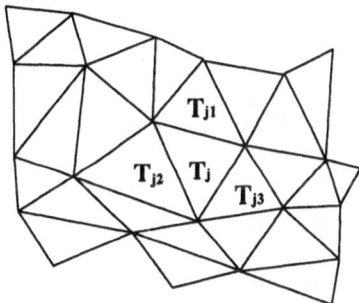

Figure 1

We describe the explicit version of the first order finite volume method. We assume that the values u^n for the time level $n \cdot \Delta t$ are given. The function u^n is assumed to be constant on each triangle with the value u_j^n on the triangle T_j. The values of u^n in the neighbouring triangles T_{j1}, T_{j2}, T_{j3} are denoted by $u_{j1}^n, u_{j2}^n, u_{j3}^n$. For this local situation we use the local numbering j_1, j_2, j_3. Then u_j^{n+1} in T_j is defined as

$$u_j^{n+1} = u_j^n - \frac{\Delta t}{|T_j|} \sum_{l=1}^3 g_{jl}(u_j^n, u_{jl}^n) \tag{4}$$

where g_{jl} is a numerical flux, satisfying

$$g_{jl}(u, u) = \nu_{jl} \begin{pmatrix} f_1(u) \\ f_2(u) \end{pmatrix}$$

$$g_{ji}(u, v) = -g_{ij}(v, u) \qquad (5)$$

$$|g_{jl}(u, v) - g_{jl}(u', v')| \le ch\left(|u - u'| + |v - v'|\right).$$

Here $j, i = 1, ..., N$ refer to global indices, and $l = 1, 2, 3$ to local ones; ν_{jl} denotes the scaled outer normal of the triangle T_j with respect to the edge l, this means $|\nu_{jl}| =$ length of the edge corresponding to l and $h := \max_j diam T_j$.

There are two famous examples for the numerical fluxes $g_{jl}(u, v)$ satisfying (5). The first one is the Lax-Friedrichs flux

$$g_{jl}(u, v) = \frac{1}{2}\nu_{jl}\left(f(u) + f(v)\right) - \frac{1}{2\mu}(v - u) \qquad (6)$$

where $f(u) = \begin{pmatrix} f_1(u) \\ f_2(u) \end{pmatrix}$ and the second one is the Engquist-Oster flux

$$g_{jl}(u, v) = f_{jl}^+(u) + f_{jl}^-(v) \qquad (7)$$

where $f_{jl}^+(s) := \int_0^s \max\left\{f_{jl}'(s), 0\right\}ds$ and $f_{jl}^-(s) := \int_0^s \min\left\{f_{jl}'(s), 0\right\}ds$. For these schemes we get the following results (see [8]).

Theorem 1 (Convergence) Let $u_0 \in L_{loc}^2(\mathbf{R}) \cap \mathbf{L}^\infty(\mathbf{R^2})$ and

$$u_j^0 := \frac{1}{|T_j|}\int_{T_j} u_0.$$

Let u_j^n be defined by (4) with g_{jl} as in (6) or (7). Then there exists a subsequence of $(u^n)_n$ such that

$$u^n \rightarrow u \qquad in \ L_{loc}^2(\Omega)$$

and u is the uniquely determined entropy solution of (3).

Remark 2 (see [3]) For systems we can still show that if $u^n \rightarrow v$ in $L_{loc}^1(\Omega)$, the limit v will be a weak solution of the system. This is a generalization of the well-known Lax-Wendroff-Theorem.

Remark 3 The local truncation error for the scheme (4) is less than or equal to one (see [3]). There are also some recent convergence results concerning higher order schemes. In [7] we haved proved convergence for the discontinuous Galerkin method as described in [1]. In this case instead of piecewise constant functions u^n we have piecewise linear ones.

Higher order methods on unstructured grids

Here we shall describe the fluctuation - distribution scheme [11], [12] which is based on piecewise linear and globally continuous ansatz-functions on the grid. There is still no convergence proof for this scheme but the numerical results are very convincing and this scheme is formally of higher order. Let us describe this algorithm first for linear scalar equations

$$\partial_t u + a\partial_x u + b\partial_y u = 0 \tag{8}$$

for constants $a, b \in \mathbf{R}$, then for nonlinear scalar conservation laws (3) and finally for nonlinear systems of conservation laws (1). We have used this scheme in combination with local mesh refinement and local mesh alignment. After describing the algorithm and the Kornhuber-Roitzsch alignment we shall show some results for numerical experiments concerning the system (1).

a) Fluctuation-distribution scheme for linear scalar conservation laws We shall use a notation which is slightly different from that one used in section 2. Let us assume that a regular triangulation $\mathcal{T}$ is given. The nodes of $\mathcal{T}$ are denoted by

$$P_j, \quad j = 1, ..., N.$$

For any $T \in \mathcal{T}$ let

$$P_T := \{j | P_j \in T\}$$

and ν_{Tj} be the scaled outer normal to ∂T with respect to the edge, opposite to P_j in the triangle T. For any $j = 1, ..., N$ let T_j be the set of neighbouring triangles of P_j.

$$T_j := \{T \in \mathcal{T} \mid P_j \in T\}.$$

Let u^0 denote the initial values, which are assumed to be linear on each $T \in \mathcal{T}$ and globally continuous. The value of u^0 in the jth-node P_j of $\mathcal{T}$ is denoted by u_j^0. Now we assume that u^n at time level is already given with values u_j^n in P_j, $j = 1, ..., N$ and that u^n is piecewise linear and globally continuous.

In order to describe the algorithm we need $\vec{a} := (a, b)^t$ and (see [11])

$$\Phi^T := \frac{1}{2} \sum_{j \in P_T} \vec{a}\nu_{Tj} u_j^n. \tag{9}$$

Then the algorithm is given by

$$u_j^{n+1} = u_j^n + \frac{\Delta t}{S_j} \sum_{T \in T_j} \alpha_j^T \Phi^T \tag{10}$$

where $S_j = \frac{1}{3}\sum_{T \in T_j} | T |$ (Veronoi-region) and α_j^T are upwind coefficients satisfying $\sum_{j \in P_T} \alpha_j^T = 1$. The explicit definition of the α_j^T is given in [11]. For instance for triangles with one inflow edge we have $\alpha_{j_1}^T = 0$, $\alpha_{j_2}^T = 1$ and $\alpha_{j_3}^T = 0$ (see Figure 2).

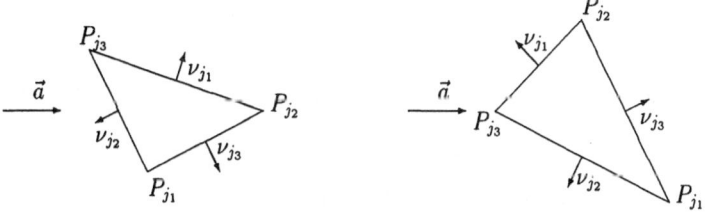

Figure 2   Figure 3

If T has two inflow edges as in Figure 3 we have $\alpha_{j_1}^T = k_1(u_{j_1} - u_{j_3})$, $\alpha_{j_2}^T = k_2(u_{j_2} - u_{j_3})$, $\alpha_{j_3}^T = 0$ where

$$k_l := -\frac{\nu_{j_l} \vec{a}}{2\Phi^T} .$$ (11)

b) Fluctuation distribution scheme for nonlinear scalar conservation laws
Now let us describe the generalization of (10) to nonlinear conservation laws as in (3). For any $T \in \mathcal{T}$ let

$$Q_T^x(f) := \sum_{j \in P_T} \nu_{Tj}^x f(u_j), \quad \text{and} \quad \vec{a}^T := (a^T, b^T)^t$$ (12)

where

$$a^T := \frac{Q_T^x(f_1)}{Q_T^x(id)}, \quad b^T := \frac{Q_T^y(f_2)}{Q_T^y(id)} .$$ (13)

Then the algorithm is given as before by

$$u_j^{n+1} = u_j^n + \frac{\Delta t}{S_j} \sum_{T \in \mathcal{T}_j} \alpha_j^T \Phi^T .$$ (14)

where

$$\Phi^T := \frac{1}{2} \sum_{j \in P_T} \vec{a}^T \nu_{Tj} u_j^n .$$ (15)

The α_j^T are chosen in the same way as in the linear case a).

c) Fluctuation distribution scheme for the nonlinear Euler equations(see [11], [12]). Now we shall describe the discretization which we have used for (1). Let us assume that the U_j^n are already given in the nodes P_j. Now we fix one j. Let $T \in \mathcal{T}_j$ be a triangle which contains P_j. All the quantities which we shall define now will depend on T and n. For $U := (\rho, u, v, p)^t$ we introduce the Roe parameter vector

$$z(U) := \sqrt{\rho}(1, u, v, H)^t$$ (16)

where $H := (e + p)/\rho$. The main property of this parameter vector z is that the conservative variables are polynomials of z of degree two. Now we define z^n and that is the main part of this scheme, such that z^n varies linearly over each triangle T and $z^n(P_j) := z(U_j^n)$. Let

$$\bar{z} := \frac{1}{3} \sum_{l \in P_T} \begin{pmatrix} \sqrt{\rho_l^n} \\ u_l^n \sqrt{\rho_l^n} \\ v_l^n \sqrt{\rho_l^n} \\ H_l^n \sqrt{\rho_l^n} \end{pmatrix} .$$ (17)

Now we consider U as a function of z and define on T

$$\bar{A} := A(U(\bar{z})), \quad \bar{B} := B(U(\bar{z})), \quad G := (\partial_z U)(\bar{z}) \nabla z^n .$$ (18)

Then instead of (1) we consider the locally linearized system

$$\partial_t U + \bar{A} \partial_x U + \bar{B} \partial_y U = 0 \text{ in } \mathbf{R}^2 \times \mathbf{R}^+$$ (19)

on T where $\bar{A}, \bar{B} \in \mathbf{R}^{(4,4)}$. The eigenvectors and eigenvalues of $\bar{A}cos\Theta + \bar{B}sin\Theta$ are denoted by $r^k = r_T^k(\Theta)$ and $\lambda^k = \lambda_T^k(\Theta)$, $k = 1, ..., 4$ respectively. Then there are some $\beta_k^T \in \mathbf{R}$, $k = 1, ..., 6$ which are uniquely defined by

$$G = \sum_{k=1}^{4} \beta_k^T p(\Theta_k) r_T^1(\Theta_k) + \beta_5^T p(\psi) r_T^4(\psi) + \beta_6^T p(\phi) r_T^3(\phi) \tag{20}$$

where $p(\Theta) := (cos\Theta, sin\Theta)^t$ and $\Theta_1 = \Theta$, $\Theta_2 = \Theta + \pi$, $\Theta_3 = \Theta + \frac{\pi}{2}$, $\Theta_4 = \Theta + \frac{3\pi}{2}$, ψ is chosen in the direction of the pressure gradient. This is a nonlinear system of eight equations and eight unknowns $\beta_1^T, \beta_2^T, ..., \beta_6^T, \Theta, \phi$. In [9] it is shown that this system is uniquely solvable for the linearized Euler equations (19). In the following we use the notation $\Theta_5 = \psi, \Theta_6 = \phi, \tilde{r}_T^1 = r_T^1(\Theta_1), ..., \tilde{r}_T^4 = r_T^1(\Theta_4), \tilde{r}_T^5 = r_T^4(\Theta_5), \tilde{r}_T^6 = r_T^3(\Theta_6), \tilde{\lambda}^1 = \lambda^1(\Theta_1), ..., \tilde{\lambda}^4 = \lambda^1(\Theta_4), \tilde{\lambda}^5 = \lambda^4(\Theta_5), \tilde{\lambda}^6 = \lambda^3(\Theta_6)$. We define for $k = 1, ..., 6$

$$w^k(x, y, t) := \beta_k^T (xcos\Theta_k + ysin\Theta_k - \tilde{\lambda}^k t) + C_k \quad w := \sum_{k=1}^{6} w^k \tilde{r}_T^k \tag{21}$$

for some constant C_k. Then we obtain

$$\nabla w = G \tag{22}$$

and w^k satisfies the linear scalar conservation law

$$\partial_t w^k + \tilde{\lambda}^k cos\Theta_k \partial_x w^k + \tilde{\lambda}^k sin\Theta_k \partial_y w^k = 0 \tag{23}$$
$$w^k(x, y, 0) = \beta_k^T (xcos\Theta_k + ysin\Theta_k) + C_k \tag{24}$$

for some constants C_k . We shall see that the algorithm will not depend on C_k. Now formally we solve this linear equation for all $T \in \mathcal{T}_j$ using (10) with $\vec{a}_k^T := \tilde{\lambda}^k(cos\Theta_k, sin\Theta_k)$ and

$$\Phi^{kT} := \frac{1}{2} \sum_{l \in P_T} \vec{a}_k^T \nu_{Tl} w^k(x_l, y_l, 0) \tag{25}$$

where $(x_l, y_l) := P_l$. Therefore Φ^{kT} is independent of the choice of C_k. Then the algorithm is given by

$$U_j^{n+1} := U_j^n + \frac{\Delta t}{S_j} \sum_{k=1}^{6} \sum_{T \in \mathcal{T}_j} \alpha_j^{kT} \Phi^{kT} \tilde{r}_T^k \tag{26}$$

where α_j^{kT} is defined as above.

Remark 4 For the $\alpha_j^{kT} \Phi^{kT} \tilde{r}_T^k$ we also get the following expression. Let $n_T = \frac{\nu_T}{|\nu_T|}$.

$$-\alpha_j^{kT} \Phi^{kT} \tilde{r}_T^k = \alpha_j^{kT} \int_{\partial T} \vec{a}_k^T n_T w^k(x, y, 0) \, dS(x, y) \tilde{r}_T^k \tag{27}$$

$$= \alpha_j^{kT} \int_T \vec{a}_k^T \nabla w^k(x, y, 0) \, dx \, dy \tilde{r}_T^k \tag{28}$$

$$= \alpha_j^{kT} \int_T \tilde{\lambda}^k p(\Theta_k) p(\Theta_k) \beta_k^T \, dx \, dy \tilde{r}_T^k \tag{29}$$

$$= \alpha_j^{kT} \beta_k^T \tilde{\lambda}^k |T| \tilde{r}_T^k \ . \tag{30}$$

If we choose the α_j^{kT} as in a) we obtain the following expressions. For the case of one inflow side (Figure 2) we have

$$\alpha_{j_1}^{kT}\Phi^{kT}\tilde{r}_T^k = 0, \ \alpha_{j_2}^{kT}\Phi^{kT}\tilde{r}_T^k = \beta_k^T\tilde{\lambda}^k|T|\tilde{r}_T^k, \ \alpha_{j_3}^{kT}\Phi^{kT}\tilde{r}_T^k = 0 \tag{31}$$

and for the case of two inflow sides (Figure 3)

$$\alpha_{j_1}^{kT}\Phi^{kT}\tilde{r}_T^k = -\frac{\nu_{Tj_1}\vec{a}_k^T\left(w^k(x_{j_1},y_{j_1},0) - w^k(x_{j_3},y_{j_3},0)\right)}{2\,\Phi^{kT}}\Phi^{kT}\tilde{r}_T^k \tag{32}$$

$$= -\frac{1}{2}\nu_{Tj_1}\vec{a}_k^T\left(w^k(x_{j_1},y_{j_1},0) - w^k(x_{j_3},y_{j_3},0)\right)\tilde{r}_T^k \tag{33}$$

$$= -\frac{1}{2}\beta_k^T\tilde{\lambda}^k\tilde{r}_T^k\,\nu_{Tj_1}p(\Theta_k)\,(P_{j_1} - P_{j_3})p(\Theta_k) \tag{34}$$

$$\alpha_{j_2}^{kT}\Phi^{kT}\tilde{r}_T^k = -\frac{1}{2}\beta_k^T\tilde{\lambda}^k\tilde{r}_T^k\,\nu_{Tj_2}p(\Theta_k)\,(P_{j_2} - P_{j_3})p(\Theta_k) \tag{35}$$

$$\alpha_{j_3}^{kT}\Phi^{kT}\tilde{r}_T^k = 0\,. \tag{36}$$

Local grid alignment

Now we shall describe the local grid alignment, which is based on the idea of Kornhuber and Roitzsch [5]. The algorithm consists of the following steps. First let us describe the alignment of the edges.

A 1) We construct a macro triangulation T^0 of Ω.

A 2) We construct a globally uniform refined mesh T^1 by refining the macro triangulation T^0.

A 3) We compute the discrete solution u_h on T^1 using the algorithm as described in c).

A 4) For all edges e of all triangles $T \in T^1$ we compute the directional derivative $D(T,e) = \frac{|m(P_i) - m(P_j)|}{|P_i - P_j|}$ of the Mach number m along the edge e . Here e is defined by its endpoints P_j, P_i.

A 5) We fix some threshold $\alpha > 0$ and we select the set T^* of all triangles for which we have an edge e such that $|D(T,e)| \geq \alpha$.

A 6) Now we consider successively all $T \in T^*$ and find the edge e_T of T such that

$$|D(T,e_T)| = min\{|D(T,e)| \,|\, e \text{ is edge of T}\}. \tag{37}$$

A 7) Consider the local situation as in Figure 4 and let $T_1, T_2 \in T^*$. Without restriction we assume $e_{T_1} := \overline{P_1P_2}$, $e_{T_2} := \overline{Q_1Q_2}$.

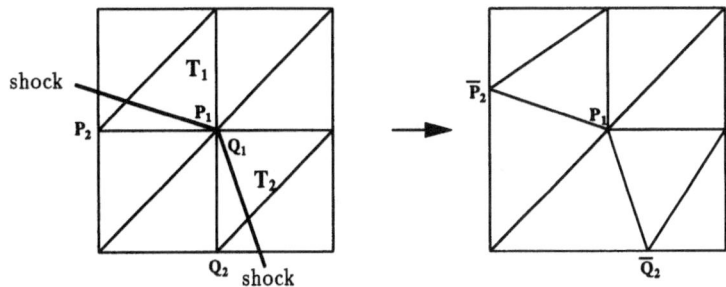

Figure 4

First case: $e_{T_1} \cap e_{T_2} \neq \emptyset$. Without restriction let $e_{T_1} \cap e_{T_2} = \{P_1\}$. Then we move P_2 and Q_2 to $\bar{P}_2$ and $\bar{Q}_2$ such that $\overline{P_1 \bar{P}_2}$ and $\overline{P_1 \bar{Q}_2}$ are nearly parallel to the front of the discontinuity. Then choose another pair of triangles from T^* .

Second case: $e_{T_1} \cap e_{T_2} = \emptyset$. Then choose another pair of triangles from T^*.
A 8) Goto A2)

In the examples this type of refinement has been repeated two or three times. Now let us describe the blue refinement and where we apply it.

B 1) We use again the refinement obtained from A on the whole triangulation. **B 2)** as A3, **B 3)** as A4, **B 4)** as A5, **B 5)** as A6.
B 6) We fix a pair T_1, T_2 of triangles.

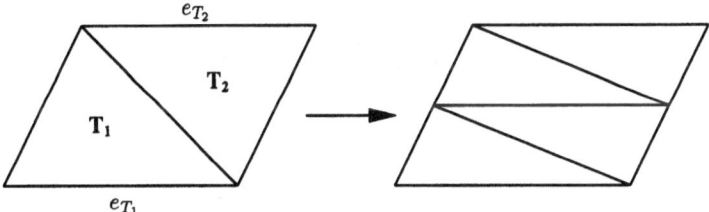

Figure 5, blue refinement

If $e_{T_1} \cap e_{T_2} \neq \emptyset$ there is no blue refinement (see Figure 5) and we choose another pair of triangles.
B 7) If
a) $e_{T_1} \cap e_{T_2} = \emptyset$ *and*
b) e_{T_1}, e_{T_2} "nearly" parallel *and*
c) an "angle condition" (see below) is satisfied then we apply the blue refinement, otherwise we choose another pair of triangles.
The angle condition means that (see Figure 6) if α, β are the largest angles in T_1 and T_2 respectively , then γ, δ are the smallest ones in T_1 and T_2 respectively.

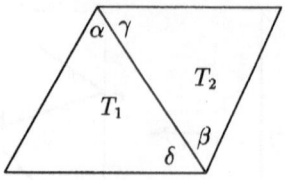

Figure 6

This refinement algorithm B is also repeated up to two or three times.

Numerical results

The first two examples are similar to those in the paper of Kornhuber, Roitzsch [5] but the algorithms are completely different. Here we use the fluctuation- distribution scheme as described before. We consider the linear scalar conservation law

$$\partial_t u + \partial_x u + \partial_y u = 0 \quad \text{in} \quad \Omega := [0,1] \times [0,1] \tag{38}$$

with initial values

$$
\begin{aligned}
u(x,y,0) &= 1 \quad \text{if} \quad y = 0 \tag{39}\\
u(x,y,0) &= 0 \quad \text{otherwise} \tag{40}
\end{aligned}
$$

and boundary conditions on the inflow part of the boundary

$$
\begin{aligned}
u(x,y,t) &= 1 \quad \text{if} \quad y = 0 \tag{41}\\
u(x,y,t) &= 0 \quad \text{otherwise.} \tag{42}
\end{aligned}
$$

The exact stationary solution is given by a straight diagonal line which separates two constant states. The numerical results are shown in Figure 7 on a grid which is not aligned and on a grid which is aligned with the shock respectively. The resolution of the shock on the aligned grid is much better than for the other one. Therefore this example shows us, that it is necessary to align the triangles in an unstructured grid with the main structures of the solutions.

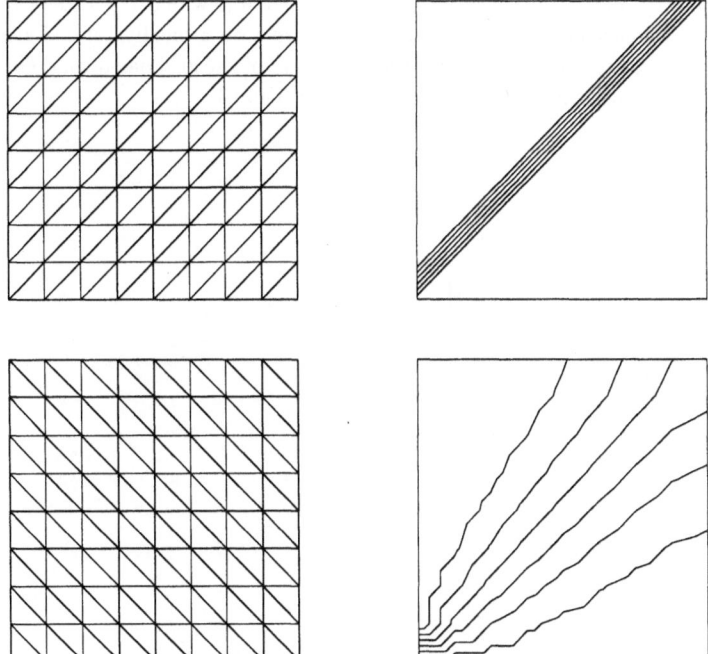

Figure 7

The second testproblem is to solve

$$\partial_t u + y\partial_x u - x\partial_y u = 0 \quad \text{in} \quad \Omega := [0,1] \times [0,1] \tag{43}$$

with initial values

$$u(x,y,0) = 1 \quad \text{if} \quad x = 0, \ y \leq \frac{2}{3} \tag{44}$$

$$u(x,y,0) = 0 \quad \text{otherwise} \tag{45}$$

and the corresponding boundary conditions on the inflow part of the boundary

$$u(x,y,t) = 1 \quad \text{if} \quad x = 0, \ y \leq \frac{2}{3} \tag{46}$$

$$u(x,y,t) = 0 \quad \text{otherwise} \ . \tag{47}$$

The exact stationary solution is given by a quarter of a circle which separates two constant states. The numerical results are shown in Figure 8 on a locally refined grid and on an aligned grid respectively. In this case the alignment of the grid has been done automatically by using the algorithm as described in section 3 and 4.

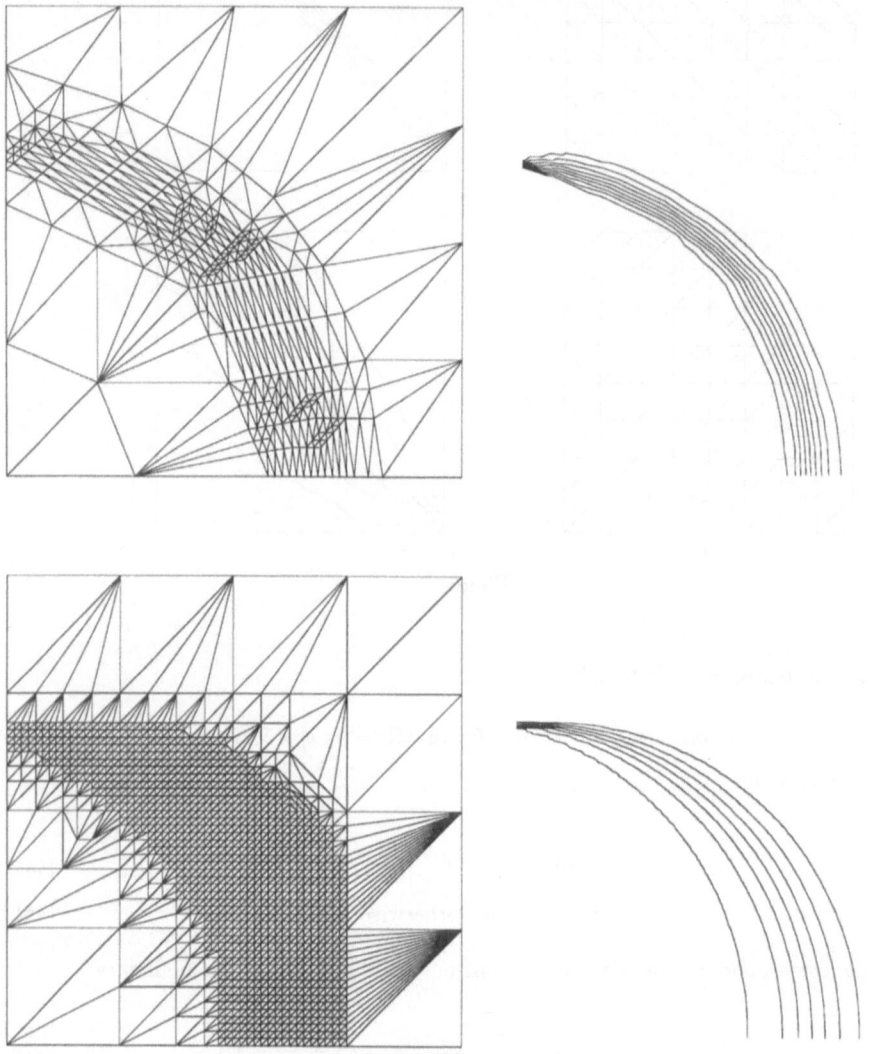

Figure 8

The third and the fourth testproblem concerns the system of the Euler equations (1). In the third one we consider the shock reflection problem in a tube (see Figure 9). The exact stationary solution concerning the density and the Mach number is given by a straight line which is reflected on the bottom of the tube and which separates constant states. The numerical calculation were performed on a mesh with

1226 nodes. In Figure 9 we can see the results concerning the density and the Mach number for the locally refined (Figure 9) and the aligned grid (see Figure 10) respectively. The resolution of the shock again is much better on the aligned grid.

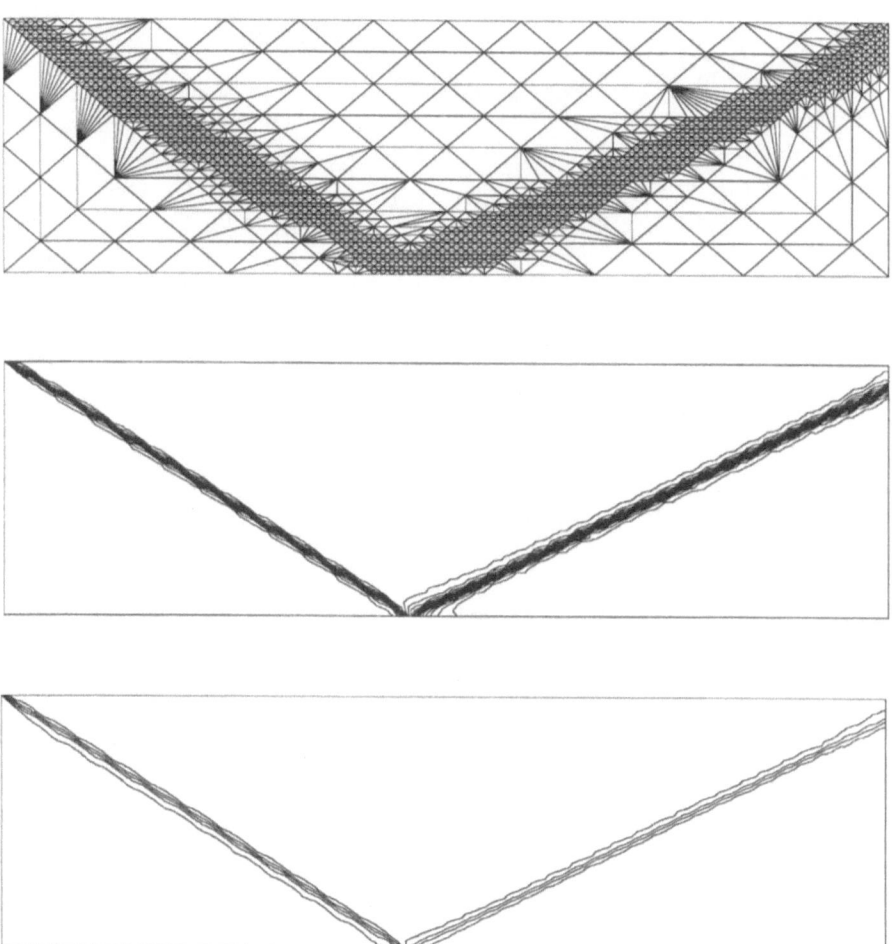

Figure 9

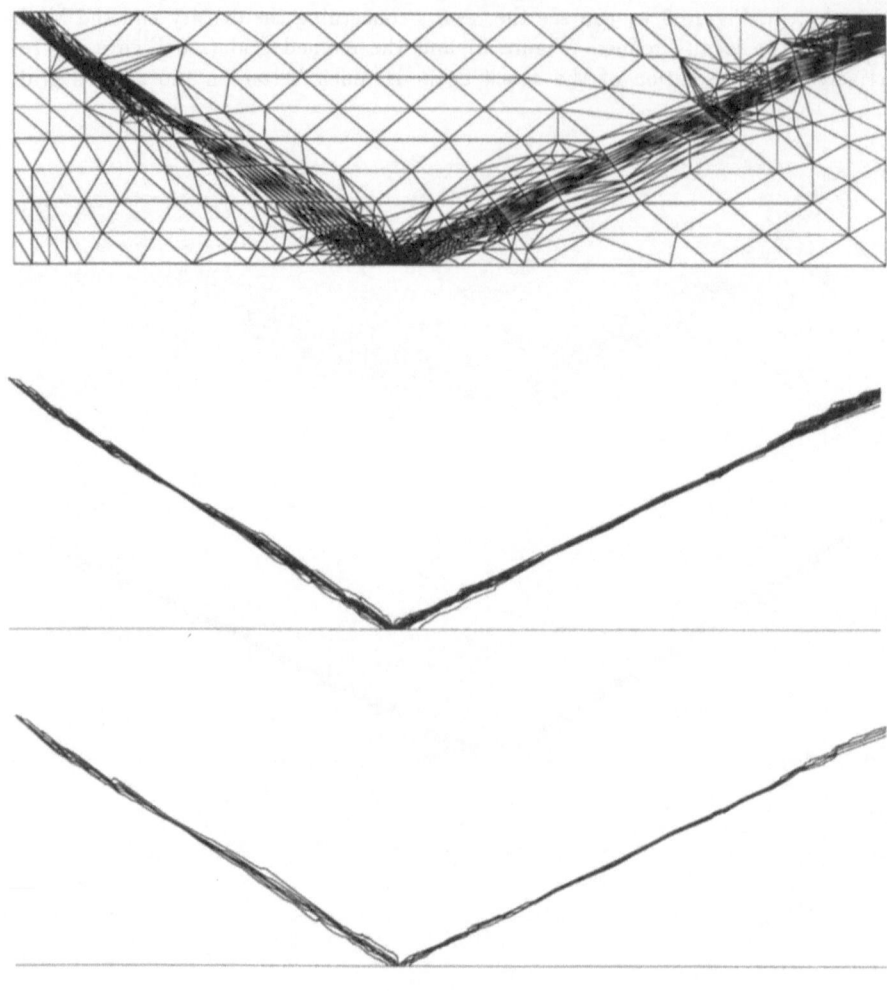

Figure 10

In the fourth example we have treated the supersonic flow around a cylinder. The results on the locally refined and locally aligned grid are shown in Figure 11, 12 and 13, 14 respectively. In Figure 11 there is the locally refined mesh with 2226 nodes and in Figure 12 the level lines of the corresponding Mach number. Also for this test problem we get some improvement of the shock resolution if we use a locally aligned grid (see Figure 13 (2436 nodes) and Figure 14) but the results can still be improved.

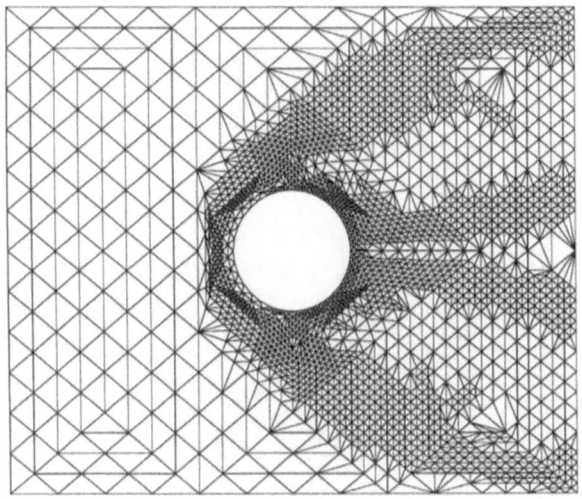

Figure 11

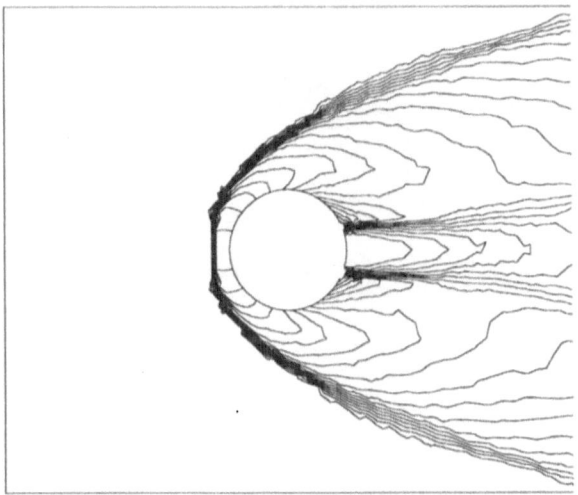

Figure 12

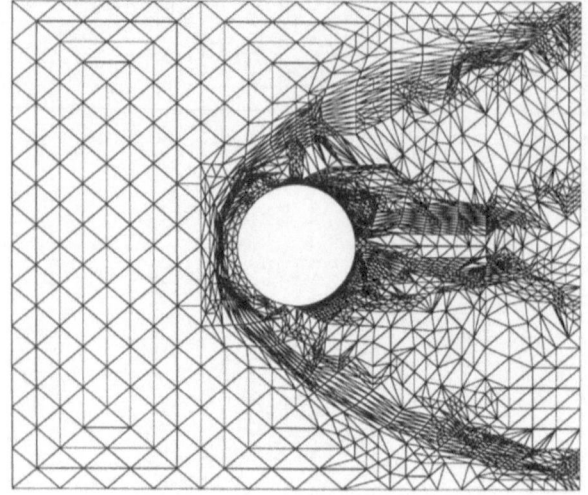

Figure 13

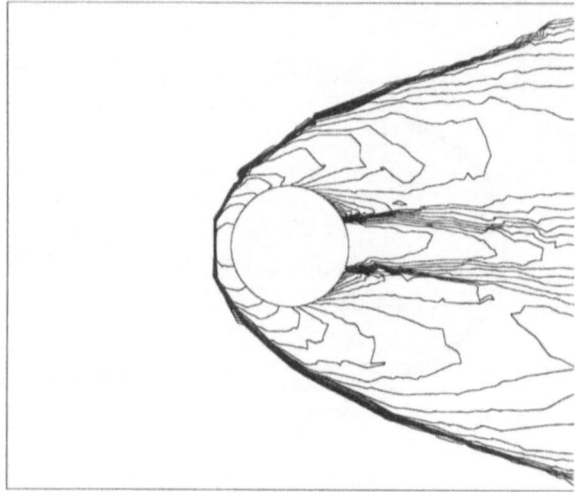

Figure 14

For the visualization of the numerical data the GRAPE system has been used [10], [13]. This graphic system has been developed at the Sonderforschungsbereich 256 of the Universität Bonn and at the Universität Freiburg.

References

[1] Cockburn,B., Shu, C.W.: TVB Runge-Kutta projection discontinuous Galerkin finite element method for conservation laws II: general framework. Math. Comp. 52 (1989),411-435.

[2] Cockburn,B. Coquel,F., LeFloch, P., Shu, C.W.: Convergence of finite volume methods. IMA Preprint #771 (1991).

[3] Geiben, M., Kröner, D., Rokyta, M.: A Lax-Wendroff type theorem for cell-centered, finite volume schemes in 2-D. Preprint 1993.

[4] Johnson, C., Szepessy, A.: Convergence of a finite element method for a nonlinear hyperbolic conservation law. Math. Comput. 49 (1987), 427-444.

[5] Kornhuber, R., Roitzsch, R.: Adaptive Finite-Element Methoden für konvektions-dominierte Randwertprobleme bei partiellen Differentialgleichungen. Konrad-Zuse-Zentrum, Preprint SC 88-9, Berlin 1988.

[6] Kröner, D.: Directionally adapted upwind schemes in 2-D for the Euler equations. Priority Research Program, Results 1986-1988. Notes on numerical fluid mechanics II., Vol.25.

[7] Kröner, D., Noelle, S., Rokyta, M.: Convergence of higher order upwind finite volume schemes on unstructured grids for scalar conservation laws in two space dimensions. Preprint 268, SFB256, Bonn 1993, eingereicht bei Num. Math.

[8] Kröner, D., Rokyta, M.: Convergence of upwind finite volume schemes for scalar conservation laws in 2-D. Preprint SFB 208 (1993), Bonn ; to appear in SIAM J. of Num. Anal.

[9] Roe, P.: Discrete models for the numerical analysis of time-dependent multi-dimensional gas dynamics. J. Comput. Phys., 63 (1986)

[10] Rumpf M., Schmidt A.: GRAPE, GRAphics Programming Environment, Report 8, SFB 256, Bonn, 1990.

[11] Struijs, R., Deconinck, H.: Multidimensional upwind schemes for the Euler equations using fluctuation distribution. VKI, 1989/90.

[12] Struijs, R., Deconinck, H., de Palma, P.: Progress on multidimensional upwind schemes for unstructured grids. AIAA paper 91-1550, 1991.

[13] Wierse A., Rumpf M.: GRAPE, Eine objektorientierte Visualisierungs- und Numerikplattform. Informatik Forschung und Entwicklung 7, 145-151, 1992.

A New Algorithm for Multi-Dimensional Adaptive Numerical Quadrature

Thomas Bonk
Institut für Informatik
Technische Universität München
Arcisstraße 21, D-80290 München, Germany

Summary

We present an algorithm for multi-dimensional quadrature that is adaptive both in the refinement of the subdomains and in the order. The method is based on an extrapolation technique using sparse grids, reducing the exponential cost explosion for increasing dimension d. The basic concepts will be introduced as a combination and generalization of the Archimedes Rule and the multi-variate extrapolation. We describe the data structures and show the asymptotic superiority of the algorithm by numerical results in comparison to a traditional algorithm.

1 Introduction

It is known from complexity theory that the problem of numerical quadrature becomes intractable in practice for high dimension [3]. The cost is

$$C = O\left(\varepsilon^{-\frac{d}{g}}\right), \tag{1}$$

so that the cost C grows exponentially with the dimension d if an accuracy ε is to be achieved. On the other hand, the cost of the problem drop, if the integrand is smooth to a degree g. So good quadrature algorithms have to exploit the smoothness of the function and use two adaptivity strategies. The first is adaptivity in the refinement of the subdomains so that regions will be partitioned into smaller ones, where local smoothness can be exploited. The second is adaptivity in the order that means that the algorithm rises the order in those subdomains, where the cost can be reduced best exploiting the smoothness of the function. It is especially desirable that the exploited smoothness grows with the dimension of the problem.

Usually smoothness is defined by

$$\left|\frac{\partial^s u}{\partial x_1^{j_1} \ldots \partial x_d^{j_d}}\right| \leq c \quad \text{where} \quad j_1 + j_2 + \ldots + j_d = s \leq g. \tag{2}$$

This means that *all* mixed derivatives of a given order of the function u have to be less than a constant c. Our algorithms need a weaker definition of smoothness, namely

$$\left|\frac{\partial^s u}{\partial x_1^{j_1} \ldots \partial x_d^{j_d}}\right| \leq c \quad \text{where} \quad j_i \leq \frac{g}{d} =: r, \ i = 1, \ldots, d. \tag{3}$$

To see the difference between this definitions, consider a multi-dimensional function composed of one-dimensional C^1 functions:

$$u(x_1,\ldots,x_d) = f_1(x_1) + \ldots + f_d(x_d) \quad \text{with} \quad f_i \in C^1.$$

$u(x_1,\ldots,x_d)$ satisfies (2) only for $g = 1$ whereas (3) is satisfied for $g = d$. So an algorithm requiring (2) may converge with much weaker rate of convergence than an algorithm using only (3). From this special example one can learn that with definition (3) the smoothness of the integrand grows proportional to the dimension.

Using (3) for our algorithms it can be shown that the cost grows as:

$$C = O\!\left(\varepsilon^{-\frac{1}{r} - \delta}\right) \quad \text{for every} \quad \delta > 0.$$

Therefore, our algorithms are only slightly worse than the theoretical bound (1).

In the following sections, we present several multi-dimensional quadrature algorithms. The first is adaptive approximatively of order $O(h^2)$. It goes back to the quadrature method used by Archimedes. The second one is nonadaptive and of higher order than $O(h^2)$, which will be achieved by extrapolation. The third one combines the advantages of the other ones and is both adaptive and of high order.

2   Archimedes Rule in the Multi-Dimensional Case

First we focus on the one-dimensional case. We want to compute the integral

$$F_1\!\left(f(\underline{x}), a, b\right) := \int_a^b f(\underline{x}) \, d\underline{x},$$

as it is shown in figure 1 (left) (In this section independent variables are underlined to enhance readability). We divide this area in two disjoint parts, a trapezoid T_1 and a function segment S_1 ranging from a to b and obtain

$$F_1\!\left(f(\underline{x}), a, b\right) = \left(f(a) + f(b)\right)\frac{b-a}{2} + S_1\!\left(f(\underline{x}), a, b\right). \tag{4}$$

As figure 1 (right) shows the remaining function segment S_1 can be divided in a triangle D_1 and two smaller function segments S_1' and S_1''

$$\begin{aligned}
S_1\!\left(f(\underline{x}), a, b\right) = {} & \left(f\!\left(\frac{a+b}{2}\right) - \frac{1}{2}\!\left(f(a) + f(b)\right)\right)\frac{b-a}{2} \\
& + S_1\!\left(f(\underline{x}), a, \frac{a+b}{2}\right) \\
& + S_1\!\left(f(\underline{x}), \frac{a+b}{2}, b\right).
\end{aligned} \tag{5}$$

(5) can be applied recursively to the remaining function segments S_1' and S_1'' so that we get four function segments, which can be divided again using (5) and so on. By this means the area to be computed will be approximated by triangles. This principle is known since the Greek time and is called according to its inventor the *Archimedes Rule*.

Both equations (4) and (5) can be used to specify an adaptive algorithm for numerical quadrature of one-dimensional functions. Now it is necessary to define a termination

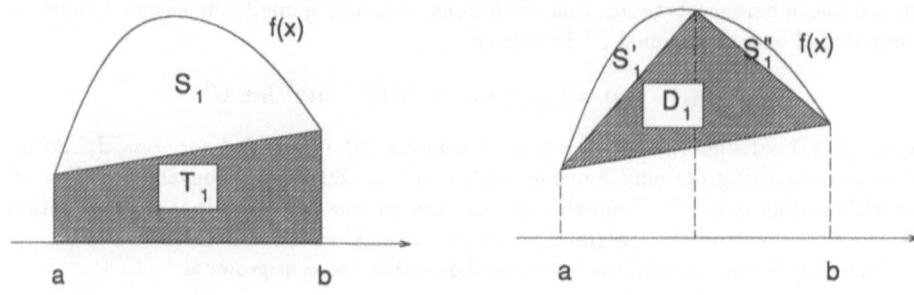

Figure 1: Partitioning a quadrature problem in trapezoidal and triangular areas

criterion for the second equation. For example we may terminate the recursion, if the area of the triangle is less than a given ε. With this enhancement we obtain a variant of the adaptive trapezoidal rule.

If we want to extend this algorithm for the two-dimensional case, we have to use the *Principle of Cavalieri*. This principle applied to our problem says that the volume to be computed can be cut into thin slices of thickness h. The integral can be computed approximatively as the product of total slice area and thickness h. For $h \to 0$ we obtain the exact integral.

Figure 2 (left) shows three slices for $\underline{x}_1 \in \{a_1, \frac{1}{2}(a_1 + b_1), b_1\}$. Since each slice can be treated as a one-dimensional integration problem, we divide each in a trapezoid and a function segment. All theses trapezoids and function segments can be summed up separately over $\underline{x}_1$ from a_1 to b_1 and yield two volumes. We get:

$$
\begin{aligned}
F_2\Big(f(\underline{x}_1, \underline{x}_2), a_1, b_1, a_2, b_2\Big) \ = \ & F_1\Big(\big(f(\underline{x}_1, a_2) + f(\underline{x}_1, b_2)\big)\frac{b_2 - a_2}{2}, a_1, b_1\Big) \\
+ \ & S_2\Big(f(\underline{x}_1, \underline{x}_2), a_1, b_1, a_2, b_2\Big).
\end{aligned}
\tag{6}
$$

As in the one-dimensional case the function segments can be split in a triangle and two smaller function segments (figure 2 right). All these areas can be summed over $\underline{x}_1$ and we get the following equation:

$$
\begin{aligned}
S_2\Big(f(\underline{x}_1, \underline{x}_2), a_1, b_1, a_2, b_2\Big) = & \\
& F_1\Big(\big(f\big(\underline{x}_1, \frac{a_2 + b_2}{2}\big) - \frac{1}{2}\big(f(\underline{x}_1, a_2) + f(\underline{x}_1, b_2)\big)\big)\frac{b_2 - a_2}{2}, a_1, b_1\Big) \\
+ \ & S_2\Big(f(\underline{x}_1, \underline{x}_2), a_1, b_1, a_2, \frac{a_2 + b_2}{2}\Big) \\
+ \ & S_2\Big(f(\underline{x}_1, \underline{x}_2), a_1, b_1, \frac{a_2 + b_2}{2}, b_2\Big).
\end{aligned}
\tag{7}
$$

All equations (4) to (7) can be transformed easily to a C or PASCAL program. It is only necessary to take care for the termination of the recursive definitions of (5) and

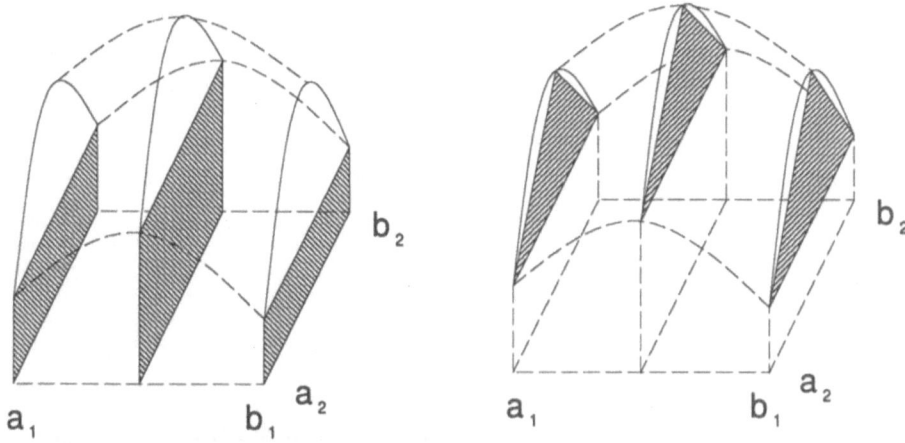

Figure 2: Two-dimensional quadrature based on the principle of Cavalieri

(7), for example by defining:

$$S_2\big(f(\underline{x}_1, \underline{x}_2), a_1, b_1, a_2, b_2\big) = 0$$
$$\text{if } \left| F_1\left(\big(f(\underline{x}_1, a_2) + f(\underline{x}_1, b_2)\big)\frac{b_2 - a_2}{2}, a_1, b_1\right) \right| \le \varepsilon$$
$$S_1(f(\underline{x}), a, b) = 0$$
$$\text{if } \left| \big(f(a) + f(b)\big)\frac{b - a}{2} \right| \le \varepsilon.$$

(8)

Recursive application of the principle of Cavalieri allows the derivation of d-dimensional quadrature algorithms ($d > 2$). The underlying formulas for F_d and S_d are similar to (6) and (7) and thus can be traced back to F_{d-1} and S_{d-1} and so on.

We apply this algorithm to a first example function

Example 1:
$$f(x_1, x_2) = x_1^2 \, x_2^2 \quad \text{on} \quad (x_1, x_2) \in [-1, 1]^2$$

and choose $\varepsilon = 10^{-3}$. Figure 3 (left) shows the grid points that will be evaluated. As we see, this is *not* the grid generated by an adaptive trapezoidal rule.

The grid of figure 3 (left) is a *sparse grid* [5]. Sparse grids have some interesting advantages over regular grids. First, the cost is significantly less if a given border step width h is required: in the d-dimensional case, a regular grid needs $O(h^{-d})$ points while a sparse grid needs only $O(h^{-1}(ld(h^{-1}))^{d-1})$ grid points. The loss of accuracy by sparse grids is comparatively small: it is known that the approximation error is of order $O(h^2)$ for regular grids. For sparse grids Bungartz [1] proved that the approximation error is $O(h^2(ld(h^{-1}))^{d-1})$, i.e. compared with regular grids the error is slightly increased by a logarithmic factor. This shows that sparse grids are much more suitable for the approximation of multi-dimensional functions, than regular grids are. Proofs for the order of convergence for numerical quadrature shall be given in a separate paper.

57

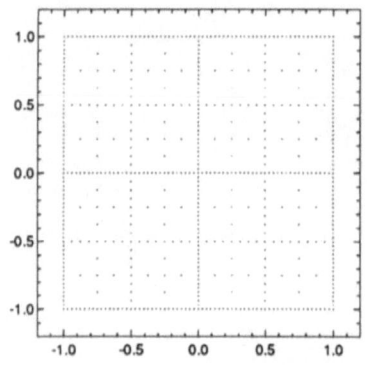 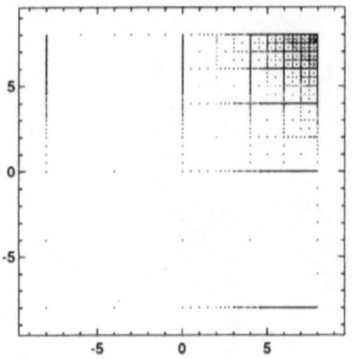

Figure 3: Archimedes Rule: the grids evaluated for example 1 (left) and 2 (right)

At this point it is to be emphasized that the grid of figure 3 (left) is the result of the algorithm derived from the equations (4) to (8). Especially no explicit means have been undertaken to evaluate just a sparse grid.

In a second example we want to demonstrate the adaptivity properties of the algorithm. We apply it to the function

Example 2:

$$f(x_1, x_2) = e^{x_1 + x_2} \quad \text{on} \quad (x_1, x_2) \in [-8, 8]^2.$$

Figure 3 (right) shows the grid for $\varepsilon = 10^3$ (corresponding to an accuracy of two decimal digits). First it can be seen that the algorithm evaluates the function on a denser grid in the nonsmooth regions (the function has a peak in the upper right corner). Second it can be observed that the sparse grid structure is preserved locally.

3 Quadrature of High Order

The *Euler-MacLaurin Summation Formula* gives the relation between a trapezoidal sum $T(h)$ and the exact integral τ_0 of a function $f \in C^{2m+2}[0,1]$

$$\tau_0 := \int_0^1 f(x)dx$$

$$T(h) := \frac{1}{2}f(0) + \sum_{i=1}^{h^{-1}-1} f(ih) + \frac{1}{2}f(1)$$

$$T(h) = \tau_0 + \tau_1 h^2 + \tau_2 h^4 + \ldots + \tau_m h^{2m} + O(h^{2m+2}), \tag{9}$$

where $\tau_j, j = 0, \ldots, m$ do not depend on h. As introduced in [2] the extrapolation technique exploits the existence of (9). Starting with a sequence of step widths $H = \{h_0, h_1, \ldots h_m\}$, the resulting trapezoidal sums $T(h_i), i = 0, \ldots, m$ are combined linearly with such weights $g_{m,i}$ that the high order error terms $\tau_i h^{2i}, i = 1, \ldots, m$ vanish

$$\hat{T}_m := \sum_{i=0}^{m} g_{m,i} \, T(h_i) = \tau_0 + O'(h^{2m+2}).$$

Introducing our quadrature technique, we start with a sequence of midpoint sums

$$M_j := M(2^{-j}) = \sum_{k=1}^{2^j} f\left(\left(i - \frac{1}{2}\right) 2^{-j}\right)$$

of geometrically decreasing step width. There also exists an expansion like (9). Let $\hat{M}_m$ be the extrapolated approximation of the integral based on $M_j, j = 0, \ldots, m$

$$\hat{M}_m = \sum_{j=0}^{m} c_{m,j} \, M_j = \tau_0 + O_m(h^{2m+2}) \text{ with } c_{m,j} = \prod_{\rho=0, \rho \neq j}^{m} \frac{4^\rho}{4^\rho - 4^j}.$$

For the multi-dimensional case we use contrary to traditional methods not an univariate but a multi-variate approach. We define $M_{i_1, \ldots, i_d} = M(2^{-i_1}, \ldots, 2^{-i_d})$ as the midpoint sum with step width 2^{-i_j} in the coordinate direction $x_j, j = 1, \ldots, d$.

Moreover we define a set of multi indices $\mathcal{I} := \{(i_1, \ldots, i_d) | \ i_j \geq 0, j = 1, \ldots, d\}$ convenient for our purposes, if it satisfies

$$(i_1, \ldots, i_j, \ldots, i_d) \in \mathcal{I} \Rightarrow (i_1, \ldots, i_j - 1, \ldots, i_d) \in \mathcal{I} \text{ for } i_j \neq 0, \ 1 \leq j \leq d. \quad (10)$$

Starting from such a set $\mathcal{I}$, we define a scheme $M_{\mathcal{I}}$ of midpoint sums as:

$$M_{\mathcal{I}} = \left\{ M_{i_1, \ldots, i_d} | \ (i_1, \ldots, i_d) \in \mathcal{I} \right\}.$$

We use this scheme $M_{\mathcal{I}}$ for the computation of an extrapolation in coordinate direction x_j using $i_j + 1$ values at the index position $(i_1, \ldots, i_d) \in \mathcal{I}$

$$\hat{M}_{i_1, \ldots, i_d}^{(j)} = \sum_{k=0}^{i_j} c_{i_j, k} \, M_{i_1, \ldots, i_{j-1}, k, i_{j+1}, \ldots, i_d} =: \Lambda_{i_1, \ldots, i_d}^{(j)} \circ M_{\mathcal{I}}, \ (i_1, \ldots, i_d) \in \mathcal{I}.$$

Applying this to all multi indices of $\mathcal{I}$ we obtain

$$\hat{M}_{\mathcal{I}}^{(j)} := \Lambda^{(j)} \circ M_{\mathcal{I}} = \left\{ \hat{M}_{i_1, \ldots, i_d}^{(j)} | \ \hat{M}_{i_1, \ldots, i_d}^{(j)} = \Lambda_{i_1, \ldots, i_d}^{(j)} \circ M_{\mathcal{I}} \right\}.$$

We define two operators $\Delta^{(j)}$ and Σ that may be applied to an arbitrary scheme $X_{\mathcal{I}} = \{X_{i_1, \ldots, i_d}\}, \mathcal{I}$ satisfying (10). $\Delta^{(j)}$ computes the differences in coordinate direction x_j:

$$\Delta^{(j)} \circ X_{\mathcal{I}} := \left\{ S_{i_1, \ldots, i_d} | \begin{array}{l} i_j > 0 : \ S_{i_1, \ldots, i_d} = X_{i_1, \ldots, i_j, \ldots, i_d} - X_{i_1, \ldots, i_{j-1}, i_j - 1, i_{j+1}, \ldots, i_d}, \\ i_j = 0 : \ S_{i_1, \ldots, i_d} = X_{i_1, \ldots, i_d} \end{array} \right\},$$

and Σ sums up the scheme:

$$\Sigma \circ X_{\mathcal{I}} := \sum_{(i_1, \ldots, i_d) \in \mathcal{I}} X_{i_1, \ldots, i_d}.$$

Our extrapolation method can now be comprised to four phases: computation of a scheme of midpoint sums $M_{\mathcal{I}}$, extrapolation in all directions ($\Lambda^{(j)}$), computation of improvements ($\Delta^{(j)}$) and the summation of all improvements (Σ) as:

$$\hat{M}_{\mathcal{I}} := \Sigma \circ \Delta^{(1)} \circ \ldots \circ \Delta^{(d)} \circ \Lambda^{(1)} \circ \ldots \circ \Lambda^{(d)} \circ M_{\mathcal{I}}. \quad (11)$$

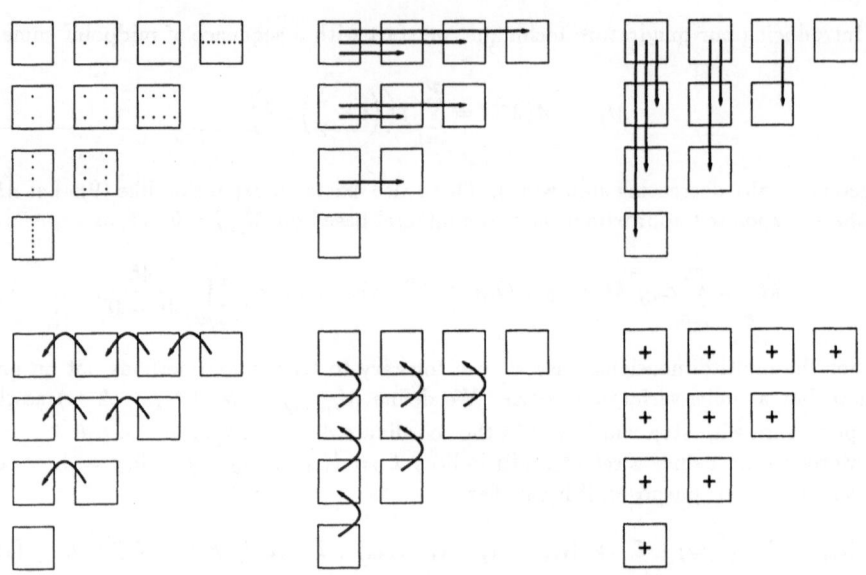

Figure 4: Quadrature of high order based on extrapolation technique and using sparse grids

As a set of multi indices we choose a d-dimensional simplex of depth c

$$\mathcal{I}_c := \big\{(i_1, \ldots, i_d) |\ i_1 + \ldots + i_d \leq c\big\}.$$

Figure 4 illustrates the extrapolation for the two-dimensional case working on a simplex scheme (in this case a triangle) of depth 3. The upper left scheme shows $M_{\mathcal{I}}$. Each square signifies a midpoint sum and the points used for its computation. It can easily be seen the multi-variate approach: going to the right the step width in direction x_1 decreases, going down the step width in direction x_2 decreases. Note that the superposition of all these grids is a sparse grid. The upper middle scheme of figure 4 illustrates how the extrapolation operator $\Lambda^{(1)}$ works. The arrows mark the values needed to compute the extrapolated value on the vertex according $\Lambda^{(1)}_{i_1,i_2}$. In the next step $\Lambda^{(2)}$ extrapolates in the direction x_2 (upper right). At this point the extrapolation phase is finished and the phase of computing improvements in all directions using the operator $\Delta^{(j)}$ starts. The lower left and lower middle scheme illustrate the behavior of $\Delta^{(1)}$ respectively $\Delta^{(2)}$. For an approximation of the integral we have to sum up all improvements (lower right).

The Euler-MacLaurin Summation Formula exists for the multi-dimensional case in multi-variate expansion in the following form:

$$M(h_1, \ldots, h_d) = \sum_{(i_1, \ldots, i_d) \in \mathcal{I}} \tau_{i_1, \ldots, i_d}\ h_1^{2i_1} \ldots h_d^{2i_d} + R_{\mathcal{I}},$$

where $R_{\mathcal{I}}$ represents the high order error term.

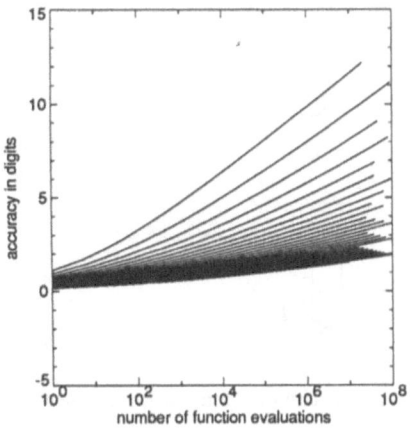 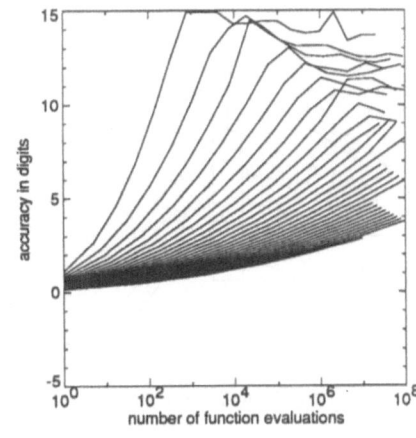

Figure 5: Numerical results of example 3 for $d = 2$ (upper line) to $d = 30$ (lower line)

It can be proved for an arbitrary set of multi indices $\mathcal{I}'$ satisfying (10) that by (11) all error terms

$$\tau_{i_1,\dots,i_d} h_1^{2i_1} \dots h_d^{2i_d}, \; (i_1,\dots,i_d) \in \mathcal{I} \setminus (0,\dots,0)$$

vanish. Especially the use of a simplex scheme $\mathcal{I}_c$ yields an approximation $\hat{M}_{\mathcal{I}_c}$ of order $O(h^{2c+2})$.

Figure 5 shows the numerical results if we apply the algorithm to the function
Example 3:

$$f(x_1,\dots,x_d) = e^{x_1+\dots+x_d} \;\; \text{on} \;\; (x_1,\dots,x_d) \in [0,1]^d.$$

The diagrams shows the number of function evaluations against the accuracy in decimal digits achieved for different dimensions ($d = 2,\dots,30$), if the simplex depth c is increased stepwise. The left one shows the results, if the extrapolation operators $\Lambda^{(j)}$ are ignored in (11). This method is in accordance to the midpoint rule on sparse grids. In the asymptotic phase it can be seen that the error decreases mainly with order $O(h^2)$ (in reality damped by a logarithmic term). The right diagram shows the results of the same experiment, but now using extrapolation (11). It can be seen that the accuracy gained by every additional extrapolation step grows more and more. We will see in the examples of the next section that this is the deciding advantage against fixed order quadrature rules.

The two quadrature algorithms introduced in the last two sections are very different. The first algorithm is adaptive but essentially of order $O(h^2)$, while the second is nonadaptive and of high order. Common to both is that they work on sparse grids. In the next section we introduce a third quadrature algorithm that combines the individual advantages of both algorithms, namely adaptivity and high order.

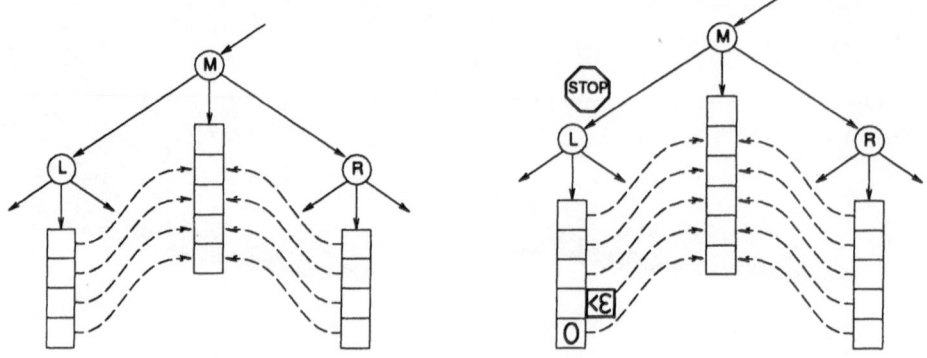

Figure 6: Adaptive quadrature of high order in the one-dimensional case

4 Adaptive Quadrature of High Order

We first consider the one-dimensional case. The algorithm computes stepwise the midpoint sums $M(0), M(\frac{1}{2}), M(\frac{1}{4}), \ldots$ and organizes the evaluated points in a binary tree. Each tree node contains an extrapolation scheme of its subtree.

First will be computed the tree node M for the midpoint of the integration domain (see figure 6 left). A first approximation of the integral can be computed using a function evaluation in the midpoint:

$$M_0 = f\left(\frac{a+b}{2}\right)(b-a).$$

$\hat{M}_0 = M_0$ will be stored in the first field of its list. In the next tree traversal we generate the subnodes L and R corresponding to the points $\frac{1}{4}(3a+b)$ respectively $\frac{1}{4}(a+3b)$ and compute an approximation of the integral in the both subdomains $[a, \frac{1}{2}(a+b)]$ and $[\frac{1}{2}(a+b), b]$ in the same way as it was done before for the midpoint. The last step of this tree traversal is the computation of an extrapolated approximation $\hat{M}_1$ of the integral on $[a, b]$ of order $O(h^4)$ based on all function values computed up to now. The improvement $\hat{M}_1 - \hat{M}_0$ will be stored in the second field of the list for node M.

In the next tree traversal a new layer of nodes will be generated. For the nodes L and R we get an $O(h^4)$ approximation of the integral on $[a, \frac{1}{2}(a+b)]$ and $[\frac{1}{2}(a+b), b]$. Their improvements will be stored in the provided fields of the lists. Now at the midpoint node, we are able to compute an $O(h^6)$ approximation for the integral.

This process of generating new layers of tree nodes and computing new approximations of higher order is repeated recursively.

In the case that the absolute of the last improvement in a subnode is less then a given ε we set a stop flag for that node (see figure 6 right). This stop flag means that the subtree belonging to this node will not be visited in further tree traversals. For the extrapolation in the father node, the improvements for all higher order approximations are set to zero. The tree traversals continues until the set of stop flags does not permit new function evaluations.

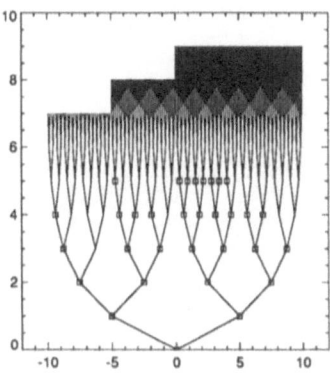
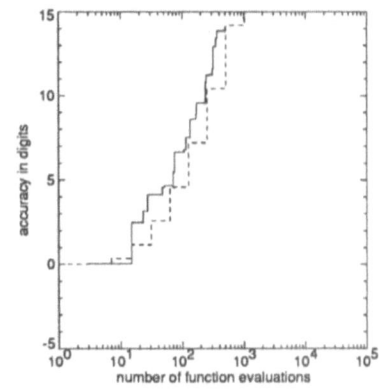

Figure 7: Example 4: tree built up by the algorithm (left), numerical results (right)

First we apply this algorithm to the function
Example 4:

$$f(x) = e^x \quad \text{on} \quad x \in [-10, 10].$$

Figure 7 (left) shows the tree that is built up by the algorithm if we choose $\varepsilon = 10^{-8}$ as accuracy limitation. The computed tree can be subdivided in three regular subtrees. Tree nodes, where the stop flag is set, are marked as squares. It can be seen that the error criterion hits at extrapolation orders higher than or equal $O(h^8)$. Nonsmooth subdomains are resolved with a finer grid. Figure 7 (right) shows the cost (measured in function evaluations) to be paid, and the accuracy in decimal digits for the method, if ε will be incremented step by step (solid line). The dashed line shows the result if the nonadaptive variant is used. The gain is comparatively small, as it can be seen in the relatively uniform shape of the tree in figure 7 (left).

Now we explain the d-dimensional quadrature for $d = 2$ as an example. Here there will be built up a two-dimensional tree as it is illustrated in figure 8. The midpoint node M has a triangular simplex scheme of improvements. For this midpoint the corresponding domain is divided up along the coordinate direction x_1 in two subdomains. The subnodes N and S represent the quadrature problems on these subdomains. The subnodes W and E belong to the one-dimensional quadrature problem on both halves of the border line between the subdomains. The extrapolation schemes of depth c of the four subnodes are compiled together to a scheme of depth $c + 1$ in the node M as shown in figure 8. With every tree traversal step, the triangular schemes will be augmented by a diagonal line of fields storing the improvement, while the linear schemes grow by an additional field. Stop flags are set in a similar way than in the one-dimensional case, namely under the condition that all absolutes of the improvements generated in one traversal step are less then a given ε.

In the following we apply the algorithm to the function
Example 5:

$$f(x_1, x_2) = e^{x_1 + x_2} \quad \text{on} \quad (x_1, x_2) \in [-10, 10]^2.$$

The points evaluated for $\varepsilon = 10$ (corresponds to eight digits accuracy) are displayed in

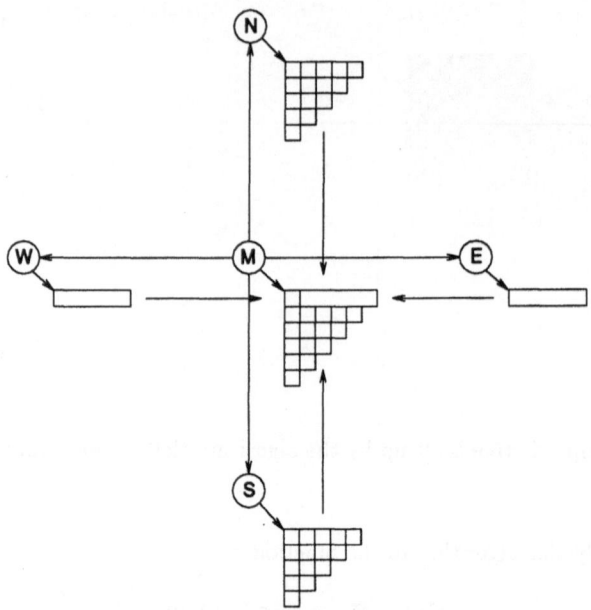

Figure 8: Adaptive quadrature of high order in the two-dimensional case

figure 9 (left). It can be seen that there are large regions of regular sparse grids. The closer to the peak in the upper right corner, the grid structure becomes finer. Figure 9 (right) shows the numerical results if ε will be decremented step by step (solid line). The dashed lines shows the results of the nonadaptive method of section 3. Adaptivity improves the results up to three digits. As it can be seen from figure 9 (left) the reduction of grid points in the lower left quadrant is very high compared with the upper right.

The dotted line shows the numerical results, if the routine d01fcf from the NAG library is used [4]. This algorithm operates by subdividing hyper-rectangular regions into smaller hyper-rectangulars. In each subregion, the integral is estimated using a seventh-degree rule.

In the starting phase this method is superior to our method, since it uses a stronger basic rule. But the increasing order of our method yields better results beginning with about $8 \cdot 10^4$ function evaluations.

5   Global Limitations

The next example should demonstrate a deficiency of our method that will be removed in this section by a modification of the algorithm.

Let's have a look to the following function defined on a circle

Example 6:

$$f(r) = \frac{1}{2\pi} e^{-\frac{r^2}{2}} \quad \text{on} \quad r \in [0,1], \ r = \sqrt{x_1^2 + x_2^2}.$$

We transform this function in direction x_2 so that the integration region becomes a square.

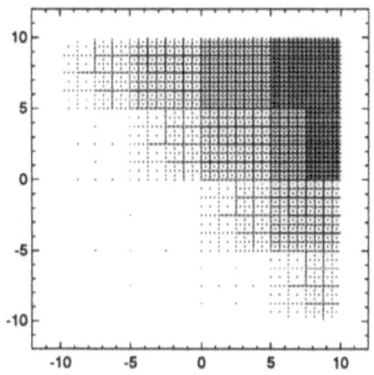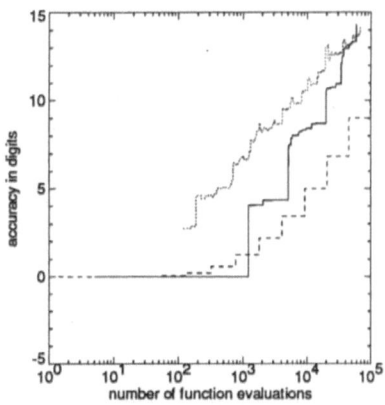

Figure 9: Example 5: grid and numerical result

Figure 10 (left) shows the points to be evaluated. At the point $(0,0)$ it is known that the resolution of the line $x_1 = 0$ is fine enough. For the extrapolation in this node, it can be expected that no higher resolution in coordinate direction x_2 is necessary. The subnodes themselves do not have this knowledge and refine deeper. The result is the dense evaluation of the integrand near the left and right border.

Figure 11 illustrates, how this deficiency is removed: at the point M we observe that the resolution in this direction is enough accurate. Thus we set a *global limitation* that will be inherited to all subnodes. This limitation signifies that the points belonging to the corresponding grids will not be evaluated, although this would be necessary seen from the view of subnodes. Improvements beyond the global limitation will be set to zero.

Figure 10 (right) shows the grid evaluated, if the algorithm is enhanced by the modification described above. Now an accumulation of grid points in the border regions as it was observed before has been inhibited.

6 Numerical Experiments

Finally we want to present some results of our algorithm and compare it with the algorithm d01fcf mentioned above. As example function we choose again the d-dimensional exponential function with arbitrary chosen integration domain

Example 7:

$$f(x_1, \ldots, x_d) = e^{x_1 + \cdots + x_d} \quad \text{on} \quad (x_1, \ldots, x_d) \in [-h, h]^d.$$

Figure 12 shows the results if the pair (d, h) is chosen as $(3, 2), (3, 3), (4, 2), (5, 2)$ (from left to right and top to bottom). The dotted line shows the results of the alternative algorithm. The straight lines result from the fixed order of this method. The slope of the line becomes lower with increasing dimension d. In the startup phase our results do

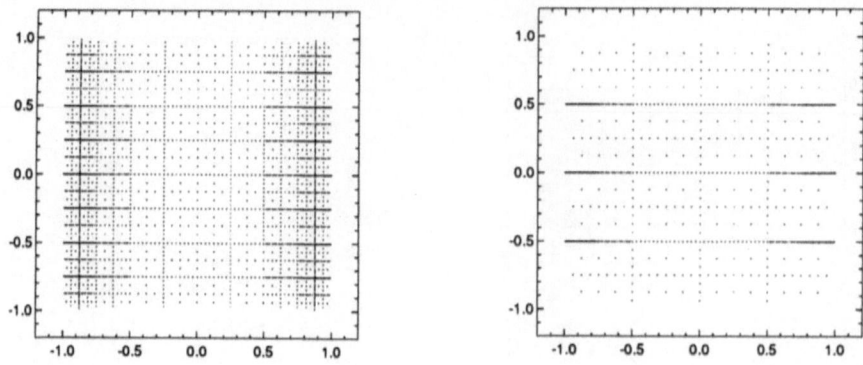

Figure 10: Example 6: without (left) and with (right) global limitation

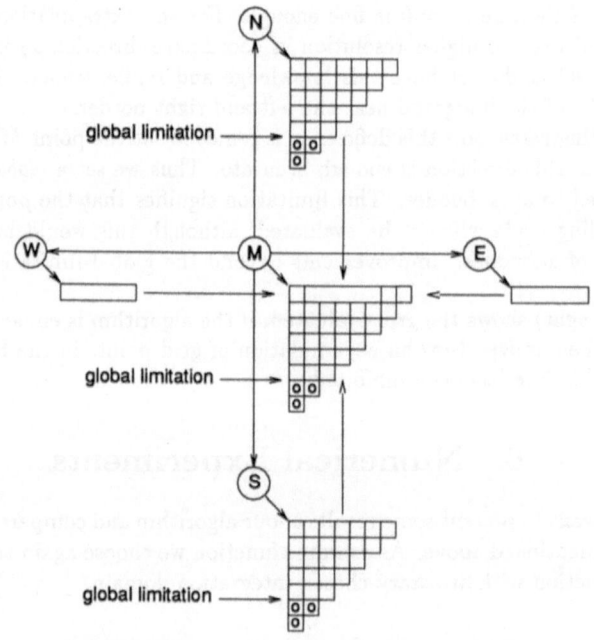

Figure 11: Global limitations will be inherited to the subnodes

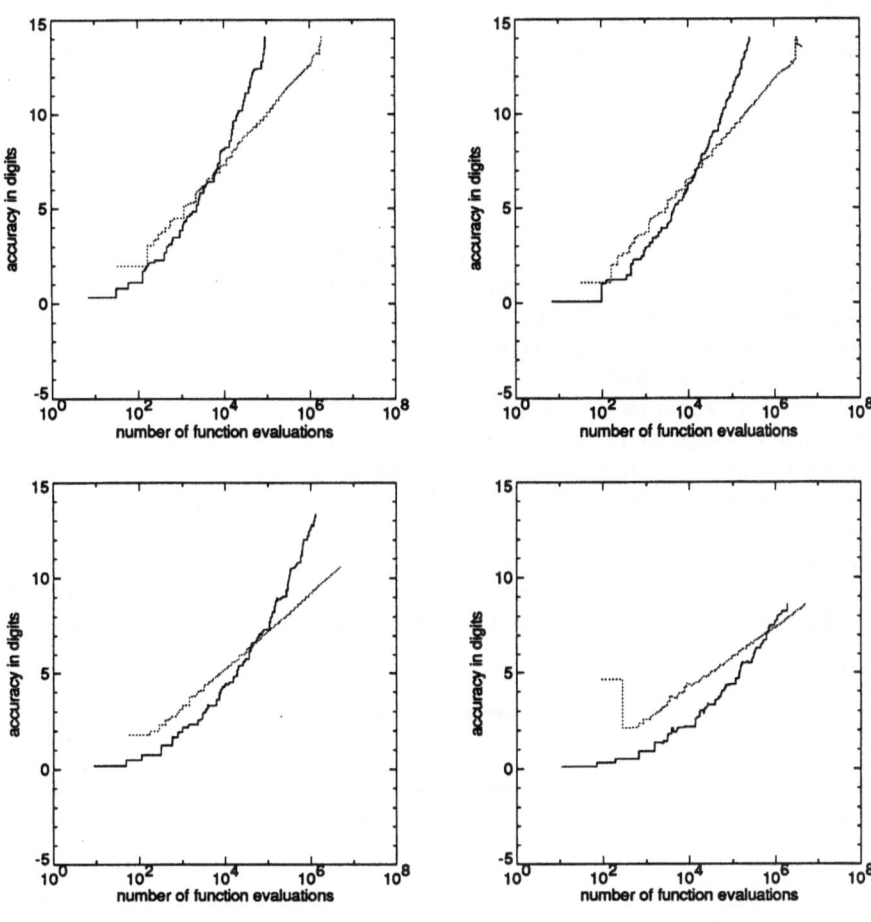

Figure 12: Example 7: algorithm 3 (solid line) against a fixed order quadrature routine (dotted line)

not achieve the accuracy, since at this point our method does not have such a high order as the competing algorithm. With increasing number of function evaluations we can rise the order as it can be seen by the over proportional increase of the accuracy. We pass the results of the routine d01fcf between 10^4 and 10^6 function evaluations.

7 Conclusion

We introduced two approaches of multi-dimensional numerical quadrature: first a generalization of the Archimedes Rule, second a multi-variate extrapolation technique. Both approaches led in a natural way to sparse grids. Starting from this knowledge we developed an algorithm preserving the advantages of both approaches, namely adaptivity and higher order than $O(h^2)$.

We saw that fixed order quadrature algorithms often yield better results in the startup phase while our algorithms are asymptotically superior. We demonstrated that the asymptotical superiority in practice comes to fruition by reasonable cost.

Further investigations have to be made to estimate the break even point for our algorithms against fixed order quadrature algorithms.

Theoretical considerations showed that our approaches are very convenient for high dimensional problems by exploiting smoothness. Multi-dimensional quadrature problems up to dimension 30 demonstrated the practical relevance of these considerations.

References

[1] H. Bungartz. *Dünne Gitter und deren Anwendung bei der adaptiven Lösung der dreidimensionalen Poisson-Gleichung*. PhD thesis, TU München, Institut für Informatik, June 1992.

[2] J. Stoer. *Einführung in die Numerische Mathematik I*, volume 1. Springer Verlag, Berlin, Heidelberg, New York, 1979.

[3] J. F. Traub, G. W. Wasilkowski, and H. Woźniakowski. *Information-Based Complexity*. Academic Press, New York, London, 1988.

[4] P. van Dooren and L. de Ridder. An Adaptive Algorithm for Numerical Integration over an N-dimensional Cube. *J. Comput. Appl. Math.*, pages 207–217, 1976.

[5] C. Zenger. Sparse Grids. SFB Bericht 342/18/90 A, Institut für Informatik, TU München, 1990.

ADAPTIVE SOLUTION OF ONE–DIMENSIONAL SCALAR CONSERVATION LAWS WITH CONVEX FLUX

Folkmar A. Bornemann

Fachbereich Mathematik, Freie Universität Berlin, Arnimallee 2–6, D-14195 Berlin, and
Konrad–Zuse–Zentrum Berlin, Heilbronner Str. 10, D-10711 Berlin, Germany
bornemann@sc.zib-berlin.de

Summary

A new adaptive approach for one–dimensional scalar conservation laws with convex flux is proposed. The initial data are approximated on an adaptive grid by a problem dependent, monotone interpolation procedure in such a way, that the multivalued problem of characteristic transport can be easily and explicitly solved. The unique entropy solution is chosen by means of a selection criterion due to HOPF and LAX. For arbitrary times, the solution is represented by an adaptive monotone spline interpolation. The spatial approximation is controlled by local L^1–error estimates. As a distinctive feature of the approach, there is no discretization in time in the traditional sense. The method is monotone on fixed grids. Numerical examples are included, to demonstrate the predicted behavior.

Key Words. method of characteristics, adaptive grids, monotone interpolation, L^1–error estimates

AMS(MOS) subject classification. 65M15, 65M25, 65M50

Introduction

The construction of an adaptive algorithm for time-dependent hyperbolic conservation laws has to face a well known problem: Explicit direct discretization in time will yield a CFL-restriction of the time step, which will be prohibitively small when shocks are resolved by an adaptive space mesh. On the other hand, an implicit direct discretization in time introduces much numerical viscosity, since the velocity of information transport is modeled the wrong way. Thus, the time derivative should not be discretized directly at all.

Instead, one should try to attack the *evolution operator* of the problem, $\mathcal{E}_t : X \to X$, which maps admissible initial data to the solution at time t. An abstract framework of promising methods is given by the so called PERU-schemes of K.W. MORTON [11], which we state in a slightly more general way: Replace the evolution operator $\mathcal{E}_t$ by the sequence $\hat{\mathcal{E}}_t = \mathcal{P}\mathcal{E}_t\mathcal{R}$,

$$ X \xrightarrow{\mathcal{R}} \hat{X} \xrightarrow{\mathcal{E}_t} X \xrightarrow{\mathcal{P}} X. $$

Here, $\mathcal{R}$ is a *(re)construction* operator, which maps the initial data in a more desirable space $\hat{X} \subset X$. The operator $\mathcal{P}$ *(re)presents* the result of the transport on $\hat{X}$ as a more appropriate element of the space X. Moreover, the exact transport $\mathcal{E}_t$ on the space $\hat{X}$ might be replaced by some simplification $\tilde{\mathcal{E}}_t$.

69

An example of this rather general approach is provided by Godunov's method and its generalizations. Here, $X = L^1$ and $\hat{X} \subset X$ consists of the piecewise constant functions on some mesh Δ. The reconstruction and representation operator are just the cell average projection onto $\hat{X}$, $\mathcal{R} = \mathcal{P}$. We have to consider three cases, depending on the Courant number $C = C(t, \Delta)$:

- $C < 1/2$. The evaluation of $\mathcal{E}_t$ on $\hat{X}$ is just the solution of non-interacting Riemann problems (classical Godunov scheme).
- $C \leq 1$. Here one can use the weak formulation of conservation laws in order to evaluate at least the product $\mathcal{P}\mathcal{E}_t$ on $\hat{X}$ by means of Riemann solvers.
- $C > 1$. There are simplifications $\tilde{\mathcal{E}}_t$ of the exact evolution operator, which approximate the now interacting Riemann problems and which are TVD. LEVEQUE [9] handles the interactions linearly in the large time-step Godunov's method. BRENIER [2] uses the transport-collapse operator, which is a certain averaging operator of the multivalued solution of characteristic transport. Both introduce errors in time.

The question which will be addressed is, whether there are any spaces $\hat{X}$ which are more powerful than piecewise constants in terms of approximation properties and which allow (at least for the most simple problems) the computation of the exact evolution operator $\mathcal{E}_t$ for arbitrary times t.

The answer given in this article is as follows: $\hat{X}$ can be chosen as the range of a certain interpolation operator $\mathcal{R}$, which depends on the right hand side of the problem. The space $\hat{X}$ will be as powerful for approximation as piecewise linears. The interpolation operator $\mathcal{R}$ is constructed in a way, that the computation of the multivalued solution of characteristic transport of initial data from $\hat{X}$ is extremely simple. Rather than averaging the multivalued solution like BRENIER [2] with the transport-collapse operator, we select the single valued entropy solution by means of the Lax-Hopf formula [6, 7]. The final representation of the solution is given by some interpolation operator $\mathcal{P}$, which could be any monotone spline interpolation. The space meshes will be constructed by an adaptive interpolation procedure, which is guided by local L^1-error estimates.

Up to now, the presented algorithm strongly relies on convexity properties of the underlying problem, so that the problem class which can be immediately handled seems to be quite restricted. However, there are strong indications that our approach will be at least applicable to the class of those systems of conservation laws, which are equivalent to Hamilton-Jacobi equations with convex Hamiltonian.

Theoretical Preparations

We are concerned with the solution of scalar conservation laws

$$u_t + f(u)_x = 0, \qquad u(\cdot, 0) = u_0, \tag{1}$$

where $u(\cdot, t)$ is a function on $\mathbb{R}$. Our general *assumptions* on the flux f will be

- $f : \mathbb{R} \to \mathbb{R}$ strictly convex and C^1.

Thus, the derivative $\alpha \equiv f' : \mathbb{R} \to]\alpha_-, \alpha_+[$ is one–one, onto and nondecreasing. The inverse of α will be denoted by $\beta \equiv \alpha^{-1} :]\alpha_-, \alpha_+[\to \mathbb{R}$.

A crucial role will be played by the *Legendre–Fenchel dual* $f^* :]\alpha_-, \alpha_+[\to \mathbb{R}$ of f, defined as

$$
\begin{aligned}
f^*(z) &= \sup_v (vz - f(v)) \\
&= u\alpha(u) - f(u) \quad \text{with } u = \beta(z).
\end{aligned}
$$

Convex analysis (e.g., TIKHOMIROV [14]) states that

$$(f^*)' = \beta.$$

Our whole approach heavily relies on the following characterization of that weak solution, which satisfies the entropy condition.

THEOREM 1. (LAX [7]). *Let $u_0 \in L^1(\mathbb{R})$ and $U_0(y) = \int_{-\infty}^y u_0(\xi)d\xi$. For $x \in \mathbb{R}$, $t > 0$ define*

$$\mathcal{E}_t u_0(x) = \beta\left(\frac{x-y}{t}\right),$$

where $y = y(x,t) \in]x - t\alpha_-, x - t\alpha_+[$ is any value which minimizes

$$U_0(y) + tf^*\left(\frac{x-y}{t}\right) = \min! . \tag{2}$$

Then $\mathcal{E}_t u_0 \in L^1(\mathbb{R})$ is the unique entropy solution at time t of the conservation law under consideration.

If there exist several different values y, which minimize (2) for a given x, then x is the position of a shock discontinuity. The limits $\mathcal{E}_t u_0(x-0)$, $\mathcal{E}_t u_0(x+0)$ exist, and

$$\mathcal{E}_t u_0(x-0) \geq \beta\left(\frac{x-y}{t}\right) \geq \mathcal{E}_t u_0(x+0)$$

holds for every such y.

Remark. For $u \in L^\infty(\mathbb{R})$, $u_- \leq u_0 \leq u_+$ a.e., the solution only depends on f restricted to the interval $[u_-, u_+]$.

Remark. This Theorem has quite some history. HOPF [6] stated it for the inviscid Burgers equation $u_t + uu_x = 0$. He obtained the result in the limit $\mu \to 0$ of his explicit solution (i.e., the Cole–Hopf transformation to the heat equation) of the viscid Burgers equation $u_t + uu_x = \mu u_{xx}$. Later LAX [7] generalized the result to arbitrary convex fluxes f. A nice interpretation as Bellman's approach to the Hamilton–Jacobi equation $v_t + f(v_x) = 0$ can be found in LAX [8] or CONWAY and HOPF [3]. In fact, $U(x,t) = \int_{-\infty}^x u(\xi,t)d\xi$ satisfies the Hamilton–Jacobi equation and

$$U(x,t) = \min_y \left(U_0(y) + tf^*\left(\frac{x-y}{t}\right)\right),$$

if f is adjusted to $f(0) = 0$. This formula is actually connected with the modern notion of *viscosity solutions* of more general Hamilton–Jacobi equations, we refer to the book of P.-L. LIONS [10]. In [4] this connection is used to propose discretizations of Hamilton–Jacobi equations for the numerical solution of conservation laws.

Our approach uses the fact that, for certain u_0, the set of values y which possibly minimize (2) can be considerably restricted. A first step in that direction is the following

COROLLARY 2. *Assumptions as in Theorem 1. Additionally let u_0 be continuous. Any value $y \in]x - t\alpha_-, x - t\alpha_+[$ which minimizes (2) satisfies*

$$y + t\alpha(u_0(y)) = x, \tag{3}$$

and allows to set $\mathcal{E}_t u_0(x) = u_0(y)$.

Proof. Differentiating relation (2) yields, by continuity of u_0,

$$u_0(y) - \beta\left(\frac{x-y}{t}\right) = 0.$$

This implies both assertions. □

The nonlinear equation (3) is just the one which allows to construct, for smooth u_0 and small t, by means of the implicit function theorem, a classical solution of the conservation law. For larger t, equation (3) does not have a unique solution y for whole intervals of x. Thus, the minimum condition (2) can be understood as a *selection principle* for the right value of y.

The following stability result allows us to change u_0 slightly, in order to obtain simpler problems of the kind (3).

THEOREM 3. (KEYFITZ [12]). *The entropy solutions of (1) form a nonlinear L^1-contractive semigroup $\mathcal{E}_t$. Thus, for $u_0, v_0 \in L^1(\mathbb{R})$, the corresponding entropy solutions satisfy the estimate*

$$\|\mathcal{E}_t u_0 - \mathcal{E}_t v_0\|_{L^1} \leq \|u_0 - v_0\|_{L^1} \tag{4}$$

for all $t \geq 0$. A proof may also be found in LAX [8].

For our further development we need some *notation*:

- The convex hull of two points u_0, u_1 will be denoted by $[u_0, u_1]$.
- For $u \in [u_0, u_1]$ the *barycentric coordinate* of u is denoted by $\lambda_{u_0,u_1}(u)$ and satisfies

$$u = (1 - \lambda_{u_0,u_1}(u))u_0 + \lambda_{u_0,u_1}(u)u_1.$$

- Let $(y_0, u_0), (y_1, u_1)$ be two points in $\mathbb{R}^2$. The β–*interpolant* of these points is given as $\mu_{u_0,u_1} : [y_0, y_1] \to [u_0, u_1]$ by

$$\mu_{u_0,u_1}\left((1-\lambda)y_0 + \lambda y_1\right) = \beta\left((1-\lambda)\alpha(u_0) + \lambda\alpha(u_1)\right), \qquad \lambda \in [0,1].$$

Our assumptions imply, that this is a monoton connection of the two points.

These β–interpolants have the very nice property that (3) may be solved uniquely, as we show now.

LEMMA 4. *For given $t > 0$ define*

$$\varphi(y) = y + t\alpha(\mu_{u_0,u_1}(y)),$$

which maps $[y_0, y_1]$ onto $[\varphi(y_0), \varphi(y_1)]$. If $\varphi(y_0) \neq \varphi(y_1)$, the equation $x = \varphi(y)$ is uniquely solved by the value y given as

$$\lambda_{y_0,y_1}(y) = \lambda_{\varphi(y_0),\varphi(y_1)}(x).$$

Proof. Simply note that

$$\varphi(y) = (1 - \lambda_{y_0,y_1}(y))(y_0 + t\alpha(u_0)) + \lambda_{y_0,y_1}(y)(y_1 + t\alpha(u_1)),$$

and that $\varphi(y_i) = (y_i + t\alpha(u_i))$ for $i = 0, 1$. $\square$

Since integrals are involved in (2), it helps a lot that integrals of β–interpolants can be computed explicitly:

LEMMA 5. *We have, for $y \in [y_0, y_1]$, that*

$$\int_{y_0}^{y} \mu_{u_0,u_1}(\eta)d\eta = \begin{cases} \dfrac{y_1 - y_0}{\alpha(u_1) - \alpha(u_0)} f^* \Big|_{\alpha(u_0)}^{\alpha(\mu_{u_0,u_1}(y))} & \text{if } u_0 \neq u_1, \\[2ex] (y - y_0)u_0 & \text{if } u_0 = u_1. \end{cases}$$

Proof. Let $u_0 \neq u_1$. The substitution $\zeta = (1 - \lambda_{y_0,y_1}(\eta))\alpha(u_0) + \lambda_{y_0,y_1}(\eta)\alpha(u_1)$ gives

$$\int_{y_0}^{y} \mu_{u_0,u_1}(\eta)d\eta = \frac{y_1 - y_0}{\alpha(u_1) - \alpha(u_0)} \int_{\alpha(u_0)}^{\alpha(\mu_{u_0,u_1}(y))} \beta(\zeta)d\zeta.$$

Thus, the assertion follows from $(f^*)' = \beta$. $\square$

Now we try to approximate u_0 by its piecewise β–interpolant, for which we have seen that problems (3) and (2) turn out to be fairly simple. For that purpose, let u_0 be a piecewise continuous function with supp $u_0 \subset\subset]a, b[$. Let $\Delta : a = y_0 < y_1 < \ldots < y_n = b$ be a subdivision of that interval, with mesh–size parameter

$$h = \max_{1 \leq j \leq n}(y_j - y_{j-1}).$$

Denote the subintervals by $I_j = [y_{j-1}, y_j]$, $j = 1, 2, \ldots, n$. The piecewise β–interpolant $\mathcal{R}_{f,\Delta}u_0$ is now defined as

$$\mathcal{R}_{f,\Delta}u_0(y) = \begin{cases} \mu_{u_0(y_{j-1}),u_0(y_j)}(y) & \text{for } y \in I_j, j = 1, 2, \ldots, n, \\ 0 & \text{elsewhere.} \end{cases}$$

Obviously we have $\mathcal{R}_{f,\Delta}u_0 \in C^0$ and $\mathcal{R}_{f,\Delta}u_0(y_j) = u_0(y_j)$, $j = 1, 2, \ldots, n$.

LEMMA 6. *Let u be a piecewise continuous function on $[a, b]$. Then the interpolation operator $\mathcal{R}_{f,\Delta}$ satisfies*

$$\|u - \mathcal{R}_{f,\Delta}u\|_{L^1[a,b]} \to 0 \quad \text{for } h \to 0.$$

Moreover, if the function u is piecewise C^2, and we make the assumptions on the flux f that $f \in C^3$ with

$$M = \|f'''/f''^3\|_{L^\infty[u_-,u_+]}\|f''\|_{L^\infty[u_-,u_+]}^2 < \infty,$$

where $u_- \leq u(x) \leq u_+$ for all $x \in [a, b]$, then there is a constant $c = c(u, M)$, which gives us the estimates

$$\|u - \mathcal{R}_{f,\Delta}u\|_{L^1[a,b]} \leq ch, \quad \text{and} \quad \|u - \mathcal{R}_{f,\Delta}u\|_{L^1(I_c)} \leq ch^2.$$

Here $I_c = \bigcup_{j \notin J_u} I_j$ with $J_u = \left\{ 1 \leq j \leq n \;\middle|\; \left(u|_{I_j}\right) \notin C^2 \right\}$.

Proof. We proof the second, smooth part. The first part follows by usual density arguments. Take any $j \notin J_u$, and denote the linear interpolation operator at the nodes y_{j-1}, y_j by $\mathcal{I}_\Delta$. Since by construction

$$\mathcal{I}_\Delta \mathcal{R}_{f,\Delta} = \mathcal{I}_\Delta$$

we estimate by the usual error expression for linear interpolation

$$\|u - \mathcal{R}_{f,\Delta} u\|_{L^\infty(I_j)} \leq \|u - \mathcal{I}_\Delta u\|_{L^\infty(I_j)} + \|\mathcal{R}_{f,\Delta} u - \mathcal{I}_\Delta \mathcal{R}_{f,\Delta} u\|_{L^\infty(I_j)}$$

$$\leq \frac{h^2}{8} \left(\|u''\|_{L^\infty[a,b]} + \|\mu''_{u(y_{j-1}),u(y_j)}\|_{L^\infty(I_j)} \right).$$

Now we compute, for $y \in I_j$, that

$$\mu''_{u(y_{j-1}),u(y_j)}(y) = -\frac{f'''\left(\mu_{u(y_{j-1}),u(y_j)}(y)\right)}{f''\left(\mu_{u(y_{j-1}),u(y_j)}(y)\right)^3} \left(\frac{\alpha(u(y_j)) - \alpha(u(y_{j-1}))}{u(y_j) - u(y_{j-1})} \cdot \frac{u(y_j) - u(y_{j-1})}{y_j - y_{j-1}} \right)^2$$

$$= -\frac{f'''\left(\mu_{u(y_{j-1}),u(y_j)}(y)\right)}{f''\left(\mu_{u(y_{j-1}),u(y_j)}(y)\right)^3} f''(\eta)^2 \, u'(\zeta)^2$$

for some $\eta \in [u(y_{j-1}), u(y_j)]$, $\zeta \in I_j$. Hence, it is $\|\mu''_{u(y_{j-1}),u(y_j)}\|_{L^\infty(I_j)} \leq M \|u'\|^2_{L^\infty[a,b]}$. For $j \in J_u$, we simply estimate

$$\|u - \mathcal{R}_{f,\Delta} u\|_{L^1(I_j)} \leq 2\|u\|_{L^\infty[a,b]} h.$$

Finally, we observe that $\#J_u \leq \nu$ as $h \to 0$, because we assumed that u is piecewise C^2. Thus, we obtain

$$\|u - \mathcal{R}_{f,\Delta} u\|_{L^1[a,b]} \leq \frac{b-a}{8} \left(\|u''\|_{L^\infty[a,b]} + M\|u'\|^2_{L^\infty[a,b]} \right) h^2 + 2\nu\|u\|_{L^\infty[a,b]} h$$

and

$$\|u - \mathcal{R}_{f,\Delta} u\|_{L^1(I_c)} \leq \frac{b-a}{8} \left(\|u''\|_{L^\infty[a,b]} + M\|u'\|^2_{L^\infty[a,b]} \right) h^2.$$

□

Note that the same result holds for the piecewise linear interpolation operator $\mathcal{I}_\Delta$, as introduced in the proof.

Remark. The value of the constant M is invariant against transformations $f \mapsto \gamma f$ with $\gamma > 0$.

Another important property of $\mathcal{R}_{f,\Delta}$ is monotonicity. This is a fairly simple consequence of the assumed monotonicity of α, β.

LEMMA 7. *Let u, v be piecewise continuous. The pointwise inequality $u \leq v$ implies that pointwise $\mathcal{R}_{f,\Delta} u \leq \mathcal{R}_{f,\Delta} v$. The same holds for the linear interpolation operator $\mathcal{I}_\Delta$.*

The Algorithm

Our algorithm solves the following problem: Given a conservation law (1), a piecewise continuous initial u_0, compute for an accuracy TOL and a time $t > 0$ an approximation $\hat{\mathcal{E}}_t u_0$ to the solution $\mathcal{E}_t u_0$ such that

$$\|\mathcal{E}_t u_0 - \hat{\mathcal{E}}_t u_0\|_{L^1} \leq \text{TOL}. \tag{5}$$

The algorithm can be stated very roughly as

A. *Construct* by an adaptive interpolation procedure a mesh Δ_0, such that the piecewise β-interpolant $\mathcal{R}_{f,\Delta_0} u_0$ of u_0 fulfills

$$\|u_0 - \mathcal{R}_{f,\Delta_0} u_0\|_{L^1} \leq \text{TOL}/2.$$

B$_x$. *Propagate* the function $\mathcal{R}_{f,\Delta_0} u_0$ with the evolution operator $\mathcal{E}_t$ to the time t, such that $\mathcal{E}_t \mathcal{R}_{f,\Delta_0} u_0(x)$ is evaluable.

C. *Represent* by an adaptive interpolation procedure (with the help of step B$_x$) the function $\mathcal{E}_t \mathcal{R}_{f,\Delta_0} u_0$ on a mesh Δ_t by its piecewise β-interpolant (resp. piecewise linear interpolant) $\hat{\mathcal{E}}_t u_0 = \mathcal{R}_{f,\Delta_t} \mathcal{E}_t \mathcal{R}_{f,\Delta_0} u_0$ (resp. $\hat{\mathcal{E}}_t u_0 = \mathcal{I}_{\Delta_t} \mathcal{E}_t \mathcal{R}_{f,\Delta_0} u_0$), such that

$$\|\mathcal{E}_t \mathcal{R}_{f,\Delta_0} u_0 - \hat{\mathcal{E}}_t u_0\|_{L^1} \leq \text{TOL}/2.$$

If these steps can be achieved, Theorem 3 guarantees for the accuracy requirement (5).

The choice of the interpolant in Step C, i.e., $\mathcal{I}_{\Delta_t}$ or $\mathcal{R}_{f,\Delta_t}$, is not really important. In fact, any adaptive monotone spline interpolation, which controls the L^1–approximation error, could be used to represent the solution for fixed times. The β–interpolant should be taken, if we intend to use the solution at a particular time as new initial data for another computation.

We now describe each step more closely. Note that Steps A and C are quite similar tasks.

Step A. Here, the choice of an appropriate mesh Δ_0 is the essential problem. This will be done in an *adaptive* way, starting with a coarse mesh Δ^0. The main loop reads as:

> **while** (estimated L^1–error > TOL$/2$)
> {
> $$\Delta^{k+1} = \text{refine}(\Delta^k);$$
> $$k = k + 1;$$
> }

Let the kth mesh be $\Delta^k : a = y_0^k < y_1^k < \ldots < y_{n_k}^k = b$. For the following, we will suppress the index k. The L^1–error of the piecewise β-interpolation on the mesh Δ is given as

$$\epsilon = \|u_0 - \mathcal{R}_{f,\Delta} u_0\|_{L^1} = \sum_{j=1}^{n} \epsilon_j,$$

with

$$\epsilon_j = \int_{y_{j-1}}^{y_j} \left| u_0(\xi) - \mu_{u_0(y_{j-1}),u_0(y_j)}(\xi) \right| d\xi.$$

The local error ϵ_j will be estimated by a trapezoidal sum, introducing the midpoint of I_j. Thus, noting the interpolation property, we obtain the local estimate

$$\epsilon_j \approx \eta_j = \frac{(y_j - y_{j-1})}{2} \left| u_0\left(\frac{y_{j-1} + y_j}{2}\right) - \mu_{u_0(y_{j-1}),u_0(y_j)}\left(\frac{y_{j-1} + y_j}{2}\right) \right|.$$

Observe that

$$\mu_{u_0(y_{j-1}),u_0(y_j)}\left(\frac{y_{j-1}+y_j}{2}\right) = \beta\left(\frac{1}{2}\left(\alpha(u_0(y_{j-1})) + \alpha(u_0(y_j))\right)\right).$$

In the case that I_j is bisected, we note that $u_0((y_{j-1}+y_j)/2)$ has already been computed for η_j. Thus, we can readily assign this value to the new node.

The proposed error estimate is sensible for accuracy and complexity reasons. The global estimate is finally given as

$$\eta = \sum_{j=1}^{n} \eta_j.$$

For actual refinement we need some *refinement strategy*, which uses the local information provided by the indicators η_j. We have implemented a strategy based on local extrapolation, which was introduced by BABUŠKA and RHEINBOLDT [1] for elliptic problems.

The actual implementation of the mesh refinement can easily be done by means of packages designed for finite element computations using tree data structures, e.g., ROITZSCH [13].

After the adaptive refinement we are provided with the final mesh Δ_0, an error estimate $\eta < \text{TOL}/2$, and each node y_j carries the interpolation information $u_0(y_j)$. For purposes of Step B_x, we should additionally store in each node the integral information

$$\int_{-\infty}^{y_j} \mathcal{R}_{f,\Delta_0} u_0(\xi) d\xi,$$

which can be computed by successive application of Lemma 5.

Step C. The piecewise β–interpolant (piecewise linear interpolant) $\hat{\mathcal{E}}_t u_0$ of the propagated function $\mathcal{E}_t \mathcal{R}_{f,\Delta_0} u_0$ and the mesh Δ_t are computed in a similar fashion as $\mathcal{R}_{f,\Delta_0} u_0$ and Δ_0. This approximation procedure only demands the possibility of evaluating the expression $\mathcal{E}_t \mathcal{R}_{f,\Delta_0} u_0$ in certain points x.

Step B_x. How do we compute $\mathcal{E}_t \mathcal{R}_{f,\Delta_0} u_0(x)$ for a given x? This question will be addressed now. In preparation of any evaluation, the following values are computed

$$\varphi(y_j) = y_j + t\alpha(\mathcal{R}_{f,\Delta_0} u_0(y_j))$$

for $j = 0, 1, \ldots, n$. These are the positions of the characteristic transport of the mesh points y_j of Δ_0. Given x, we first determine the set J_x of indices j, such that

$$x = y + t\alpha(\mathcal{R}_{f,\Delta_0} u_0(y)) \tag{6}$$

possesses a solution $y \in I_j$. By construction of $\mathcal{R}_{f,\Delta_0} u_0$ and Lemma 4, this set is exactly given by

$$J_x = \{1 \le j \le n \mid x \in [\varphi(y_{j-1}), \varphi(y_j)]\}.$$

For $j \in J_x$ with $\varphi(y_{j-1}) \neq \varphi(y_j)$, we compute the barycentric coordinate

$$\lambda_j = \lambda_{\varphi(y_{j-1}),\varphi(y_j)}(x),$$

which is, by Lemma 4, also the barycentric coordinate of the unique solution $\bar{y}_j \in I_j$, i.e., $\bar{y}_j = (1 - \lambda_j)y_{j-1} + \lambda_j y_j$. In the exceptional case $j \in J_x$, $\varphi(y_{j-1}) = \varphi(y_j)$, all values $y_{j-1} \leq y \leq y_j$ satisfy (6). Thus, the value of the expression in (2) remains constant on the whole interval $[y_{j-1}, y_j]$. Hence, we may take $\bar{y}_j = y_{j-1}$ as representative candidate for the minimizing value of (2). Summarizing, our construction of $\mathcal{R}_{f,\Delta_0} u_0$ allows us to compute a set of critical points of (2) with cardinality $\#J_x$, in which a minimizing value is included.

In view of Theorem 1 and its Corollary we choose the smallest value $\bar{y}_\ell$ among those $\bar{y}_j, j \in J_x$, which minimize (2). Our desired value of $\mathcal{E}_t \mathcal{R}_{f,\Delta_0} u_0(x)$ is given by

$$\mathcal{E}_t \mathcal{R}_{f,\Delta_0} u_0(x) = \mathcal{R}_{f,\Delta_0} u_0(\bar{y}_\ell) = \beta \left((1 - \lambda_\ell)\alpha(u_0(y_{\ell-1})) + \lambda_\ell \alpha(u_0(y_\ell)) \right).$$

For the evaluation of (2) it is necessary to rely on Lemma 5.

If there are several values $\bar{y}_k$ which minimize (2) among the $\bar{y}_j, j \in J_x$, we are allowed, due to Theorem 1, to take any of them: In this case, x is exactly the position of a shock of $\mathcal{E}_t \mathcal{R}_{f,\Delta_0} u_0$. All minimizing $\bar{y}_k$ produce values for $\mathcal{E}_t \mathcal{R}_{f,\Delta_0} u_0(x)$ which are between the left and the right shock value. In fact, since we choose the smallest $\bar{y}_k$, which minimizes (2), we can be more specific. We obtain

$$\mathcal{E}_t \mathcal{R}_{f,\Delta_0} u_0(x) = \mathcal{E}_t \mathcal{R}_{f,\Delta_0} u_0(x - 0) \tag{7}$$

for any shock position x. Note that the specification $\bar{y}_j = y_{j-1}$ in the case $\varphi(y_{j-1}) = \varphi(y_j)$ also served this purpose: It guarantees, that the smallest minimizing value of the $\bar{y}_j$ is really the smallest value of all minimizing values for (2).

A simple implication of the monotonicity property of $\mathcal{R}_{f,\Delta}$ and $\mathcal{I}_\Delta$ (Lemma 7), together with the monotonicity of the semigroup, is the monotonicity of our algorithm as long as we fix the meshes Δ_0, Δ_t.

LEMMA 8. *Let u_0, v_0 be piecewise continuous functions, such that pointwise $u_0 \leq v_0$. If we use for both functions the same meshes Δ_0, Δ_t, we obtain that pointwise $\hat{\mathcal{E}}_t u_0 \leq \hat{\mathcal{E}}_t v_0$.*

Proof. Care should be taken, if x is a shock position of both $\mathcal{E}_t \mathcal{R}_{f,\Delta_0} u_0$ and $\mathcal{E}_t \mathcal{R}_{f,\Delta_0} v_0$. Here, one has to rely on (7). If (7) wouldn't hold, one would have to exclude a neighborhood of x. □

In order to run our algorithm, we need procedures for evaluating f, α and β. If β is not given analytically, we may compute it by Newton's method.

Numerical Examples

Important: Since we can evaluate $\mathcal{E}_t \mathcal{R}_{f,\Delta_0} u_0(x)$ exactly for any (x, t), $t > 0$, it should be clear, that the time–steps of the examples have been solely introduced for graphical reasons. They are completely arbitrary and independent, and we work for all times with the same $\mathcal{R}_{f,\Delta_0} u_0$. Thus, *there is no discretization in time!* Once more, we remark, that any adaptive monotone spline interpolation, which controls the L^1-approximation error, could be used to represent the function $\mathcal{E}_t \mathcal{R}_{f,\Delta_0} u_0$. For simplicity, we have chosen piecewise linear interpolation whenever displacing our approximate solution graphically.

Example 1. Here, we consider the nonlinear conservation law

$$u_t + \left(\frac{u^4}{4} \right)_x = 0$$

with initial data

$$u_0(x) = \begin{cases} 1 & \text{for } 0 \leq x \leq 1, \\ 0 & \text{elsewhere.} \end{cases}$$

The inverse of the flux derivative, $\beta = (\cdot)^{1/3}$, is quite different from a linear function, giving the β–interpolant a distinguishable shape.

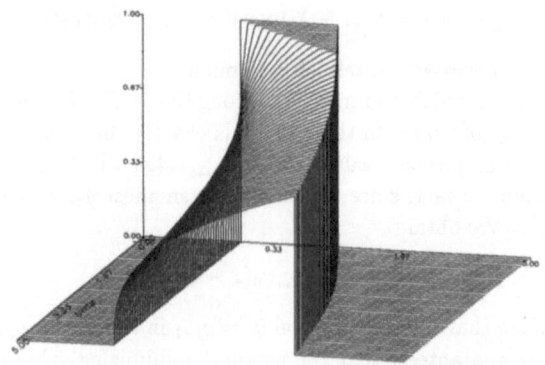

FIG. 1. *Example 1. Evolution of the computed solution.*

The exact solution is given by

$$u(x,t) = \begin{cases} \begin{cases} \left(\frac{x}{t}\right)^{1/3} & 0 \leq x \leq t \\ 1 & t \leq x \leq 1 + t/4 \\ 0 & \text{elsewhere} \end{cases} & \text{for } 0 < t \leq 4/3, \\[2em] \begin{cases} \left(\frac{x}{t}\right)^{1/3} & 0 \leq x \leq \left(\frac{4}{3}\right)^{3/4} t^{1/4} \\ 0 & \text{elsewhere} \end{cases} & \text{for } t > 4/3. \end{cases}$$

The computed solution for $0 \leq t \leq 5$, using a time–step $\tau = 0.1$ *for graphical reasons*, can be seen in Fig. 1. If not stated otherwise, we choose as accuracy $\mathsf{TOL} = 10^{-4}$. The solution was represented by the adaptive linear interpolation of Step C.

The solution for the particular time $t = 1.0$ is shown in Fig. 2, represented by the adaptive linear interpolation. We observe that, as a result of our construction (7), there is no mesh point with a value between the left and the right shock value.

The development of the interpolation mesh in time, here with time–step $\tau = 0.05$, can be seen in Fig. 3. We can observe nicely, how the rarefaction wave runs into the shock and slows it down.

Using the adaptive β–interpolation to represent the solution, we get much less mesh points. This is precisely what should have been expected for this example: Rarefaction waves are exactly represented by the β–interpolant of the left and right value. For the same accuracy as above, the corresponding mesh ($\tau = 0.05$) is shown in Fig. 4.

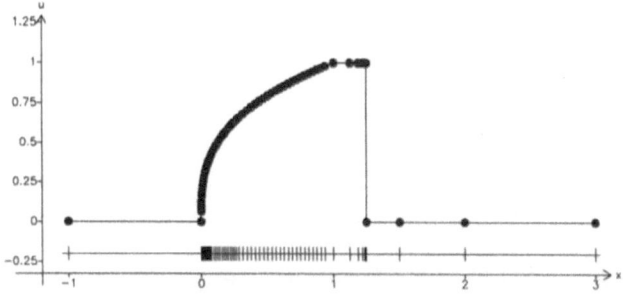

FIG. 2. *Example 1. Solution for t = 1.0, represented by adaptive linear interpolation.*

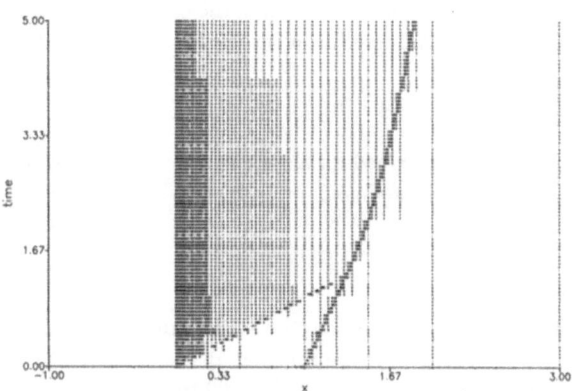

FIG. 3. *Example 1. Grid, using adaptive linear interpolation.*

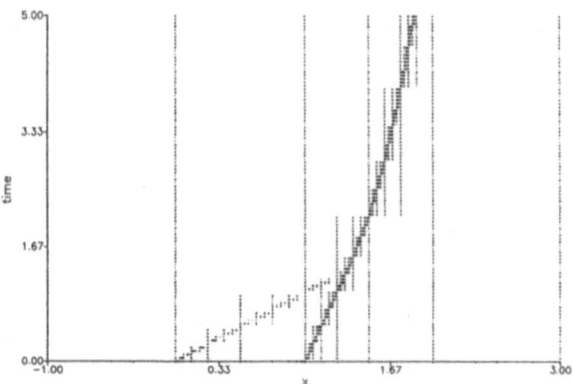

FIG. 4. *Example 1. Grid, using adaptive β–interpolation.*

79

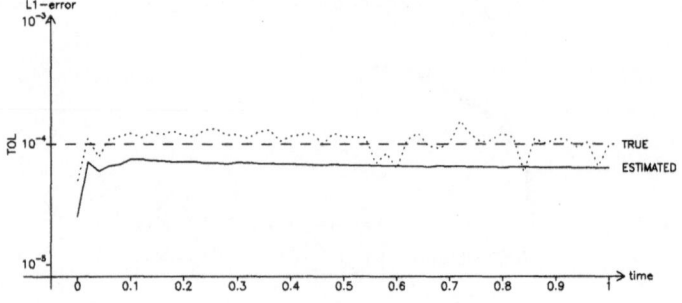

FIG. 5. *Example 1. Evolution of the error, true (...) and estimated (—).*

The quality of our error estimator can be seen in Fig. 5. We observe a slight error underestimation. Our estimated error η at time t is the *sum* of the estimated β–interpolation error η_0 of the initial data and of the estimated linear interpolation error η_t of $\hat{u}$,

$$\eta = \eta_0 + \eta_t.$$

Compared with the true L^1–error ϵ, we obtained for all of our experiments (i.e., $0 \leq t \leq 5$, $\tau = 0.05$, $\mathsf{TOL} = 10^{-1}, \ldots, 10^{-8}$, linear as well as β–interpolation of the solutions), that

$$0.33 \leq \frac{\eta}{\epsilon} \leq 1.97 \ .$$

Finally, we show in Fig. 6 the dependence of the CPU–time (in seconds) on the accuracy TOL, for the case, that we represent the solution at each time by the adaptive linear interpolation. The comparison has been made using 100 time–steps of size $\tau = 0.05$ for each accuracy. The dotted line in the double–logarithmic scale has slope $-1/2$. We observe, that asymptotically

$$\text{CPU-time} \propto \mathsf{TOL}^{-1/2} \ . \tag{8}$$

This is an optimal result, since, for the set $\mathcal{S}_{2,n}$ of piecewise linear functions with no more

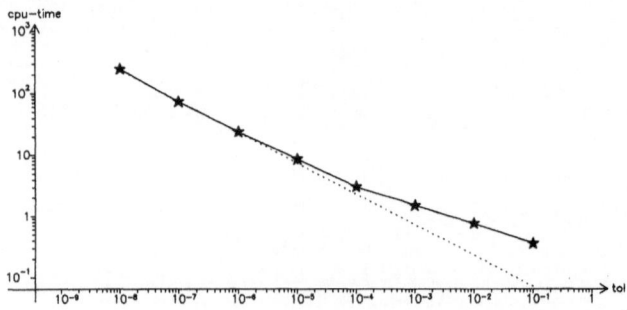

FIG. 6. *Example 1. Computing time vs.* TOL.

than n breaks in the first derivative, we obtain

$$\text{dist}(\mathcal{E}_t u_0, \mathcal{S}_{2,n}) = \mathcal{O}\left(n^{-2}\right),$$

a result, which can be found in DE BOOR [5, Theorem III.2]. Thus, the behavior (8) shows two things: First, that our mesh was chosen nearly optimal, second, that we realized our algorithm with an complexity of $\mathcal{O}(\#\text{nodes})$.

Example 2. The problem of this example is given by the inviscid Burgers equation

$$u_t + \left(\frac{u^2}{2}\right)_x = 0$$

with initial data

$$u_0(x) = \begin{cases} 2.4 + \sin(\pi(x - 0.5)) & \text{for } 0.5 \le x \le 2.5, \\ 2.4 & \text{elsewhere.} \end{cases}$$

This initial data does not have a compact support, but we can obviously modify our algorithm to handle this kind of problems.

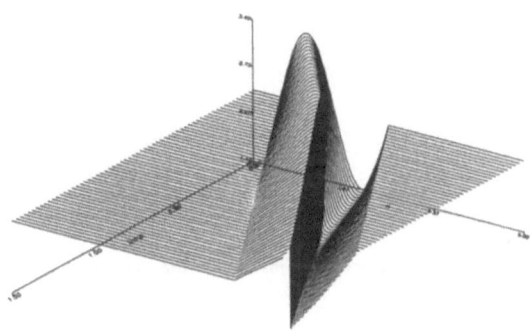

FIG. 7. *Example 2. Evolution of the computed solution.*

The continuous initial u_0 develops a shock at time $t = 1/\pi \approx 0.318$. The computed solution can be seen in Fig. 7. It was computed with accuracy TOL $= 10^{-4}$ in the time interval $[0, 1.5]$, using a time–step $\tau = 0.025$.

The corresponding mesh is shown in Fig. 8. The number of mesh points varies between 330 at the beginning and 11 at the end.

Figs. 9 and 10 show a zoom into the computed solution represented by piecewise linear interpolation just before and just after the shock formation. In both cases we have taken the position $x_s = 1.5 + 2.4t$, and have shown the computed solution in the interval $[x_s - 0.01, x_s + 0.01]$.

Note that in this problem there is no difference between piecewise linear and β-interpolation: $\mathcal{I}_\Delta = \mathcal{R}_{f,\Delta}$.

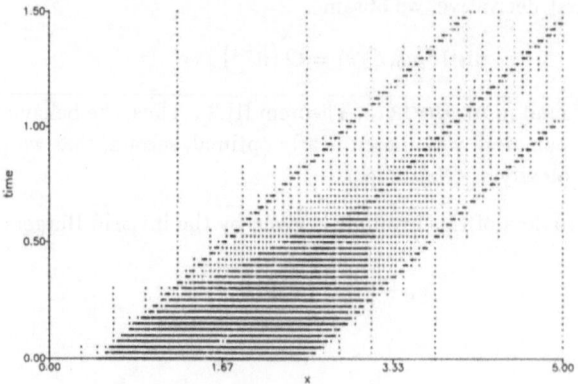

FIG. 8. *Example 2. Evolution of the mesh.*

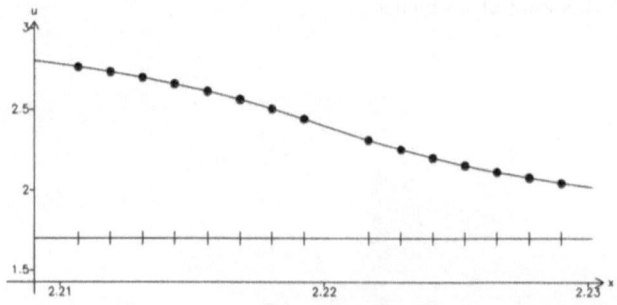

FIG. 9. *Example 2. Zoom into solution, just before the shock (t = 0.3).*

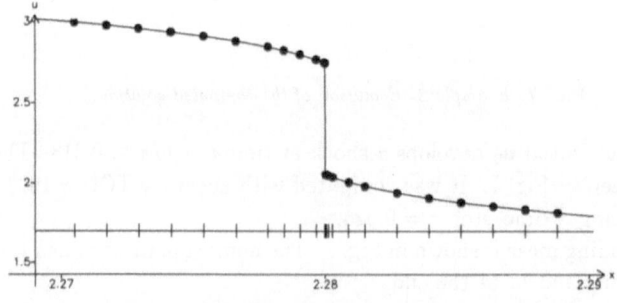

FIG. 10. *Example 2. Zoom into solution, just after the shock (t = 0.325).*

REFERENCES

[1] I. Babuška and W. C. Rheinboldt, *Error estimates for adaptive finite element computations*, SIAM J. Numer. Anal., 15 (1978), pp. 736–754.

[2] Y. Brenier, *Averaged multivalued solutions for scalar conservation laws*, SIAM J. Numer. Anal., 21 (1984), pp. 1013–1037.

[3] E. Conway and E. Hopf, *Hamilton's theory and generalized solutions of the Hamilton–Jacobi equation*, J. Math. Mech., 13 (1964), pp. 939–986.

[4] L. Corrias, M. Falcone, and R. Natalini, *Numerical schemes for conservation laws via Hamilton-Jacobi equations*, Istituti per le applicazioni del calcolo, Roma, 1992.

[5] C. de Boor, *A Practical Guide to Splines*, Springer–Verlag, New York, Heidelberg, Berlin, 1978.

[6] E. Hopf, *The partial differential equation $u_t + uu_x = \mu u_{xx}$*, Comm. Pure Appl. Math., 3 (1950), pp. 201–230.

[7] P. D. Lax, *Hyperbolic systems of conservation laws II*, Comm. Pure Appl. Math., 7 (1957), pp. 537–566.

[8] ———, *Hyperbolic Systems of Conservation Laws and the Mathematical Theory of Shock Waves*, Conf. Board Math. Sci. 11, SIAM, Philadelphia, 1973.

[9] R. J. LeVeque, *A large time step generalization of Godunov's method for systems of conservation laws*, SIAM J. Numer. Anal., 22 (1985), pp. 1051–1073.

[10] P. L. Lions, *Generalized Solutions of Hamilton–Jacobi Equations*, Research Notes in Mathematics 69, Pitman, Boston, London, Melbourne, 1982.

[11] K. W. Morton, *Private communication*.

[12] B. L. Quinn-Keyfitz, *Solutions with shocks, an example of an L_1-contractive semigroup*, Comm. Pure Appl. Math., 24 (1971), pp. 125–132.

[13] R. Roitzsch, *KASKADE Programmer's Manual*, Konrad–Zuse–Zentrum, Berlin, 1989. Technical Report TR89-5.

[14] V. M. Tikhomirov, *Convex analysis*, in Analysis II, Encyclopaedia of Mathematical Sciences Vol. 14, R. Gamkrelidze, ed., Springer–Verlag, Berlin, Heidelberg, New York, 1990.

ADAPTIVE, BLOCK–STRUCTURED MULTIGRID ON LOCAL MEMORY MACHINES

Joël Canu
Simulog
1, rue J. Joule, F-78182 Saint Quentin en Yvelines, France

Hubert Ritzdorf
GMD, Institut für Methodische Grundlagen
Schloß Birlinghoven, D-5205 Sankt Augustin 1, Germany

SUMMARY

While the general acceptance of parallel computers is accelerating, an important step to its wider employment is the demonstration that highly efficient numerical algorithms may also be efficiently implemented on· such architectures. This is particularly true of adaptive algorithms, especially in conjunction with MIMD systems with distributed memory, on which we concentrate in this paper. Our aim is to investigate the parallel performance which may be obtained when employing an adaptive multigrid algorithm, whose numerical efficiency has been successfully demonstrated on serial computers.

The multigrid approach used here is the multi-level adaptive technique. We will consider its combination with the use of two-dimensional, block-structured grids which allow geometrical flexibility while maintaining simple, logically rectangular grid structures. Communication and re-mapping strategies which minimize the overhead of the parallel implementation will be introduced. Computational results will be presented for model problems, whereby exact error properties can be used to illustrate the effectiveness of the approach, and for fluid flow applications, including the compressible Euler equations.

INTRODUCTION

Different approaches have been proposed to link multigrid solvers and locally refined meshes, starting with stretched or unstructured but nevertheless static grids. Adaptive grids are much more attractive, and have led to methods such as MLAT (Multi–Level Adaptive Technique) proposed by Brandt [2] or FAC (Fast Adaptive Composite grid method) proposed by McCormick [9] whose effectiveness have been proved for sequential machines.

Good parallel efficiencies have already been obtained when solving PDE systems with the multigrid method using static grids [8]. The use of adaptive grids changes the nature of the problem. Indeed, with static grids, the user has a full control of the grid. He can therefore optimize the parallel application with regard to load balancing, which is the most crucial parameter for the effectiveness of the parallel application. Adaptive grids, however, are a result of the computation. As a consequence, achieving load balance is a task which has to be performed at run time. The amount of data which has to be exchanged between processors increases drastically. Therefore, careful strategies, such as those we want to present here, must be employed in order not to destroy the parallel efficiency through communication overhead.

An approach to link multigrid solvers and locally refined meshes on parallel machines has already been proposed with AFAC (Asynchronous Fast Adaptive Composite grid method) developed by McCormick [10]. However, this approach has the disadvantage that performance of the multigrid method decreases (the convergence rates of AFAC are the square root of the convergence rates of FAC) and thus the maximal efficiency of AFAC is 50 percent.

PARALLEL MLAT ON BLOCK–STRUCTURED GRIDS

Block–Structured Grids

If a partial differential equation or a system of partial differential equations has to be solved on a general domain, the question arises: which grid should be used for the numerical calculations ? Certainly, the answer to this question depends on the numerical methods to be applied, the incorporation of parallelization and/or vectorization, the programming overhead. A compromise solution between geometrical flexibility and simplicity is provided by the use of block–structured grids, composed of *logically* rectangular subgrids. In order to construct such a block–structured grid on a given general domain, the domain is divided into several pieces (blocks). Logically rectangular, boundary–fitted grids are then generated on all blocks. Block–structured grids have several advantageous features:

- The use of block–structured grids allows for a great geometrical flexibility.
- Due to their logically rectangular parts, grid functions on one block can easily be described by a Fortran array, enabling easy vectorization without pointers or indirect addressing.
- The use of line relaxations, which are often much more efficient than point relaxations, is possible at least within blocks.
- The block structure provides a natural basis for the parallelization of partial differential equation solvers. Each block is mapped to one process (processor) and there is, between blocks, an overlap region whose width depends on the order of accuracy expected for the solution.

An example of such a block–structured grid is given in Figure 5. Block–structured grids are used in industrial codes in the aircraft industry for example.

MLAT

We will not give here an exhaustive description of MLAT, but only state the principle and refer to [2] and [3] for details. Essentially, the idea is to use the same sequence of grids and the same interaction between them, as in the standard multigrid method; simply, some of the finest levels may only cover a part of the domain covered by the coarser level. Relaxation sweeps are performed only on the points of the same level.

It should be emphasized that locally refined grids are not stored as composite grids, because the locally refined levels are managed exactly as normal, non–refined levels; i.e. if the area of a refined grid is logically rectangular, the refined grids can be stored as a block.

Combining MLAT and Block–Structured Grids

The aims of introducing locally refined meshes were to combine the efficiency of MLAT (already proved on serial machines) and the advantages provided by block–structured grids and to achieve a good parallel performance on local memory machines. These aims imply the following requirements:

- The local refinements must form a block–structure, in order that the advantages of block–structured grids can also be maintained on refinement levels.
  This implies, that each local refinement area of a block must be logically rectangular.
- The work must be "nearly optimally" load balanced on all levels because the main work (relaxation sweeps, etc.) is performed only on the points of the same level.
  This implies, that the refinement blocks, including grid functions, of a level must be distributed to all processes (processors).

The advantage of this concept is that the programmer works only on logically rectangular grids and that all the advantageous features mentioned above are preserved for locally refined grids. The disadvantages are that this concept is not as flexible as unstructured grids and that the load balancing may be bad if the number of refinement areas is approximately the same as the number of processes (processors).

Mapping and Communication Strategies

All the communication and mappings tasks, which are required by 2–dimensional block–structured application, are performed by the GMD Communications Library for grid–oriented problems. This is a library of Fortran routines and provides a user interface fully independent of the target machine, thus ensuring this way portability as soon as the Communications Library, which uses the portable message passing interface PARMACS, is installed on different computers.

The adaption to locally refined grids requires new functionalities of the Communications Library. From the user's point of view, a locally refined level is introduced as follows: In each block, the user determines the logically rectangular areas in which a locally refined grid should be introduced. The user passes the index ranges of these refinement areas to a routine of the Communications Library. This routine also recognizes the refinement areas of neighbouring blocks, finds out the neighbourhood relations between the refinement blocks and creates a new, temporary block–structure which covers the part of the domain where the grid should be refined. The work must then be load balanced within this new block–structure. The load balancing is implemented by the following algorithm, whereby it is assumed that the work a block has to perform is proportional to the number of grids cells. Thus, the algorithm tries to minimize the maximal number of cells which a block has in the new distributed block structure (it should be noticed, that the load balancing algorithm assumes that number of processes in the new block–structure must not be larger than in the originally created block–structure):

1. The "optimal" number of grid cells a block should have is determined.
   The optimal number of grid cells is the total number of grid cells in the new block–structure divided by the number of processes (processors) to which the block–structure should be distributed. However, there may be refinement blocks whose number of grid cells is less than the optimal one. It doesn't make sense to subdivide and distribute such a refinement block to several processes. Thus, these blocks are excluded from the computation of the "optimal" number of grid cells which is then repeated and the number of processes (processors) to which the remaining blocks of the new block–structure should be distributed is decremented correspondingly.

2. Each block B_i, whose number of grid cells is greater than the optimal one, is subdivided into at most n_i sub–blocks, where

$$n_i := \frac{\text{number of grid cells of block } B_i}{\text{"optimal" number of grid cells computed in Step 1}} \ .$$

It is possible, that a block B_i is subdivided into less than n_i sub–blocks because geometrical reasons or user–options don't allow a subdivision into n_i sub–blocks.

3. If there are free processes, i.e. if the total number of sub–blocks is smaller than the number of available processes, the block B_j which has generated the largest sub–block with C_j^{max} grid cells is subdivided into N_j, where $N_j > n_j$ is the smallest number into which the block B_j can be sub–divided. If $N_j - n_j$ is larger than the number of free processes, an attempt is made to to subdivide other blocks B_k into a smaller number of M_k sub–blocks, where $M_k < n_k$ and the number of grid cells in the sub–blocks is smaller than C_j^{max}.
   This step is continued as long as free processes are available and blocks can be subdivided.
   Note : This step looks sequential. But this step can be performed simultaneously for many blocks, so that this step, in general applications, is executed at most twice.

Note : It is not claimed, that this algorithm generates the optimally load balanced, distributed block–structure. However, it rapidly generates a relatively good load balanced block–structure. Of interest must be the form of the dependence of the time spent in the algorithm on the number of processes. Because the communication within the algorithm is performed between neighbour blocks or along an embedded tree, the dependence on the number of processes is

$$\text{time spent} = \mathcal{O} \ (\log \ (\text{number of processes})).$$

And, of course, the algorithm can be further optimized. One possible optimization is to join, if possible, the refinement areas a user passes to the Library routine before the new block structure is subdivided and distributed.

Let us return to the user's point of view. After the load balancing algorithm has finished, the Library routine returns the block size and some further block information of the refinement block on which the calling process has to work on within the refined block–structure. The user can then allocate space for the storage of grid functions within

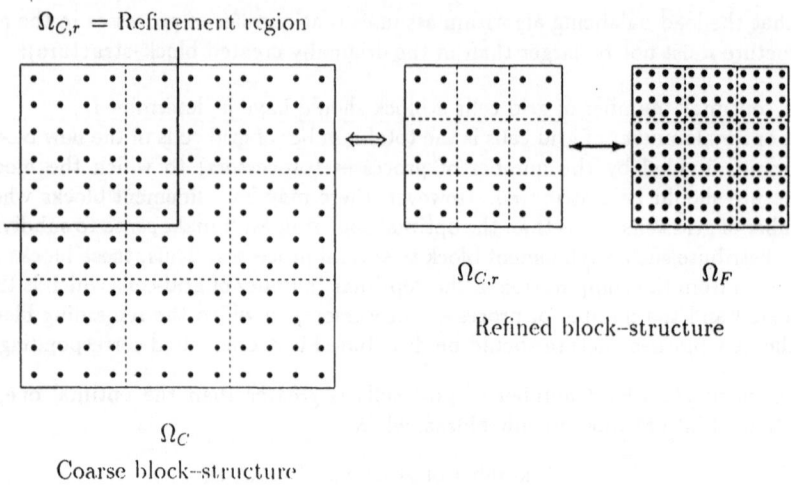

$\Omega_{C,r}$ = Refinement region

$\Omega_{C,r}$ Ω_F

Refined block–structure

Ω_C

Coarse block–structure

Figure 1: Communication strategy between coarse and refined block–structure

the refined block–structure. However, the grid functions values must be transferred between the coarse and the refined block–structure. This task is performed from two routines; a routine distributes grid function values from the coarse block structure into the refined block structure and the other collects the grid function values back to the original process. These both routines are integrated in our multigrid solver, which is part of the L_iSS environment, as follows (cf. Figure 1):

- Relaxation on the coarser grid and exchange of the overlap values within the coarse block–structure.

⇒ The Library routine which distributes the grid function values into the refined block–structure is called. Thereby, grid function values of the <u>coarse, not refined</u> grid are transferred to the refined block–structure.

- <u>Parallel and load balanced</u> execution within the refined block–structure :

  → The coarse grid corrections are computed, interpolated and added to the refined grid. The grid function values at the inner refinement boundaries are interpolated and the overlap regions are exchanged within the refined block–structure.

  - Relaxation on the refined grid and the overlap regions are exchanged within the refined block–structure.

  ← The residuals are computed and together with the grid function values restricted back to coarse grid of the refined block–structure.

⇐ The Library routine which collects the grid function values transfers the <u>coarse</u> grid values from the refined block–structure into the coarse block–structure. The overlap regions of the coarse block–structure are updated.

- The new right hand side on the coarse grid is computed and exchanged in the overlap regions.

- Relaxation on the coarser grid and exchange of the overlap values within the coarse block–structure.

In this implementation, only <u>coarse</u> grid function values are transferred between the coarse and refined block–structures. The advantages over the other possible implementation, in which coarse grid corrections and residuals are computed in the coarse block–structure, are obvious; only a quarter of the must be transferred between the block–structures and the coarse grid corrections and residuals can be computed in a load balanced manner.

PARALLEL PERFORMANCE

The achieved *parallel efficiencies*, shown in the following tables and figures, were computed using the formula (as used in [8])

$$E := \frac{\sum_{i=1}^{p} a_i}{p \max_i (a_i + c_i)}$$

where a_i denotes the (wall–clock) time for arithmetic computations in processor i, c_i the corresponding total communication time including idle time and p the number of processors. This term is used instead of the *real efficiency*

$$E_{REAL} := \frac{\text{solution time using 1 processor}}{p \times \text{solution time using } p \text{ processors}}$$

because storage limitations on distributed memory machines do not permit the solution of the complete problem on one processor. The two definitions for efficiency are equivalent under the assumptions that the processes are synchronized when the parallel application and the time measurements are begun, one process (block) is mapped to each processor, the algorithm on p processors does not involve additional arithmetic work, and the time for floating point operations does not depend on the number of processes involved.

The parallel performance is studied for pre–adaptive refinements as well as for self–adaptive refinements. For pre–adaptive refinements, the refinement zones are defined in advance (i.e. before the solution process starts) by the user; for self–adaptive refinements, the refinement zones are determined within the solution process. These proceedings have different effects when computing the parallel efficiencies: for pre–adaptive refinements, the time spent in collecting the refinement areas, creation and load balancing of the refined block–structures (see above) is not contained in the communication time c_i, because theses tasks are performed in the initialization phase before the solution and the timing starts; for self–adaptive refinements, the time spent within these tasks is contained in the communication time c_i, because these are parts of the solution process.

The results presented are obtained using the L_iSS package into which we implemented the adaptive algorithm. L_iSS is an environment for the multigrid solution of

PDE systems on 2–dimensional block–structured grids. The core PDE–solver of *LiSS* includes both serial and parallel version in the code, employs the GMD Communications Library and supports vertex–centred discretization schemes. What isn't yet implemented is the λ–FMG algorithm (cf. [4], [11]). It must be pointed at this because the computational cost of a normal FMG algorithm isn't longer $O(\mathcal{N})$, where $\mathcal{N}$ is the number of grid points, for a locally refined grid.

The first application, which should demonstrate the performance of the parallelization concept, is the solution of the Poisson's equation on a domain with a re–entrant corner. It might be objected that it is well–known how to solve the Poisson's equation on such a domain. However, the solution of the Poisson's equation is the "worst case" for the solution of a PDE on a local memory machine because the ratio of computation to communication is small.

Application to the Poisson's Equation

The first Test problem considered is the solution of the Poisson's equation

$$\Delta U = F$$

on a square with a re–entrant corner (cf. Figure 2). The boundary conditions are Dirichlet conditions which are given by the solution, $U = r^{1/2} sin(\phi/2)$, which is taken. This solution satisfies the Laplace's equation ($F = 0$).

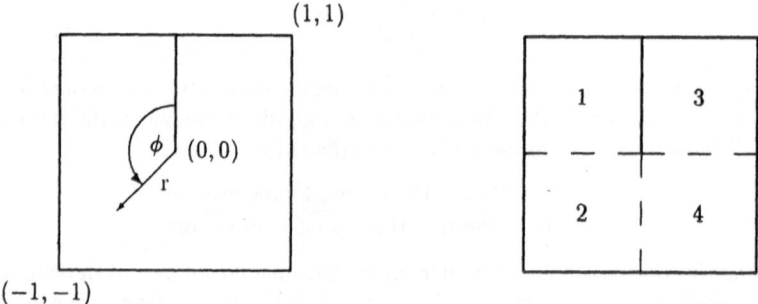

Figure 2: Geometry of the Test problem (1) and the blocking into 4 blocks

The singularity near the re–entrant corner (0,0) results only from the structure of the boundary and it was shown (cf. [1]), that the discretization error (using the standard 5–point discretization) on a uniform mesh is

$$||U - u_h||_\infty = O(h^{1/2-\epsilon}) \qquad\qquad \forall \epsilon > 0 \ .$$

Pre–adaptive as well as self–adaptive refinement algorithms are known for the solution of the Poisson's equation on such a domain (see [7], [4], [11]). The pre–adaptive refinement algorithm which is used in the computations is the algorithm of Kaspar and Remke [7]. In this algorithm, each grid point whose distance to the re-entrant corner is smaller than or equal to r_i with

$$r_i := 2^{-2i/(2-\alpha)} \qquad\qquad (\alpha = 0.5)$$

has to be discretized with mesh size $h_i := h/2^i$ where h is the finest global meshsize.. Thus, the square which contains all grid points satisfying the condition above as inner grid points and whose boundary lies on a grid line of grid h_{i-1}, is selected as i-th refinement region. The refinement process stops if $r_i \leq h_i$.

The self–adaptive algorithm which is used is the algorithm of Brandt [3]. This algorithm uses, very roughly described, the ratio of improving accuracy per work

$$Q(x) = \frac{G(x) \, |\tau_h^{2h}(x)|}{h(x)^2}$$

where

$$\tau_h^{2h} = I_h^{2h} L_h(u_h) - L_{2h}(I_h^{2h} u_h),$$

is the magnitude of the estimation of the local truncation error, I_h^{2h} is the fine-to-coarse restriction operator, L_h the discrete operator, u_h the current solution on the fine grid and $G(x)$ is the "error–weighting function". For the test problem, the error–weighting function

$$G(x) = r^{1/2}$$

is taken (cf. [4]). The global control parameter λ is selected so, that the same accuracy as in the pre–adaptive refinement algorithm is obtained.

Test Problem (1) was solved on square Cartesian meshes using standard multigrid components (i.e. FMG, V (2,1)–Cycles, Red–Black point Gauss–Seidel Relaxation) up to a final residual less than 10^{-4}. The grid function values at the inner boundaries of refinement areas were interpolated by cubic interpolation. L_iSS uses a 9-point finite volume discretization which is equivalent to the standard 5-point stencil for square Cartesian grids.

In Table 1, the solution on a non-refined grid using 16 processors is compared with the solution on locally refined grids using 4 processors, where h denotes the meshsize of the

Table 1: Parallel results for Test problem (1)

h	m	m_{loc}	Number of grid points	$\|U - u_h\|_\infty$	iPSC/2 Time [secs]	iPSC/2 Parallel efficiency	iPSC/860 Time [secs]	iPSC/860 Parallel efficiency
uniform grid; no local refinement					16 processors			
1/128	7	0	66177	1.70D-2	178.67	89.96 %	18.28	72.79 %
pre-adaptive refinement					4 processors			
1/4	7	5	375	1.77D-2	23.37	55.95 %	4.43	39.99 %
1/8	11	8	1045	4.52D-3	65.35	61.80 %	11.31	43.46 %
1/16	15	11	3263	1.13D-3	166.67	71.30 %	24.49	51.15 %
1/32	19	14	11032	2.84D-4	–	–	55.30	63.74 %
1/64	23	17	40672	7.10D-5	–	–	152.78	79.58 %
self-adaptive refinement					4 processors			
1/8	7	4	1641	1.70D-2	50.22	71.26 %	10.26	32.33 %
1/16	11	7	6581	4.28D-3	196.43	76.91 %	26.01	49.63 %
1/32	15	10	25541	1.07D-3	767.46	84.93 %	75.98	68.49 %
1/64	19	13	98891	2.67D-4	–	–	270.22	78.25 %

finest grid which covers the entire domain, m the total number of grids, m_{loc} the number of locally refined grids, *time* is the solution (wall–clock) time ($= \max_i (a_i + c_i)$). The table demonstrates that the pre– and self–adaptive algorithms achieve the same accuracy as on the corresponding non–refined grid. The parallel efficiencies of the adaptive algorithms are quite good already for small problems. However, it must be highlighted that the transfer of grid functions between processors isn't necessary if Test problem (1) is solved on 4 processors, because each block always refines in 1 region and thus the refinement block is located on the same processor.

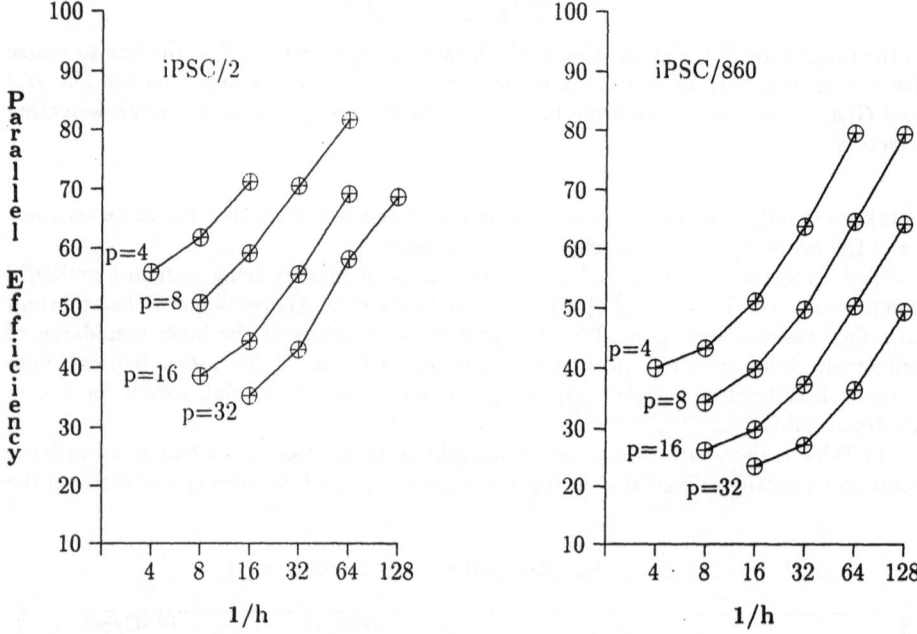

Figure 3: Parallel efficiencies for the pre–adaptive refinement on iPSC/2 and iPSC/860

In Figures 3 and 4, the parallel efficiencies are shown for different meshsizes h (h again denotes the finest global meshsize) and different number of processors p used. The parallel efficiencies of the self–adaptive refinement algorithm are worse than the efficiencies of the pre–adaptive algorithm. As already mentioned above, the reason for this is the different timing of creation and load balancing of the refined block–structures. The figures show that the difference in the parallel efficiency decreases if the problems become larger. The figures also show that already for the Poisson's equation more than 50 percent efficiency (the maximal efficiency of AFAC) is achieved. However, it must be noted that solving the Poisson's Equation using the L_iSS package isn't really the "worst case" because the environment introduces some overhead (for example, the 9–point stencil is used instead of the 5–point stencil) and a vectorized Red-Black relaxation was used which is really a 4 colour relaxation and somewhat more expensive than the non–vectorized relaxation.

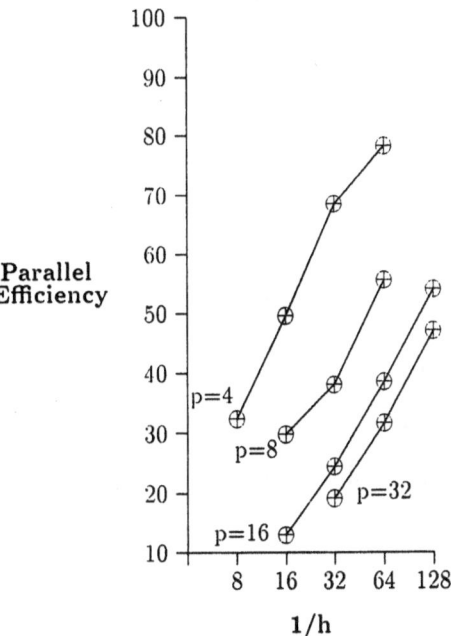

Figure 4: Parallel efficiencies for the self-adaptive refinement on iPSC/860

Application to the Euler Equations

The Test problem (2) is to solve the Euler equations

$$\frac{\partial f}{\partial x} + \frac{\partial g}{\partial y} = 0, \tag{1}$$

around the well-known NACA0012–airfoil with a far–field Mach number $M = 0.85$ and an angle of attack $\alpha = 1°$ (cf. [12]). In Equation 1, we have

$$f = \begin{bmatrix} \rho u \\ \rho u^2 + p \\ \rho u v \\ (E + p)u \end{bmatrix} \quad \text{and} \quad g = \begin{bmatrix} \rho v \\ \rho u v \\ \rho v^2 + p \\ (E + p)v \end{bmatrix}$$

where ρ denotes the density, u, v the Cartesian velocity components, E the total energy, and p the pressure. The state equation allows the closure of the previous system:

$$p = (\gamma - 1)(E - \frac{1}{2}\rho(u^2 + v^2)). .$$

The peculiarity of the Euler equations, namely their hyperbolicity, has led to the development of specific numerical methods which now allow the employment of effective multigrid solvers ([6],[5]). Following [6], we use a finite volume discretization combined with the well-known Osher's flux–difference–splitter, but we prefer the vertex centered scheme to the cell centered one. The characteristics of the computation are as follows:

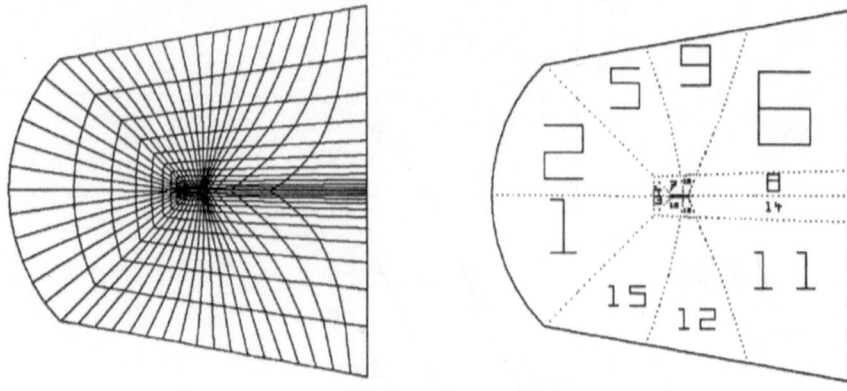

Figure 5: Finest global grid and block–structure of Test problem (2)

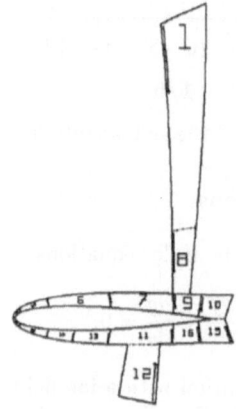

Figure 6: Block–structure of the finest refinement level of Test problem (2)

- A C-grid is built around the airfoil with a far–field boundary at a distance of approximatively 10 chord lengths (cf. Figure 5),
- the global domain is divided into 16 blocks (cf. Figure 5),
- 2 refinement levels are defined around the airfoil (cf. Figure 6).
- the grid function values at inner boundaries of refinement areas are interpolated by cubic interpolation. This is a non–conservative treatment of the inner boundaries.

As these equations admit discontinuous solutions (shocks or contact discontinuities), an important gain, in terms of solution time, is expected from the use of an adaptive technique. Unfortunately, we did not have enough time to investigate the development of a suitable refinement criterion. That is the reason why only pre–adaptive results will be presented for this application. Therefore, we will concentrate on comparison of the parallel performance, and the accuracy obtained with and without local refinements, since the solution time gain may be much greater with a proper refinement criterion.

Fluid dynamics results

Concerning the fluid dynamics results, reference solutions can be found in [12]. For each of our computation, the residual is reduced to 10^{-3}. To measure the accuracy of our solutions, we consider the two following parameters:

- the pressure coefficient ($C_p = \frac{p-p_\infty}{1/2\rho_\infty V_\infty^2}$) along the upper and lower surface of the airfoil,
- the position of the upper and lower shocks (x_s).

We compare the solution on a grid with two local refinement levels (cf. Figure 6), and the solution on a uniform grid built by refining the finest global grid of the previous computation, twice. The goal of this proceeding is to obtain in both cases the same definition (i.e. the same grid coordinates) of the profile of the wing, in order to be able to compare the C_p, since the evolution of the pressure along the airfoil depends on its curvature (grid coordinates). This proceeding is currently necessary because the grid coordinates of a local refinement region including the physical boundary are interpolated by bi-linear interpolation, so that refining a curved boundary does not give more accuracy in the definition of the boundary. That explains the oscillations which can be seen in the C_p curves (cf. Figure 7 and 8). That result shows that, as the adaptive grid is a result of the computation, the data describing the domain must be somehow included in the solution process.

The results show a slight shifting of the shocks and a little underestimation of the minima of the pressure coefficients (cf. Figures 7 and 8 and Table 2 - Adaptive grid (1)). In our opinion, the slight deterioration of accuracy is due to the non-conservative treatment at the boundary of the refinement area. To verify it, we have enlarged the refinement area in such a way that its boundary is in a part of the domain where the solution is smoother than in the first case (cf. Figure 9 - Adaptive grid (2)). Improvements of the accuracy of the results can then be seen, particularly concerning the level of the C_p minima. It has also to be mentioned that, much more important than the accuracy deterioration is the deterioration of the convergence rate on the final level of the Adaptive grid (1). An acceptable convergence rate is retrieved with the Adaptive grid (2) (cf. Table 2), but the number of points does not allow us to expect an important gain in terms of computation time. These results prove that a conservative treatment of the boundaries of the local refinement zones can not be avoided.

Table 2: Convergence rates and position of the shocks and minimal values of pressure coefficients with uniform and adaptive grids

	Uniform grid		Adaptive grid (1)		Adaptive grid (2)	
	upp. surf.	low. surf.	upp. surf.	low. surf.	upp. surf.	low. surf.
x_s	0.825	0.65	0.815	0.625	0.82	0.635
C_{p_m}	-1.00	-0.75	-0.95	-0.70	-1.00	-0.75
Conv. rate	0.42		0.79		0.51	
# grid points	49967		11313		29438	

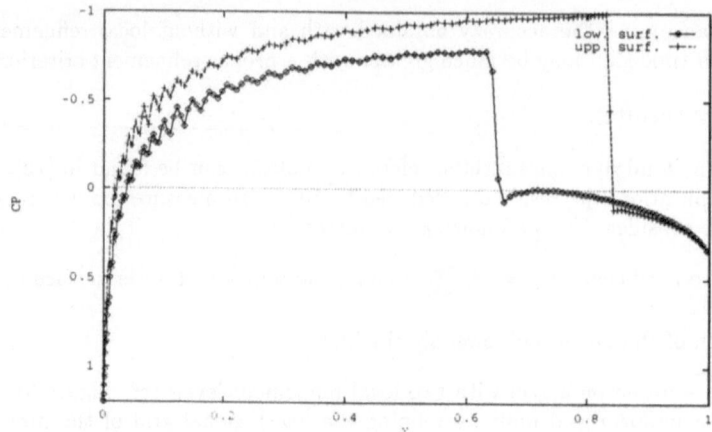

Figure 7: Lower and upper surface pressure coefficient on the uniform grid

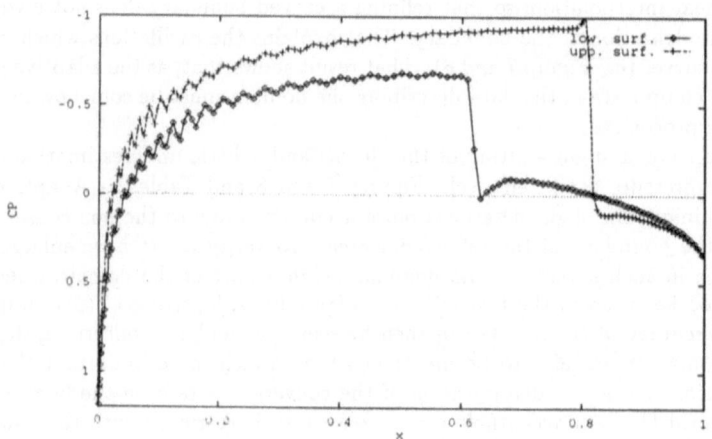

Figure 8: Lower and upper surface C_p obtained with Adaptive grid (1)

Parallel results

We are interested in the parallel performance (cf. Table 3) obtained with the config-uration shown Figure 6 (Adaptive grid (1)), which is supposed to be representative of what could be obtained with a self–adaptive procedure, and with an other configuration (Adaptive grid (3)), where the refinement areas are chosen such a way that a perfect load balance is obtained on all levels. The loss of efficiency when using the adaptive technique is moderate, 10%. The main reason of this loss is certainly the decrease in the number of points pro block in the local refinement levels. Indeed, a computation on a uniform grid with only 4 levels (9432 points on the composite grid) yields a parallel efficiency of 77 %. Another reason is the non–optimal load balancing that explains the

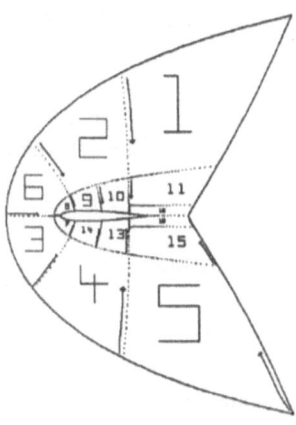

Figure 9: Block–structure of the finest refinement level of the Adaptive grid (2)

Table 3: Parallel efficiencies for the uniform and adaptive grids on the iPSC/2

	Number of levels	Nb. of points on the comp. grid	Parallel efficiency
Uniform grid	5	49667	87 %
Adaptive grid (1)	5	11313	77 %
Adaptive grid (3)	5	17348	81 %

difference of efficiency between the two adaptive configurations (4 %). The last reason is the additional communication work performed for the local refinement, which is much less important than in the two other cases, at least when the computation work per point is great, as it is in the case of the Euler equation.

CONCLUSIONS

It has been shown that MLAT and block–structured grids are an effective combination for the adaptive solution of PDE systems on general domains. The current investigations indicate that the parallelization concept presented can be implemented on MIMD systems without severe loss in parallel efficiency and demonstrate that parallel efficiencies of more than 50 percent can be achieved for the number of processors used. Further investigations must be the conservative treatment of inner boundaries of refinement regions, the correct generation of the physical boundary coordinates on refined grids and a further optimization of the load balancing algorithm.

ACKNOWLEDGEMENTS

We would like to thank the Zentralinstitut für Angewandte Mathematik of the Research Centre Jülich for providing the computing time on the Intel iPSC/860.

REFERENCES

[1] Bramble, J.H., Hubbard, B.E., Zlamal, M.: "Discrete Analogues of the Direichlet Problem with Isolated Singularities", SIAM J. Numerical Analysis, 5, 1968, pp. 1–25.

[2] Brandt, A.: "Multi–Level Adaptive Solutions to Boundary–Value Problems", Math. Comp., Vol. 31, (1977), pp. 333–390.

[3] Brandt, A.: "Multi–Level Adaptive Techniques for Singular Pertubation Problems", In "Numerical Analysis of Singular Pertubation Problems", (P.W.Hemker and J.J.M. Miller eds.), Academic Press, 1979, pp. 53–142.

[4] Brandt, A.; Bai, D.: "Local Mesh Refinement Multilevel Techniques", SIAM J. Sci. Statist. Comput. 8 (1987), pp. 109–134.

[5] Dick, E.: "Multigrid Formulation of Polynomial Flux-Difference Splitting for Steady Euler Equations", J. Comp. Phys., 91, 161 (1990)

[6] Hemker, P.W.; Koren, B.: "A Non–Linear Multigrid Method for the Steady Euler Equations", CWI Report NM-R8621, (1986).

[7] Kaspar, W.; Remke, R.: "Die numerische Behandlung der Poisson Gleichung auf einem Gebiet mit einspringenden Ecken", Computing, Vol. 22, 1979, pp. 141–151.

[8] Lonsdale, G.; Schüller, A.: "Multigrid Efficiency for Complex Flow Simulations on Distributed Memory Machines", Parallel Computing, Vol. 19, 1993, pp. 23–32.

[9] McCormick, S; Thomas, J.: "The Fast Adaptive Composite Grid (FAC) Method for Elliptic Equations", Math. Comp., Vol. 46, (1986), pp. 439–456.

[10] McCormick, S; Quinlan, D.: "Asynchronous Multilevel Adaptive Methods for Solving Partial Differential Equations on Multiprocessors: Performance Results", Parallel Computing, Vol. 12, 1989, pp. 145–156.

[11] Ritzdorf, H.: "Lokal verfeinerte Mehrgitter-Methoden für Gebiete mit einspringenden Ecken", Diplomarbeit, Universität Bonn, 1984.

[12] Rizzi, A.; Viviand, H.: "Numericals Methods for the Computation of Inviscid Transonic Flows with Shock Waves", Notes on Numerical Fluid Mechanics, Vol. 3.

Biorthogonal Wavelets and Multigrid

Stephan Dahlke

Institut für Geometrie und Praktische Mathematik, RWTH Aachen
52056 Aachen, Germany

Angela Kunoth[*]

FB Mathematik und Informatik, Institut für Informatik, FU Berlin
14195 Berlin, Germany

Abstract

We will be concerned with the solution of an elliptic boundary value problem in one dimension with polynomial coefficients. In a Galerkin approach, we employ biorthogonal wavelets adapted to a differential operator with constant coefficients, and use the refinement equations to set up the system of linear equations with exact entries (up to round-off). For the solution of the linear equation, we construct a biorthogonal two-grid method with intergrid operators stemming from wavelet-type operators adapted to the problem.

Key words: Adapted biorthogonal wavelets, boundary value problems, refinement equations, two-grid (multi-grid) methods.

AMS subject classification: 65L10, 65L60, 65F10, 41A30.

1   Introduction

When solving elliptic boundary value problems by means of a Galerkin approach, the trial functions for the approximation spaces are usually chosen with compact support such that the resulting stiffness matrices are as sparse as possible. But even then, the resulting system of equations is so large that one might want to apply a multigrid method to obtain the solution much faster than with just using a standard iterative method.

These two aspects will be our starting point to employ the concept of shift-invariant spaces. We will be concerned with (elliptic) boundary value problems with variable coefficients, and a first attempt is to use special biorthogonal wavelets adapted to a corresponding differential operator with constant coefficients as trial functions in a Galerkin approach. This will lead to stiffness and mass matrices with band-block structure, and in the case of polynomials as variable coefficients, the entries of the matrices can be determined exactly (up to round-off) by solving an eigenvector problem. After setting up the system of linear equations, we develop intergrid transfer operators originating from biorthogonal wavelet operators to be used in a two-(multi-)grid method. The flexibility

---

[*]The work of this author is supported by the Deutsche Forschungsgemeinschaft in the Graduiertenkolleg "Algorithmische Diskrete Mathematik" at the Freie Universität Berlin.

within the concept of biorthogonal wavelets will then be utilized to accelerate the rate of convergence.

It should be noted, however, that using any kind of wavelets causes trouble with satisfying boundary conditions. Since our aim is to indicate two relevant advantages of the shift-invariant concept, we adhere to periodic boundary conditions here. Furthermore, we will restrict ourselves to a one-dimensional problem, in particular, we will consider a Sturm's boundary value problem with polynomial coefficients. For more general coefficients, one could use an approximation in terms of shifts of some refinable function as indicated in [DK2]. Generalizing our attempts to higher dimensions will cause problems only for the task of constructing corresponding biorthogonal wavelets as it was done e.g. in [DW2], which we do not use here in order to avoid complicating matters.

To fix the setup, we will treat a general Sturm's boundary value problem in self-adjoint form

$$
\begin{aligned}
-(p(x)u(x)')' \;+\; q(x)u(x) &= g(x) \quad \text{on } (a,b) \subset \mathbb{R}, \\
u(a) &= u(b), \qquad u'(a) = u'(b).
\end{aligned}
\tag{1.1}
$$

where $p(x) \geq 0$ on (a,b), $p(a) = p(b)$ is required.

Since the boundary terms vanish due to periodicity, the Galerkin formulation of (1.1) is to find some u in the Sobolev space $H^1(a,b)$ satisfying

$$
\int_a^b (p(x)u'(x)v'(x) + q(x)u(x)v(x))\, dx = \int_a^b g(x)v(x)\, dx
\tag{1.2}
$$

for all $v \in H^1(a,b)$.

The remainder of this paper is organized as follows. It consists of two parts that can essentially be seen independent of each other, the first part containing Section 2 and 3, and the second part consisting of the remaining two sections. In Section 2, we construct special biorthogonal wavelets adapted to the corresponding differential operator with constant coefficients which we will use in Section 3 as trial funcions in a Galerkin approach for setting up the system of linear equations. In particular, special care will be devoted to the exact determination of the entries of the matrices and the right hand side. While this material and the corresponding software programs are pretty developed, the remaining two sections are of a more heuristic and less rigorous nature. In Section 4, we derive a biorthogonal two-grid method generalizing the approach in [R]. The so-gained flexibility will be used to adapt the two-grid method appropriately to the problem which will be the topic in Section 5. A more detailed and thorough investigation of the latter points will be taken up in a forthcoming paper, and an extended version of the results presented here can be found in [DK1].

2 Biorthogonal Wavelets Adapted to a Differential Operator

For the solution of Sturm's boundary value problem in weak formulation (1.2), we attempt to use biorthogonal wavelets adapted to the differential operator $-\frac{d^2}{dx^2}$. We briefly recall the construction as carried out in [DW1].

In general, a function ψ is called an (orthonormal) **wavelet** if the scaled and translated versions form an (orthonormal) basis of $L^2(\mathbb{R})$. The most important tool for the construction of wavelets is the multiresolution analysis approximation of functions introduced by Mallat [Ma], see also [Mey]. A **multiresolution approximation** of $L^2(\mathbb{R})$ is a nested sequence $\{V_j\}_{j\in\mathbb{Z}}$ of closed shift-invariant subspaces of $L^2(\mathbb{R})$ such that their union is dense in $L^2(\mathbb{R})$ while their intersection is $\{0\}$. In addition, one requires $f(\cdot) \in V_j$ iff $f(2\cdot) \in V_{j+1}$, and that V_0 is spanned by the integer translates of one function φ called the **generator** of the multiresolution analysis.

A natural way to construct an orthonormal basis is to find a function ψ in the orthogonal complement W_0 of V_0 in V_1 whose translates are orthonormal i.e.,

$$\langle \psi(\cdot), \psi(\cdot - \ell)\rangle := \int_R \psi(x)\overline{\psi(x-\ell)}\,dx = \delta_{0l}, \quad \langle \varphi(\cdot), \psi(\cdot-\ell)\rangle = 0\ , \quad \ell \in \mathbb{Z}, \qquad (2.1)$$

and whose translates span W_0. Hence, defining $W_j := \{f \in L^2(\mathbb{R}) : f(2^{-j}\cdot) \in W_0\}$, one has $V_{j+1} = V_j \oplus W_j$, $L^2(\mathbb{R}) = \oplus_{j\in\mathbb{Z}} W_j$, and the functions $\psi_{jk} := 2^{j/2}\psi(2^j\cdot - k)$ form a complete orthonormal system for $L^2(\mathbb{R})$.

ψ is called the **orthonormal wavelet** associated with φ. Orthonormal wavelet bases can be generalized a little bit further. In many applications, it is convenient to deal with **biorthogonal wavelet bases**. In this case, one has two hierarchical sequences of approximation spaces $\{V_j\}_{j\in\mathbb{Z}}$, $\{\tilde{V}_j\}_{j\in\mathbb{Z}}$. Now the space W_0 will be a complement to V_0 in V_1, but not necessarily the orthogonal one. The orthogonality condition from before is now replaced by the conditions $\tilde{W}_0 \perp V_0$, $W_0 \perp \tilde{V}_0$. As above, one looks for functions $\psi, \tilde{\psi}$ whose translates span $W_0, \tilde{W}_0$ and moreover satisfy the biorthogonality condition

$$\langle \psi_{jk}, \tilde{\psi}_{i\ell}\rangle = \delta_{ij}\delta_{\ell k} . \qquad (2.2)$$

For the construction of such a biorthogonal wavelet basis, it is necessary that the generators φ and $\tilde{\varphi}$ form a **dual pair**, i.e.,

$$\langle \varphi(\cdot), \tilde{\varphi}(\cdot - k)\rangle = \delta_{0k} \quad \forall k \in \mathbb{Z}. \qquad (2.3)$$

The idea in [DW1] was to find a biorthogonal wavelet ψ such that

$$\langle \varphi'(\cdot), \psi'(\cdot - k)\rangle = 0 \quad \forall k \in \mathbb{Z} \qquad (2.4)$$

holds. It was shown in [DW1] that the stiffness matrix originating from the differential operator $-\frac{d^2}{dx^2}$ is of block-diagonal form when one uses trial functions in a Galerkin approach consisting of dilates and translates of this wavelet ψ.

Since the sequences $\{V_j\}$, $\{\tilde{V}_j\}$ are nested, it can be shown that φ and $\tilde{\varphi}$ satisfy the two-scale-equations

$$\varphi(x) = \sum_{k\in\mathbb{Z}} a_k\varphi(2x - k)\ , \qquad \tilde{\varphi}(x) = \sum_{k\in\mathbb{Z}} b_k\tilde{\varphi}(2x - k)\ , \quad x \in \mathbb{R}, \qquad (2.5)$$

with **masks a,b**. Fourier-transforming (2.5) leads to

$$\hat{\varphi}(\xi) = \tfrac{1}{2}a(z)\hat{\varphi}\left(\tfrac{\xi}{2}\right)\ , \qquad \hat{\tilde{\varphi}}(\xi) = \tfrac{1}{2}b(z)\hat{\tilde{\varphi}}\left(\tfrac{\xi}{2}\right)\ , \quad z = e^{-i\frac{\xi}{2}}\ , \qquad (2.6)$$

where the **symbol** $a(z)$ corresponding to some mask **a** is defined by $a(z) := \sum_{k\in\mathbb{Z}} a_k z^k$. The aim is to determine $b(z)$ corresponding to some given $a(z)$ such that (2.4) holds. Some calculation shows that, for arbitrary $a(z), b(z)$ must satisfy

$$b(z) = \frac{4a(z)g(\xi)}{|a(z)|^2 g(\xi) + |a(-z)|^2 g(\xi + 2\pi)}\ , \quad g(\xi) := \sum_{n\in\mathbb{Z}}(\xi + 4\pi n)^2 \left|\hat{\varphi}\left(\tfrac{\xi}{2} + 2\pi n\right)\right|^2 . \tag{2.7}$$

One observes that $b(z)$ given in (2.7) is in general not a Laurent polynomial implying that its mask will not have finite support. We can circumvent this trouble when starting with a function φ associated with an appropriate symbol. In particular, choose $a(z)$ of the form

$$a(z) := \tfrac{1}{2}(1+z)q_N(z), \tag{2.8}$$

where $q_N(z)$ is the symbol of the Daubechies' refinable function η corresponding to the parameter N, which induces compactly supported orthonormal wavelets ([Dau]). In this case, we have

$$\varphi' = \eta - \eta(\cdot - 1) \tag{2.9}$$

(see [CDM], Chapter 8, or [DW1], Theorem 2.3). As it was shown in [Dau], the symbol of η can be written as

$$q_N(z) = (1+z)^{N-1}Q_N(z) , \tag{2.10}$$

where $Q_N(z)$ has some "nice" behaviour. Employing (2.9), the function g can easily be computed, and the general formula (2.7) leads to

$$b(z) = 2z(1+z)^{N-1}Q_N(z). \tag{2.11}$$

Reiteration of (2.6) results in

$$\hat{\varphi}(\xi) = \prod_{j=1}^{\infty} \tfrac{1}{2}b(e^{-i\xi/2^j}) \tag{2.12}$$

which converges in L_2 for $N \geq 4$. To be on the safe side, we choose $N = 5$ here (see as well the remarks at the end of this chapter). The coefficients for q_N can be found in [Dau] p.980. Note that we have used a different normalization so that we have to multiply all coefficients by $2^{1/2}$, i.e.,

$$q_N(z) = 2^{1/2} \sum_{n \in \mathbb{Z}} h_n^N z^n , \quad z = e^{-i\frac{\xi}{2}} , \tag{2.13}$$

where h_N are Daubechies' coefficients.

For the symbol of φ defined in (2.8), we obtain by (2.13)

$$a(z) = 2^{-1/2}(1+z) \sum_{n \in \mathbb{Z}} h_n^N z^n = \sum_{n \in \mathbb{Z}} a_n z^n , \tag{2.14}$$

where the coefficients are given by

$$a_n = 2^{-1/2}(h_n^N + h_{n-1}^N). \tag{2.15}$$

The symbol for $\tilde{\varphi}$ will be computed analogously. $b(z)$ must satisfy

$$b(z) = 2\frac{z}{1+z} q_N(z) \tag{2.16}$$

which is equivalent to

$$b(z)(1+z) = 2z q_N(z).$$

Inserting (2.13) leads to

$$\sum_{n \in \mathbb{Z}} (b_n + b_{n-1})z^n = 2^{3/2} \sum_{n \in \mathbb{Z}} h_{n-1}^N z^n,$$

and comparing coefficients yields

$$b_{n-1} = 2^{3/2} h_{n-1}^N - b_n , \quad \text{for } n < 10. \tag{2.17}$$

The coefficients vanish for $n \geq 10$ since $N = 5$, and $q_5(z) = \sum_{n=0}^{9} h_n^5 z^n$. We finally arrive at the following table where the coefficients indeed satisfy the necessary requirement so that the corresponding subdivision schemes converge (cf. [CDM]),

$$\sum_{n \in \mathbb{Z}} a_n = \sum_{n \in \mathbb{Z}} b_n = 2 . \tag{2.18}$$

Table: Mask coefficients of refinable functions

n	h_n^5	a_n	b_n
0	0.160102397974	0.11320949	0
1	0.603829269797	0.54018125	0.45283796
2	0.724308528438	0.93913523	1.2550491
3	0.138428145901	0.61004694	0.79360475
4	-0.242294887066	-0.07344488	-0.40207084
5	-0.032244869585	-0.19412891	-0.28324257
6	0.077571493840	0.03205077	0.19204033
7	-0.006241490213	0.05043793	0.02736498
8	-0.012580751999	-0.01330933	-0.04501858
9	0.003335725285	-0.00653722	0.00943484
10	0	0.00235871	0

The corresponding wavelets mentioned in (2.2) then satisfy the equations

$$\psi(x) = \sum_{k \in \mathbb{Z}} (-1)^k b_{1-k} \varphi(2x - k) , \quad \tilde{\psi}(x) = \sum_{k \in \mathbb{Z}} (-1)^k a_{1-k} \tilde{\varphi}(2x - k) , \quad x \in \mathbb{R}. \tag{2.19}$$

We recall that the number of nonvanishing mask coefficients for $\varphi, \tilde{\varphi}$ guides their smoothness, e.g., it was shown in [Dau], p.984, that $\eta \in C^{1.6}$ for $N = 5$. The construction of biorthogonal wavelets with respect to the energy inner product therefore yields here, by (2.9), $\varphi \in C^{2.6}$, $\tilde{\varphi} \in C^{0.6}$, implying the same smoothness for their corresponding wavelets $\psi, \tilde{\psi}$. The smoothness is intimately connected with the amount of vanishing moment conditions. More precisely, it was shown in [CDM] p.157 that a refinable $\varphi \in C_0^d(\mathbb{R}^s)$ implies $\Pi_d \subseteq \{\varphi(\cdot - \alpha) : \alpha \in \mathbb{Z}^s\}$ where Π_d are the polynomials of degree at most d. In the biorthogonal setting dealed with here, we can therefore represent polynomials up to degree two only by shifts of φ (see Chapter 3).

In practice, it might be possible to choose the parameter N smaller than 5. But in such a case, there might not exist a dual basis implying that one might not be able to estimate the condition number of the resulting system of linear equations, and find a preconditioner so that the condition number is uniformly bounded with respect to the refinement level, see [DK2] and [DW1].

3   Generation of System of Linear Equations

In the following, w.l.o.g., the domain of the boundary value problem shall be the interval $(-1, 1)$. We will consider the problem

$$\int\limits_{-1}^{1} (p(x)u'(x)v'(x) + q(x)u(x)v(x))\, dx = \int\limits_{-1}^{1} g(x)v(x)\, dx ,$$ (3.1)

where $p(x) \geq 0$, $p(-1) = p(1)$ and $p, q, g \in \Pi_k(-1, 1)$, the polynomials of degree at most k on $(-1, 1)$.

Our aim is to find the two-periodic function u that solves (3.1) for all two-periodic test functions v. For setting up the system of linear equations corresponding to (3.1) using biorthogonal wavelets, we have to determine inner products of periodic functions. But, as we will see, we can trace these terms to inner products of functions on $\mathbb{R}$. Let us first introduce an appropriate two-periodic setting.

We define the two-periodic periodization operator by

$$[f](x) := \sum_{k \in \mathbb{Z}} f(x + 2k) .$$ (3.2)

Let the space $[V_j]$ be defined as the span of the functions

$$\varphi_\ell^j(x) := 2^{j/2}[\varphi(2^j \cdot -\ell)] = 2^{j/2} \sum_{k \in \mathbb{Z}} \varphi(2^j(x + 2k) - \ell)$$ (3.3)

with $[\varphi(r(\cdot))](x) = \sum_{k \in \mathbb{Z}} \varphi(r(x+2k))$. Then a basis of $[V_j]$ consists of 2^{j+1} basis functions φ_i^j, $i = 0, \ldots, 2^{j+1} - 1$ (cf. [Mey] p.112).

Furthermore, we define an inner product on $L_2(\tilde{T})$, $\tilde{T} = \mathbb{R}/2\mathbb{Z}$ the one-dimensional torus, by $[u, v] := \int\limits_{-1}^{1} u(x)\overline{v(x)}\, dx$. Then trivially the inclusion $[V_j] \subset L_2(\tilde{T})$ holds. Moreover, it is easy to check that the direct sum decompositions carry over to the periodic setting, i.e., $[V_j] = [V_{j-1}] \oplus [W_{j-1}]$ so that $L_2(\tilde{T}) = [V_0] \oplus_{j \geq 0} [W_j]$. The proof of the following remark can be found in [DK1], see also [DPS].

Remark 3.1 *If the biorthogonal wavelet system satisfies (2.3), it follows, for every $j \in \mathbb{N}_0$,*

$$[\varphi_n^j, \tilde{\varphi}_m^j] = \begin{cases} 1 , & m = n + 2^{j+1}r \ \text{for some } r \in \mathbb{Z} \\ 0 , & \text{else} \end{cases} , \qquad n, m \in \mathbb{Z}.$$ (3.4)

After these preliminaries, we can start with the setup of the linear equations originating from (3.1). But instead of working with a multilevel basis which requires the determination of many integrals of various basis functions on different refinement levels, we will only use the nodal basis functions on a fixed finest level m. We calculate the respective inner products, and perform a change of basis, as indicated in [DK2], afterwards.

To this end, we fix $m \in \mathbb{N}$ as finest level so that $[V_m]$ is spanned by 2^{m+1} basis functions,

$$[V_m] = \text{span}\{\varphi_0^m, \ldots, \varphi_{2^{m+1}-1}^m\} .$$ (3.5)

Representing $u \in [V_m]$ in terms of these nodal basis functions and choosing $v(x)$ consecutively as basis functions, the equation (3.1) reads

$$\sum_{\ell=0}^{2^{m+1}-1} u_\ell^m \int_{-1}^{1} p(x)\left(\tfrac{d}{dx}\varphi_\ell^m(x)\right)\left(\tfrac{d}{dx}\varphi_r^m(x)\right) dx + \sum_{\ell=0}^{2^{m+1}-1} u_\ell^m \int_{-1}^{1} q(x)\varphi_\ell^m(x)\varphi_r^m(x)\, dx \quad (3.6)$$

$$= \int_{-1}^{1} g(x)\varphi_r^m(x)\, dx \ , \quad \text{for} \quad r = 0,\dots,2^{m+1}-1.$$

Therefore, we need to determine

$$a_{\ell r} \ := \ \int_{-1}^{1} p(x)\left(\tfrac{d}{dx}\varphi_\ell^m(x)\right)\left(\tfrac{d}{dx}\varphi_r^m(x)\right)\, dx\ ,\ \ell,r = 0,\dots,2^{m+1}-1, \qquad (3.7)$$

$$b_{\ell r} \ := \ \int_{-1}^{1} q(x)\varphi_\ell^m(x)\varphi_r^m(x)\, dx\ , \qquad \ell,r = 0,\dots,2^{m+1}-1, \qquad (3.8)$$

$$f_r \ := \ \int_{-1}^{1} g(x)\varphi_r^m(x)\, dx, \qquad r = 0,\dots,2^{m+1}-1. \qquad (3.9)$$

Note that $\tfrac{d}{dx}\varphi_\ell^m(x) = 2^m(\varphi_\ell^m)'(x)$. Treating (3.7) first, inserting (3.3) leads to

$$a_{\ell r} = \sum_{k,t \in \mathbb{Z}} 2^{2m} \int_{-2^m}^{2^m} p(2^{-m}y)\varphi'(y + 2^{m+1}k - \ell)\varphi'(y + 2^{m+1}t - r)\, dy. \qquad (3.10)$$

Similarly, we obtain for $(b_{\ell r})$ in (3.8)

$$b_{\ell r} = \sum_{k,t \in \mathbb{Z}} \int_{-2^m}^{2^m} q(2^{-m}y)\varphi(y + 2^{m+1}k - \ell)\varphi(y + 2^{m+1}t - r)\, dy\ , \qquad (3.11)$$

and for (f_r) in (3.9)

$$f_r = \sum_{k \in \mathbb{Z}} 2^{-\frac{m}{2}} \int_{-2^m}^{2^m} g(2^{-m}y)\varphi(y + 2^{m+1}k - r)\, dy\ . \qquad (3.12)$$

The next step in the setup process is the representation of the polynomials p, q, and g by shifts of a refinable function.

To formulate the following observation which is proved in [DM] or [DK1], let ϕ denote any function satisfying a refinement equation with finite mask $\mathbf{f}$ satisfying (2.18),

$$\phi(x) = \sum_{l \in \mathbb{Z}} f_l \phi(2x - l)\ , \quad x \in \mathbb{R}\ . \qquad (3.13)$$

Lemma 3.1 *For any refinable function ϕ satisfying (3.13) and $k \in \mathbb{N}$, the identity*

$$\int_{\mathbb{R}} y^k \phi(y)\, dy = \frac{1}{2(2^k-1)} \sum_{r=0}^{k-1} \binom{k}{r} \Big(\sum_{l\in\mathbb{Z}} f_l l^{k-r}\Big) \int_{\mathbb{R}} y^r \phi(y)\, dy . \tag{3.14}$$

holds.

One can obtain explicit expressions upon recalling the normalization for $k = 0$, $\int \phi(y)\, dy = 1$ (see [DK1] for $k = 1, \ldots, 4$).

We now employ Lemma 3.1 as a remedy to represent polynomials by means of certain refinable functions.

Proposition 3.1 *Let $\phi, \tilde{\phi}$ be a compactly supported dual pair, and let $\phi \in C_0^k$. Then one can represent any polynomial $p \in \Pi_k$, $p(x) = \sum_{i=0}^{k} p_i x^i$, as*

$$p(x) = \sum_{n\in\mathbb{Z}} \tilde{p}_n \phi(x-n) \tag{3.15}$$

where the coefficients $\tilde{p}_n$ are given by

$$\tilde{p}_n = \sum_{i=0}^{k} p_i \sum_{j=0}^{i} \binom{i}{j} n^{i-j} \int_{\mathbb{R}} y^j \tilde{\phi}(y)\, dy \tag{3.16}$$

and the integrals can be determined by (3.14).

Proof: It was shown in [CDM] p.157 that, for refinable $\phi \in C_0^k$, $\Pi_k \subseteq \mathrm{span}\, \{\phi(\cdot - j),\, j \in \mathbb{Z}\}$ implying that we can represent any polynomial $p(x) = \sum_{i=0}^{k} p_i x^i$ in the form (3.15). Since $\phi, \tilde{\phi}$ form a dual pair, the coefficients are given by

$$\tilde{p}_n := \langle p, \tilde{\phi}(\cdot - n)\rangle \tag{3.17}$$

for all $n \in \mathbb{Z}$. Inserting the polynomial and expanding yields the assertion. ∎

We have seen in Proposition 3.1 how any polynomial coefficients can be represented in terms of integer shifts of ϕ. One should note here that one need not take the trial functions in the Galerkin approach for the representation of the coefficients, one could take a different dual pair or an orthogonal wavelet instead. For simplicity, we restrict ourselves to polynomial coefficients of degree less than three so that we can stick with the representation (3.15) with the same φ employed in the Galerkin approach since $\varphi \in C^2$.

Applying Proposition 3.1, we obtain in particular for any quadratic polynomial $p(x) = \alpha + \beta x + \gamma x^2$, $\alpha, \beta, \gamma \in \mathbb{R}$,

$$p(x) = \sum_{n\in\mathbb{Z}} \Big(\alpha + \beta\Big(\tfrac{1}{2}\sum_{l\in\mathbb{Z}} l b_l + n\Big) + \gamma\Big(\tfrac{1}{6}\Big(\sum_{l\in\mathbb{Z}} l b_l\Big)^2 + \tfrac{1}{6}\sum_{l\in\mathbb{Z}} l^2 b_l + n\sum_{l\in\mathbb{Z}} l b_l + n^2\Big)\Big)\varphi(x-n). \tag{3.18}$$

For the polynomial coefficients in (3.6), we therefore get the representations

$$p(\tfrac{y}{2^m}) = \sum_{n\in\mathbb{Z}} p_n \varphi(y-n), \quad q(\tfrac{y}{2^m}) = \sum_{n\in\mathbb{Z}} q_n \varphi(y-n), \quad g(\tfrac{y}{2^m}) = \sum_{n\in\mathbb{Z}} g_n \varphi(y-n) \tag{3.19}$$

with coefficients according to (3.18).

106

Up to this point, we have not yet specified the range of the sums in the (3.10)-(3.12) but will do so in the sequel. In fact, since φ and $\tilde{\varphi}$ are assumed to be compactly supported with finite masks, all sums will be finite.

Let **a** be supported in $[k_0, k_1]$, and **b** in $[\ell_0, \ell_1]$. With denoting by $[\alpha]$ the smallest integer bigger or equal to α if $\alpha < 0$, and the biggest integer less or equal to α if $\alpha > 0$, after some calculation, we end up with the task to determine the entries of the stiffness and mass matrices and the right hand side for $\ell, r = 0, \ldots, 2^{m+1} - 1$ as

$$a_{\ell r} = 2^{2m} \sum_{k,t=m_0}^{m_1} \sum_{n=-2^m-k_1+1}^{2^m-k_0-1} p_n \int_{-2^m}^{2^m} \varphi(y-n)\varphi'(y+2^{m+1}k-\ell)\varphi'(y+2^{m+1}t-r)\,dy \quad (3.20)$$

$$b_{\ell r} = \sum_{k,t=m_0}^{m_1} \sum_{n=-2^m-k_1+1}^{2^m-k_0-1} q_n \int_{-2^m}^{2^m} \varphi(y-n)\varphi(y+2^{m+1}k-\ell)\varphi(y+2^{m+1}t-r)\,dy \quad (3.21)$$

$$f_r = 2^{-m/2} \sum_{k=m_0}^{m_1} \sum_{n=-2^m-k_1+1}^{2^m-k_0-1} g_n \int_{-2^m}^{2^m} \varphi(y-n)\varphi(y+2^{m+1}k-r)\,dy \ , \quad (3.22)$$

where $m_0 := \left[-\frac{1}{2} + \frac{k_0}{2^{m+1}}\right]$ and $m_1 := \left[\frac{3}{2} + \frac{k_1-1}{2^{m+1}}\right]$. Therefore, setting up the system of linear equations boils down to the computation of terms

$$\int_{-2^m}^{2^m} \varphi(x-\alpha_1)\varphi'(x-\alpha_2)\varphi'(x-\alpha_3)\,dx = \sum_{i=-2^m}^{2^m-1} \int_0^1 \varphi(y+i-\alpha_1)\varphi'(y+i-\alpha_2)\varphi'(y+i-\alpha_3)\,dy$$

$$=: \sum_{i=-2^m}^{2^m-1} \tilde{A}(\alpha_1-i, \alpha_2-i, \alpha_3-i), \quad (3.23)$$

$$\sum_{i=-2^m}^{2^m-1} \int_0^1 \varphi(y+i-\alpha_1)\varphi(y+i-\alpha_2)\varphi(y+i-\alpha_3)\,dy =: \sum_{i=-2^m}^{2^m-1} \tilde{B}(\alpha_1-i, \alpha_2-i, \alpha_3-i), \quad (3.24)$$

$$\sum_{i=-2^m}^{2^m-1} \int_0^1 \varphi(y+i-\alpha_1)\varphi(y+i-\alpha_2)\,dy =: \sum_{i=-2^m}^{2^m-1} \tilde{F}(\alpha_1-i, \alpha_2-i). \quad (3.25)$$

Each of the new terms is of the form

$$\int_R \chi_{[0,1)}(x)(D^{\nu_1}\varphi_1)(x-\alpha^1)(D^{\nu_2}\varphi_2)(x-\alpha^2)(D^{\nu_3}\varphi_3)(x-\alpha^3)\,dx \quad (3.26)$$

with characteristic function $\chi_{[0,1)}(x) =: \chi(x)$ satisfying a refinement equation with coefficients $a_0^0 = a_1^0 = 1$,

$$\chi(x) = \chi(2x) + \chi(2x-1). \quad (3.27)$$

By construction of the biorthogonal system, the subdivision scheme corresponding to **a** converges so that we can apply the results in [DM]. We derive refinement equations for $\tilde{A}, \tilde{B}$ and $\tilde{F}$ and solve the resulting eigenvector problems.

Inserting the refinement equations (2.5) and (3.27) into $\tilde{A}$ leads to (cf. [DM], (2.13), $(\varphi_0 = \chi, \quad \varphi_1 = \varphi_2 = \varphi_3 = \varphi, \quad \deg Q = 2)$),

$$2^{-2}\tilde{A}(\alpha) = \sum_{\mu \in \mathbb{Z}^3} c_\mu \tilde{A}(2\alpha - \mu), \quad \alpha \in \mathbb{Z}^3 \quad (3.28)$$

where

$$c_{(\mu_1,\mu_2,\mu_3)} = \tfrac{1}{2}\left(a_{-\mu_1}a_{-\mu_2}a_{-\mu_3} + a_{1-\mu_1}a_{1-\mu_2}a_{1-\mu_3}\right).$$ (3.29)

The range of indices where a_k is not vanishing, $a_k \neq 0$ for $k = k_0, \ldots, k_1$, determines the range of the indices of c_μ such that $c_\mu \neq 0$ for $-k_1 \leq \mu_1, \mu_2, \mu_3 \leq -k_0 + 1$. Similarly, the refinement equation for $\tilde{B}$ is

$$\tilde{B}(\alpha) = \sum_{\mu \in \mathbb{Z}^3} c_\mu \tilde{B}(2\alpha - \mu), \quad \alpha \in \mathbb{Z}^3,$$ (3.30)

with the same coefficients as for $\tilde{A}$.

Finally, the refinement equation for $\tilde{F}$ reads

$$\tilde{F}(\alpha) = \sum_{\mu \in \mathbb{Z}^2} d_\mu \tilde{F}(2\alpha - \mu), \quad \alpha \in \mathbb{Z}^2,$$ (3.31)

with coefficients

$$d_{(\mu_1,\mu_2)} = \tfrac{1}{2}\left(a_{-\mu_1}a_{-\mu_2} + a_{1-\mu_1}a_{1-\mu_2}\right),$$ (3.32)

and $d_\mu \neq 0$ for $-k_1 \leq \mu_1, \mu_2 \leq -k_0 + 1$. Substituting variables in (3.28), (3.30) and (3.31) leads to the eigenvector problems

$$2^{-2}\tilde{A}(\alpha) = \sum_{\mu \in \mathbb{Z}^3} c_{2\alpha-\mu}\tilde{A}(\mu), \quad \tilde{B}(\alpha) = \sum_{\mu \in \mathbb{Z}^3} c_{2\alpha-\mu}\tilde{B}(\mu), \quad \alpha \in \mathbb{Z}^3,$$ (3.33)

$$\tilde{F}(\alpha) = \sum_{\mu \in \mathbb{Z}^2} d_{2\alpha-\mu}\tilde{F}(\mu), \quad \alpha \in \mathbb{Z}^2.$$ (3.34)

The problems for $\tilde{B}$ and $\tilde{F}$ are uniquely determined since no derivatives occur in the integral whereas for the unique computation of $\tilde{A}$, one has to pose additional conditions (cf. [DM]). The size of these eigenvector problems depends on the support of φ. For $-k_1 + 1 \leq n \leq -k_0$, the support of $\varphi(\cdot - n)$ intersects $[0,1)$ so that the corresponding values of $\tilde{A}, \tilde{B}$ and $\tilde{F}$ are not trivially zero, i.e., we have to determine $\tilde{A}(\alpha), \tilde{B}(\alpha), \tilde{F}(\alpha)$ such that $-k_1 + 1 \leq \alpha_i \leq -k_0$. The eigenvector problems are then given by

$$2^{-2}\tilde{A}(\alpha) = \sum_{\mu_i=-k_1+1, i=1,2,3}^{-k_0} c_{2\alpha-\mu}\tilde{A}(\mu), \quad -k_1 + 1 \leq \alpha_i \leq -k_0, \ i = 1,2,3,$$ (3.35)

$$\tilde{B}(\alpha) = \sum_{\mu_i=-k_1+1, i=1,2,3}^{-k_0} c_{2\alpha-\mu}\tilde{B}(\mu), \quad -k_1 + 1 \leq \alpha_i \leq -k_0, \ i = 1,2,3,$$ (3.36)

$$\tilde{F}(\alpha) = \sum_{\mu_i=-k_1+1, i=1,2}^{-k_0} d_{2\alpha-\mu}\tilde{F}(\mu), \quad -k_1 + 1 \leq \alpha_i \leq -k_0, \ i = 1,2.$$ (3.37)

Note that since they can be solved exactly, all entries of the stiffness and mass matrices and the right hand side can be computed without any approximation via (3.20)-(3.22). The number of basis functions for $[V_m]$ and therefore the size of the matrices does not depend on the size of the support of φ, but on the periodization interval. For refinement level m, A, B are of size $2^{m+1} \times 2^{m+1}$, and F with entries f_r has length 2^{m+1}.

The system of linear equations derived to solve (3.1) corresponding to the nodal basis of $[V_m]$ can now be formulated as

$$A_m u^m = F_m$$ (3.38)

where

$$A_m := A + B \quad , \qquad A, B \in I\!R^{2^{m+1} \times 2^{m+1}} \, ,$$
$$F_m := F \qquad , \qquad F \in I\!R^{2^{m+1}} \, , \tag{3.39}$$

and

$$(A)_{r\ell} = a_{r\ell} = a_{\ell r} \, , \quad (B)_{r\ell} = b_{r\ell} = b_{\ell r} \, , \quad (F)_r = f_r \tag{3.40}$$

given by (3.20)-(3.22), and $u^m = (u_0^m, \ldots, u_{2^{m+1}-1}^m)^T$ are the coefficients of the solution represented in the nodal basis of $[V_m]$.

As mentioned at the beginning of this section, we now want to perform a change of basis since we are ultimately interested in the coefficients with respect to the multilevel basis. We aim to determine the matrix that performs the change of basis from the nodal to the multilevel basis acting as a preconditioner, and its inverse.

To this end, let M_m denote the stiffness matrix with respect to the multilevel basis

$$\Psi^m = \{\varphi_j^0, \, j = 0, 1, \quad \psi_j^i, \, j = 0, \ldots, 2^{i+1} - 1, \, 1 \le i \le m - 1\}. \tag{3.41}$$

We want to determine L_m such that

$$L_m^{-1} A_m L_m = M_m. \tag{3.42}$$

Note that (3.38) is equivalent to

$$L_m^{-1} A_m L_m L_m^{-1} u^m = L_m^{-1} F_m \tag{3.43}$$

which can be written as

$$M_m v^m = G_m, \tag{3.44}$$

by using (3.42) and setting $G_m := L_m^{-1} F_m$. Then

$$v^m := L_m^{-1} u^m \tag{3.45}$$

is the coefficient vector of the solution with respect to the multilevel basis Ψ^m.

Starting with the nodal basis in $[V_m]$, we will step by step determine the transformation matrix with respect to $[V_{m-1}] \bigoplus [W_m]$ and vice versa, and then replace $[V_m]$ by $[V_{m-1}]$ and repeat the procedure.

Let $v^{m,(1)} \in [V_m]$ be represented with respect to the basis of $[V_m] \bigoplus [W_m]$, $\{\varphi_j^{m-1}, \, j = 0, \ldots, 2^m - 1, \quad \psi_j^{m-1}, \, j = 0, \ldots, 2^m - 1\}$,

$$v^{m,(1)}(\cdot) = \sum_{j=0}^{2^m - 1} u_j^{m-1} \varphi_j^{m-1}(\cdot) + \sum_{j=0}^{2^m - 1} \tilde{v}_j^{m-1} \psi_j^{m-1}(\cdot). \tag{3.46}$$

Inserting the refinement equation for φ_j^{m-1} and the functional equation for ψ_j^{m-1} yields a representation of $v^{m,(1)}$ with respect to the nodal basis of $[V_m]$. To determine the functional equations for $\varphi_j^{m-1}, \psi_j^{m-1}$, recall definition (3.3) and the functional equations for nonperiodic φ, ψ, (2.5) and (2.19). For level s, we have

$$\begin{aligned}
\varphi_\ell^s(x) &= 2^{s/2} \sum_{k \in \mathbb{Z}} \varphi(2^s(x + 2k) - \ell) \\
&= 2^{s/2} \sum_{r=k_0}^{k_1} a_r \sum_{k \in \mathbb{Z}} \varphi(2^{s+1} x + 2^{s+2} k - 2\ell - r) \\
&= 2^{-1/2} \sum_{r=2\ell+k_0}^{2\ell+k_1} a_{r-2\ell} \varphi_r^{s+1}(x) \, , \tag{3.47}
\end{aligned}$$

and

$$\psi_\ell^s(x) := 2^{s/2} \sum_{k \in \mathbb{Z}} \psi(2^s(x+2k) - \ell)$$

$$= 2^{-1/2} \sum_{r=2\ell-\ell_1+1}^{2\ell-\ell_0+1} (-1)^r b_{1-r+2\ell} \varphi_r^{s+1}(x), \tag{3.48}$$

where **a** is the mask for φ and **b** the mask for $\tilde{\varphi}$. Inserting (3.47) and (3.48) into (3.46) yields

$$v^{m,(1)}(\cdot) = 2^{-1/2} \sum_r \sum_j a_{r-2j} u_j^{m-1} \varphi_r^m(\cdot) + 2^{-1/2} \sum_r \sum_j (-1)^r b_{1-r+2j} \tilde{v}_j^{m-1} \varphi_r^m(\cdot). \tag{3.49}$$

Because of periodicity, $\varphi_{r+2^{m+1}k}^m = \varphi_r^m$, $r = 0, \ldots, 2^{m+1} - 1$, and therefore

$$\sum_r a_{r-2j} \varphi_r^m(\cdot) = \sum_{r=0}^{2^{m+1}-1} \sum_k a_{r-2j+2^{m+1}k} \varphi_r^m(\cdot), \tag{3.50}$$

$$\sum_r (-1)^r b_{1-r+2j} \varphi_r^m(\cdot) = \sum_{r=0}^{2^{m+1}-1} \sum_k (-1)^r b_{1-r+2j+2^{m+1}k} \varphi_r^m(\cdot),$$

i.e., there are in fact 2^{m+1} different nodal basis functions. Putting (3.50) into (3.49) yields

$$v^{m,(1)} = \sum_{r=0}^{2^{m+1}-1} 2^{-1/2} \left(\sum_{j=0}^{2^m-1} \left(\sum_k a_{r-2j+2^{m+1}k} \right) u_j^{m-1} + \sum_{j=0}^{2^m-1} \left((-1)^r \sum_k b_{1-r+2j+2^{m+1}k} \right) \tilde{v}_j^{m-1} \right) \varphi_r^m. \tag{3.51}$$

On the other hand, $v^{m,(1)}$ can be written in terms of the nodal basis functions,

$$v^{m,(1)}(\cdot) = u^m(\cdot) = \sum_{r=0}^{2^{m+1}-1} u_r^m \varphi_r^m(\cdot). \tag{3.52}$$

Comparing coefficients in (3.51) and (3.52) for $r = 0, \ldots, 2^{m+1} - 1$ results in

$$u_r^m = \sum_{j=0}^{2^m-1} \left(2^{-1/2} \sum_k a_{r-2j+2^{m+1}k} \right) u_j^{m-1} + \sum_{j=0}^{2^m-1} \left(2^{-1/2}(-1)^r \sum_k b_{1-r+2j+2^{m+1}k} \right) \tilde{v}_j^{m-1}. \tag{3.53}$$

Defining $L_m^{(1,1)}, L_m^{(1,2)} \in \mathbb{R}^{2^{m+1} \times 2^m}$ by

$$\left(L_m^{(1,1)} \right)_{rj} = 2^{-1/2} \sum_{k \in \mathbb{Z}} a_{r-2j+2^{m+1}k}$$

$$\left(L_m^{(1,2)} \right)_{rj} = 2^{-1/2}(-1)^r \sum_{k \in \mathbb{Z}} b_{1-r+2j+2^{m+1}k}, \tag{3.54}$$

$$u^m = (u_0^m, \ldots, u_{2^{m+1}-1}^m)^T \in \mathbb{R}^{2^{m+1}}$$

$$v^{m,(1)} = (u_0^{m-1}, \ldots, u_{2^m-1}^{m-1}, \tilde{v}_0^{m-1}, \ldots, \tilde{v}_{2^m-1}^{m-1})^T \in \mathbb{R}^{2^{m+1}}$$

$$=: (u^{m-1}, \tilde{v}^{m-1}) \in \mathbb{R}^{2^m + 2^m},$$

the change of basis from $[V_{m-1}] \bigoplus [W_m]$ to $[V_m]$, (3.53), can be written in matrix-vector-form as

$$u^m = \left(L_m^{(1,1)} \quad L_m^{(1,2)} \right) \begin{pmatrix} u^{m-1} \\ \tilde{v}^{m-1} \end{pmatrix} =: L_m^{(1)} v^{m,(1)}, \tag{3.55}$$

where

$$\left(L_m^{(1)} \right)_{nr} = \begin{cases} \left(L_m^{(1,1)} \right)_{nr}, & r = 0, \ldots, 2^m - 1 \\ \left(L_m^{(1,2)} \right)_{n,r-2^m}, & r = 2^m, \ldots, 2^{m+1} - 1, \end{cases} \quad n = 0, \ldots, 2^{m+1} - 1 . \tag{3.56}$$

Repeating this procedure with u^{m-1} starting with (3.54) gives $L_m^{(2)}$ in (3.55) instead of $L_m^{(1)}$ but of half of the size of $L_m^{(1)}$. Completing $L_m^{(2)}$ with an identity matrix to the size of $L_m^{(1)}$ and repeating the process finally leads to

$$u^m = L_m^{(1)} \begin{pmatrix} L_m^{(2)} & 0 \\ \hline 0 & I_{2^m \times 2^m} \end{pmatrix} \begin{pmatrix} L_m^{(3)} & & 0 \\ \hline & I_{2^{m-1} \times 2^{m-1}} & \\ 0 & & I_{2^m \times 2^m} \end{pmatrix}$$

$$\cdots \begin{pmatrix} L_m^{(m)} & & & \\ \hline & I_{2^2 \times 2^2} & & \\ & & \ddots & \\ & & & I_{2^{m-1} \times 2^{m-1}} \\ & & & & I_{2^m \times 2^m} \end{pmatrix} \begin{pmatrix} u^0 \\ \tilde{v}^1 \\ \vdots \\ \tilde{v}^{m-1} \end{pmatrix} \tag{3.57}$$

which is abbreviated as

$$u^m =: L_m v^m . \tag{3.58}$$

That is, the matrix consisting of the product of the transformation matrices with respect to two consecutive levels performs the change of basis from multilevel to nodal basis representation. Note that the determination of L_m is not very costly: firstly, the masks $\mathbf{a}$ and $\mathbf{b}$ are known and of finite support so that the $L_m^{(j)}$'s can be set up quickly, and secondly, the amount of entries that must be calculated decreases geometrically.

For (3.45), we must know L_m^{-1} as well. Of course, one could invert L_m directly. Instead, we can derive L_m^{-1} much cheaper, similar to the procedure for obtaining (3.53) by using biorthogonality of the basis functions (cf. [DK1]).

Defining $R_m^{(1,1)}, R_m^{(1,2)} \in I\!\!R^{2^m \times 2^{m+1}}$ by

$$\left(R_m^{(1,1)} \right)_{jr} := 2^{-1/2} \sum_{k \in \mathbb{Z}} b_{r-2j+2^{m+1}k}$$

$$\left(R_m^{(1,2)} \right)_{jr} := 2^{-1/2} (-1)^j \sum_{k \in \mathbb{Z}} a_{1-r+2j+2^{m+1}k} , \tag{3.59}$$

and with $u^m, v^{m,(1)}, u^{m-1}, \tilde{v}^{m-1}$ as in (3.54), the change of basis from $[V_m]$ to $[V_{m-1}] \bigoplus [W_m]$ is performed via

$$v^{m,(1)} = \begin{pmatrix} u^{m-1} \\ \tilde{v}^m_1 \end{pmatrix} = \begin{pmatrix} R_m^{(1,1)} \\ R_m^{(1,2)} \end{pmatrix} u^m =: R_m^{(1)} u^m, \tag{3.60}$$

where

$$\left(R_m^{(1)} \right)_{rn} = \begin{cases} \left(R_m^{(1,1)} \right)_{rn}, & r = 0, \ldots, 2^m - 1 \\ \left(R_m^{(1,2)} \right)_{r-2^m,n}, & r = 2^m, \ldots, 2^{m+1} - 1, \end{cases} \quad n = 0, \ldots, 2^{m+1} - 1. \tag{3.61}$$

By construction,
$$L_m^{(1)} R_m^{(1)} = R_m^{(1)} L_m^{(1)} = I_{2^{m+1} \times 2^{m+1}}, \tag{3.62}$$

i.e., $R_m^{(1)} = \left(L_m^{(1)}\right)^{-1}$. Repeating the procedure for determining $R_m^{(2)} \ldots, R_m^{(m)}$ as before leads to a representation of L_m^{-1} as a product. Computing the stiffness matrix M_m according to (3.42), and the right hand side G_m as $G_m = L_m^{-1} F_m$, the system (3.38) is finally transformed to
$$M_m v^m = G_m \tag{3.63}$$

where its solution vector v^m gives the coefficients in terms of the multilevel basis Ψ^m.

4   The Wavelet Two-Grid Method

The preceeding chapters were devoted to the setup of the system of linear equations when employing adapted biorthogonal wavelets in a Galerkin method. In the case of polynomial coefficients, we were able to compute all terms involved exactly up to round-off, but the resulting linear systems are usually too large to be solved by a direct method. To accelerate standard iterative methods, it seems natural to connect the multilevel structure of the trial spaces with its representative in iterative methods, namely, the multigrid methods (see e.g. [Ha]), which has been done as well in [BV]. Quite recently, in [R], a wavelet multigrid method for the numerical treatment of linear equations was presented which main difference to standard multigrid algorithms is that the intergrid operators stem from the masks of refinable functions. We do not attempt to give a complete description of multigrid methods here but will refer to e.g. [Ha] and [R] instead. The restriction and prolongation operators between different refinement levels defining the multigrid method were constructed in [R] from the mask coefficients $\{a_k\}_{k \in \mathbb{Z}}$ corresponding to the two-scale equations
$$\varphi(x) = \sum_{k \in \mathbb{Z}} a_k \varphi(2x - k) , \quad \psi(x) = \sum_{k \in \mathbb{Z}} (-1)^k a_{1-k} \varphi(2x - k). \tag{4.1}$$

For the "classical" orthogonal Daubechies–wavelets of compact support, the so-defined intergrid transfer operators could be seen to be unique up to the parameter N from Chapter 2 controlling the smoothness of the wavelet and the size of its support.

In a biorthogonal setting, one has obviously more flexibility for selecting the mask **b** for the dual refinable function $\tilde{\varphi}$, once the mask **a** for φ in (2.5) is chosen. The additional degrees of freedom can be used to adapt the integrid operators in such a way that the convergence rate of the multigrid method is improved, see Section 5.

In this chapter, we will now construct a biorthogonal wavelet multigrid method according to the biorthogonal wavelet basis from Chapter 2. Based on the equations (2.5) and (2.19),
$$\varphi(x) = \sum_{k \in \mathbb{Z}} a_k \varphi(2x - k) , \quad \tilde{\varphi}(x) = \sum_{k \in \mathbb{Z}} b_k \tilde{\varphi}(2x - k) , \tag{4.2}$$
$$\psi(x) = \sum_{k \in \mathbb{Z}} c_k \varphi(2x - k) = \sum_{k \in \mathbb{Z}} (-1)^k b_{1-k} \varphi(2x - k) ,$$
$$\tilde{\psi}(x) = \sum_{k \in \mathbb{Z}} d_k \tilde{\varphi}(2x - k) = \sum_{k \in \mathbb{Z}} (-1)^k a_{1-k} \tilde{\varphi}(2x - k) ,$$

we define biinfinite matrices A_0, A_1, B_0, B_1 by

$$A_0 = (a_{\ell-2k})_{k,\ell \in \mathbb{Z}}, \qquad A_1 = (c_{\ell-2k})_{k,\ell \in \mathbb{Z}}, \tag{4.3}$$
$$B_0 = (b_{\ell-2k})_{k,\ell \in \mathbb{Z}}, \qquad B_1 = (d_{\ell-2k})_{k,\ell \in \mathbb{Z}}$$

which are related by

$$B_0^T A_0 + B_1^T A_1 = I, \tag{4.4}$$

(cf. [CDF](2.3)) which follows from the biorthogonality. In a general biorthogonal setting, it is possible to classify all matrices B_0, B_1 and A_1 satisfying (4.4) for a given biinfinite matrix A_0. This in turn means that, for a given refinable function φ, one is able to construct all possible dual generators $\tilde{\varphi}$ and the corresponding wavelet basis. This concept was studied in [CDP] and [D]. For later use, we briefly recall some of the basic facts. Let (L, K) be a pair of biinfinite matrices that satisfy the following condition (C):

    ○ L takes $\ell_2(2\mathbb{Z})$ boundedly into $\ell_2(2\mathbb{Z})$, and

    ○ K possesses an inverse K^{-1}, and both K and K^{-1} are bounded linear operators taking $\ell_2(2\mathbb{Z})$ into itself.

The following proposition proved in [CDP] states that all biorthogonal matrices are ultimately equivalent.

Proposition 4.1 *Let (L, K) be a pair of biinfinite matrices such that condition (C) holds. Furthermore, let A_0, A_1, B_0, B_1 satisfy (4.4). Then $\tilde{A}_1, \tilde{B}_0, \tilde{B}_1$ defined by*

$$\tilde{A}_1 = LA_0 + KA_1, \quad \tilde{B}_1 = (K^{-1})^T B_1, \quad \tilde{B}_0 = B_0 - L^T B_1 \tag{4.5}$$

satisfy (4.4). Conversely, for any two sets $(A_0, \tilde{A}_1, \tilde{B}_0, \tilde{B}_1)$ and (A_0, A_1, B_0, B_1) both satisfying (4.4), there exists a pair (L, K) satisfying condition (C) such that (4.5) holds.

We are now ready to define finite-dimensional **restriction operators** by

$$H := P_{N+[\frac{n-1}{2}]}^T S_{N-1} A_0 P_n : \mathbb{R}^n \to \mathbb{R}^{N+[\frac{n-1}{2}]},$$
$$\tilde{H} := P_{N+[\frac{n-1}{2}]}^T S_{N-1} B_0 P_n : \mathbb{R}^n \to \mathbb{R}^{N+[\frac{n-1}{2}]}, \tag{4.6}$$

where a shift operator S_{N-1} and a projection operator P_n are given by

$$S_{N-1} : \ell^2(\mathbb{Z}) \to \ell^2(\mathbb{Z}), \qquad (S_{N-1}\lambda)_k := \lambda_{k-(N-1)} \tag{4.7}$$

$$P_n : \mathbb{R}^n \to \ell^2(\mathbb{Z}), \qquad (P_n x)_k = \begin{cases} x_k, & 0 \le k \le n-1 \\ 0, & \text{else} \end{cases}, \tag{4.8}$$

and the parameter N depends on the support of the masks on consideration. G and $\tilde{G}$ are defined correspondingly with A_1, B_1. The corresponding **prolongation operators** are obtained by just taking the adjoints of $H, \tilde{H}, G$ and $\tilde{G}$. Similar to the investigations in [R], it is easy to check that (4.4) implies

$$H^T \tilde{H} + G^T \tilde{G} = I. \tag{4.9}$$

After these preliminaries, we can now define a biorthogonal two-grid method. We want to solve some system of linear equations of the form (3.38) or (3.63). Since we aim at presenting the basic idea here rather than giving a throrough description of the method, we confine ourselves to some fixed finer level so that we will omit the reference parameter m. For the iterative solution of the equation

$$Ax = F \qquad (4.10)$$

with some positive definite matrix A, we use the operators (4.6) to define a **biorthogonal two-grid method** by

$$
\begin{aligned}
x^{j+1} &= (I - H^T(\tilde{H}AH^T)^{-1}\tilde{H}A)(S^\nu x^j + \sum_{i=0}^{\nu-1} S^i Lb) + H^T(\tilde{H}AH^T)^{-1}\tilde{H}b \\
&= (I - H^T(\tilde{H}AH^T)^{-1}\tilde{H}A)S^\nu x^j \\
&\quad + \left((I - H^T(\tilde{H}AH^T)^{-1}\tilde{H}A)\sum_{i=0}^{\nu-1} S^i L + H^T(\tilde{H}AH^T)^{-1}\tilde{H} \right) b \\
&=: M(\nu)x^j + N(\nu)b, \qquad j \geq 0,
\end{aligned}
\qquad (4.11)
$$

where S and L originate from a standard iterative method

$$x^{j+1} = Sx^j + Lb,$$

such as Jacobi or Richardson iteration, cf. e.g. [V] or [DH].

In the following, we will always assume that ν indicating the number of smoothing steps is equal to 1 since the value of ν is not relevant for our analysis here. Then $M := M(1)$ may be written as

$$M = TS, \quad T := (I - H^T(\tilde{H}AH^T)^{-1}\tilde{H}A) \qquad (4.12)$$

where T denotes the coarse grid correction operator. Note that, in contrary to classical multigrid methods, the restriction and prolongation operators employed in the definition of T, H and $\tilde{H}$, are not adjoints of each other. Since therefore the operator

$$T_A = I - A^{1/2}H^T(\tilde{H}AH^T)^{-1}\tilde{H}A^{1/2} \qquad (4.13)$$

is no longer an orthogonal projector, one has no a-priori convergence estimates as in [R].

However, for stiffness matrices as in Chapter 3, first numerical tests indicate that (4.11) converges quite well which motivates the investigations in the following chapter.

5 Adapting a Biorthogonal Two-Grid Method

According to the flexibility arising from the biorthogonal setting, we attempt to adapt the intergrid operators to a given problem such that the method converges as fast as possible. To simplify matters, we confine ourselves to the Richardson method, i.e. $S = I - \vartheta A$ where ϑ is some relaxation parameter.

Let the Euklidian norm on $I\!\!R^n$ be defined by $\|x\| := (\sum_{i=1}^n x_i^2)^{1/2}$, and denote $\|x\|_A := \|A^{1/2}x\|$ the energy norm with matrix norm $\|B\|_A = \|A^{1/2}BA^{-1/2}\|$.

The investigations in this chapter are based on the following proposition (compare [R], Theorem 4.3, and [DK1] for the proof).

Proposition 5.1 *Let $A \in \mathbb{R}^{n \times n}$ be a positive definite matrix with eigenvalues $0 < \lambda_1 \leq \lambda_2 \leq \ldots \leq \lambda_n$ and corresponding normalized eigenvectors u_i. Then, for the Richardson method $S = 1 - \vartheta A$ such that $0 < \vartheta \leq \|A\|^{-1}$, it follows that*

$$\frac{\|M\|_A}{\|S\|_A} \leq \left(\sum_{i=1}^{n} \frac{\lambda_n}{\lambda_i} \|T u_i\|^2 \right)^{1/2}, \tag{5.1}$$

where T is the coarse grid correction operator defined in (4.12).

Although the estimate (5.1) is pretty rough, it will be sufficient for our purposes. The sum on the right-hand side is dominated by its first terms

$$\frac{\lambda_n}{\lambda_i} \|T u_i\|^2 \qquad \text{for } i = 1, \ldots, s,$$

and some s small relative to n, so it seems advantageous to construct an operator T such that $\|T u_i\|^2$ is small. In principle, this would require the knowledge of the eigenvectors of A which is equivalent to the solution of the system of linear equations on consideration. However, on a low refinement level, the stiffness matrix is small, and accordingly, the calculation of the eigenvectors is cheap. When using the nodal basis instead of a multilevel basis, the stiffness matrix on a higher refinement level is obtained by a simple scaling process implying that a coarse grid correction operator T constructed almost optimally on a low refinement level might work satisfactorily on higher scales as well. For that reason, let A now denote a stiffness matrix relative to a nodal basis so that here the basis transformation in Chapter 3 does not have to be performed. Let T be some coarse grid correction operator constructed with the aid of some biinfinite matrix $\tilde{B}_0$, i.e.,

$$T = I - H^T (JAH^T)^{-1} JA, \tag{5.2}$$

where J is defined by (4.6),

$$J = P_{N + \left[\frac{n-1}{2} \right]}^T S_{N-1} \tilde{B}_0 P_n. \tag{5.3}$$

We want to minimize, for i small,

$$\|T u_i\| = \|(I - H^T (JAH^T)^{-1} JA) u_i\|, \tag{5.4}$$

where A denotes the stiffness matrix with respect to a coarse level. Our aim is to choose $\tilde{B}_0$ in such a way that J defined by (5.3) induces a coarse grid correction operator T in (5.2) which satisfies even $T u_i = 0$ for small i. Instead of calculating the term $(JAH^T)^{-1}$ in (5.4) exactly, it makes sense to replace it simply by the inverse of the stiffness matrix $\hat{A}$ on a coarser level. Note that for the standard intergrid operators I_1, I_2 used in algebraic multigrid methods, one has $I_1^T \hat{A} I_2 = A$ (cf. [RS] p.77). In practice, one would even approximate the inverse of $\hat{A}$ by its preconditioner since the task is only to force the terms in (5.4) to be sufficiently small but not to vanish exactly.

Now we are able to adapt the biorthogonal two-grid method to the problem (4.10) as follows. Define for some integer $m \in \mathbb{N}$ the matrix

$$C = ((B_1)_{ij})_{i=-m,\ldots,m, \ j=0,\ldots,n}, \tag{5.5}$$

and define for a set of eigenvectors $u^1, \ldots, u^s$, $s \leq n$, of A the sequences $\tilde{y}^i = (\tilde{y}_k^i)_{k \in \mathbb{Z}}$ by

$$\tilde{y}_k^i := \begin{cases} (Cu^i)_k, & k = -m, \ldots, m \\ 0 & \text{else.} \end{cases} \tag{5.6}$$

Furthermore, define for some pair of operators (L, K) satisfying condition (C) the vector $v \in \mathbb{R}^{(N+[\frac{n-1}{2}]) \cdot (2m+1)}$ by

$$v := (L_{(2m+1)i+j}^T)_{i=-(N-1),\ldots,[\frac{n-1}{2}],\ j=-m,\ldots,m} \ . \tag{5.7}$$

Out of all possible pairs (L, K) that transform the biorthogonal system A_0, A_1, B_0, B_1 into the system $A_0, \tilde{A}_1, \tilde{B}_0, \tilde{B}_1$, we now introduce conditions on L so that the coarse grid correction operator of the two-grid method has the desired properties.

Theorem 5.1 *Let* A_0, A_1, B_0, B_1 *satisfy (4.4), and let* $u^1, \ldots, u^s$ *be eigenvectors of* A *with respect to the eigenvalues* $\lambda_1, \ldots, \lambda_s$ *for some* $s \leq n$. *Furthermore, let* $\tilde{A}_1, \tilde{B}_0, \tilde{B}_1$ *be constructed by Proposition 4.1, and let* J *be defined by*

$$J = P_{N+[\frac{n-1}{2}]}^T S_{N-1} \tilde{B}_0 P_n \ , \tag{5.8}$$

and $\tilde{J}$ *and* $\tilde{\tilde{J}}$ *are defined correspondingly with* $\tilde{A}_1$ *resp.* $\tilde{B}_1$.

Then the biorthogonal two-grid method with coarse grid correction operator T *defined by*

$$T := I - H^T \hat{A}^{-1} J A \tag{5.9}$$

satisfies

$$Tu^i = 0, \qquad i = 1, \ldots, s, \tag{5.10}$$

if and only if the vector v *defined by (5.7) is a solution of the linear equation*

$$\begin{pmatrix} D^1 \\ \vdots \\ D^s \end{pmatrix} v = \begin{pmatrix} (H^T \hat{A}^{-1} \tilde{H}^{-1} - \lambda_1^{-1}) u^1 \\ \vdots \\ (H^T \hat{A}^{-1} \tilde{H}^{-1} - \lambda_s^{-1}) u^s \end{pmatrix} \tag{5.11}$$

where, for $i = 1, \ldots, s$, *the matrix* D^i *is defined by*

$$(D^i) = (d_{k,r}^i)_{k=1,\ldots,n,\ r=-(N-1)(2m+1)-m,\ldots,[\frac{n-1}{2}](2m+1)+m} \tag{5.12}$$

$$d_{k,r}^i = \sum_{\ell=1}^{N+[\frac{n-1}{2}]} (H^T \hat{A}^{-1})_{k,\ell}\ \tilde{y}_{r-(2m+1)\ell}^i \ .$$

Proof: According to (4.5) and (4.6), we may write

$$\begin{aligned} J &= P_{N+[\frac{n-1}{2}]}^T S_{N-1} \tilde{B}_0 P_n \\ &= P_{N+[\frac{n-1}{2}]}^T S_{N-1} B_0 P_n - P_{N+[\frac{n-1}{2}]}^T S_{N-1} L^T B_1 P_n =: \tilde{H} - \hat{H} \end{aligned}$$

with

$$\hat{H} := P_{N+[\frac{n-1}{2}]}^T S_{N-1} L^T B_1 P_n \ . \tag{5.13}$$

We want to choose $\hat{H}$ in such a way that

$$Tu^i = (I - H^T(\hat{A}^{-1})JA)u^i = 0, \quad i = 1,\ldots,s, \tag{5.14}$$

holds. Reformulating the last equation leads to

$$u_i = \lambda_i H^T \hat{A}^{-1} \hat{H} u^i - \lambda_i H^T \hat{A}^{-1} \hat{H} u^i.$$

Therefore, $\hat{H}$ has to be a solution of the linear equations

$$(H^T \hat{A}^{-1} \hat{H} - \lambda_i^{-1} I)u^i = (H^T \hat{A}^{-1})\hat{H} u^i, \quad i = 1,\ldots,s. \tag{5.15}$$

To derive conditions for the operator L guaranteeing that (5.15) holds for the corresponding matrix $\hat{H}$ constructed by (5.13), we have to determine the matrix associated with $\hat{H}$. By definition of P_n (4.8) and applying L^T and S_{N-1} yields, for some $m \in \mathbb{N}$,

$$\hat{H}(x) = \left(P^T_{N+[\frac{n-1}{2}]} S_{N-1} L^T B_1 P_n\right) x = \begin{pmatrix} \sum_{j=0}^{n-1} \sum_{k=-m}^{m} L^T_{-(N-1),k}(B_1)_{kj}x_j \\ \vdots \\ \sum_{j=0}^{n-1} \sum_{k=-m}^{m} L^T_{[\frac{n-1}{2}],k}(B_1)_{kj}x_j \end{pmatrix}, \tag{5.16}$$

since the mask $\mathbf{d}$ according to B_1 is finite (cf. (4.2)). Denoting by $\tilde{L}^T$ the matrix

$$\tilde{L}^T := (L^T_{ij})_{i=-(N-1),\ldots,[\frac{n-1}{2}],\ j=-m,\ldots,m},$$

and recalling (5.5), we obtain

$$\hat{H}(x) = \tilde{L}^T Cx. \tag{5.17}$$

Inserting (5.17) into (5.15) finally yields an equation for $\tilde{L}$,

$$(H^T \hat{A}^{-1} \hat{H} - \lambda_i^{-1} I)u^i = (H^T \hat{A}^{-1})\tilde{L}^T Cu^i, \quad i = 1,\ldots,s. \tag{5.18}$$

Now it remains to derive an explicit expression for the right hand side of (5.18). Using the abbreviation $Cu^i =: y^i$ and identifying each vector in $\mathbb{R}^{(N+[\frac{n-1}{2}])(2m+1)}$ with some $(N+[\frac{n-1}{2}]) \times (2m+1)$-matrix by employing the canonical mapping $(Ih)_{i,j} = h_{(2m+1)i+j}$, we obtain for the canonical basis vector e^r such that $(e^r)_j = \delta_{rj}$,

$$(H^T \hat{A}^{-1}(Ie^r)y^i)_k = \sum_{\ell=1}^{N+[\frac{n-1}{2}]} (H^T \hat{A}^{-1})_{k,\ell}\, \tilde{y}^i_{r-(2m+1)\ell} \tag{5.19}$$

where $\tilde{y}^i$ is defined by (5.6). This means that the matrix D^i associated with (5.18) is given by (5.12). Each eigenvector u^i, $i = 1,\ldots,s$, generates an equation of the form (5.18) so that combining them to one large linear system yields the result. ∎

Remark 5.1 *Theorem 5.1 indicates that increasing the number of eigenvectors under consideration speeds up the convergence of the method. But increasing s leads to a rapid increase of the dimension of the problem (5.11) as well, so one will choose s small relative to n. In addition, for the construction, one has to know the first s eigenvalues and eigenvectors. A procedure to calculate these terms quickly was proposed e.g. in [BP]. Numerical results elaborating this point will be published elsewhere.*

Remark 5.2 *If the number of eigenvectors is small, then (5.11) has a lot of solutions. One can use this additional freedom to find a solution leading to an L so that the matrix $\hat{H}$ is as sparse as possible, e.g. one might require that $\hat{H}$ is some band matrix. We will investigate this question as well in a forthcoming paper.*

Acknowledgement: We thank Wolfgang Dahmen very much for his valuable comments.

References

[BP]   J.H.Bramble, and J.E.Pasciak, A preconditioned algorithm for eigenvector/-eigenvalue computation, Manuscript, 1992.

[BV]   W.L.Briggs, and Van Emden Henson, Wavelets and multigrid, Siam J. Scient. Comp. **14**(2), 506-510, 1993.

[CDP]   J.M.Carnicer, W.Dahmen, and J.M.Peña, Locally finite decompositions of nested spaces, in preparation.

[CDM]   A.S.Cavaretta, W.Dahmen, and C.A.Micchelli, Stationary Subdivision, Memoirs of Amer. Math. Soc. Vol. 93, #453, 1991.

[CDF]   A.Cohen, I.Daubechies, and J.C.Feauveau, Biorthogonal bases of compactly supported wavelets, Comm. Pur. Appl. Math. **45**, 485-560, 1992.

[DW1]   S.Dahlke, and I.Weinreich, Wavelet galerkin methods: an adapted biorthogonal wavelet basis. Constr. Approx. **9**(2), 237-262, 1993.

[DW2]   S.Dahlke, and I.Weinreich, Wavelet bases adapted to pseudo-differential operators, Preprint FU Berlin, 1992.

[DK1]   S.Dahlke, and A.Kunoth, A biorthogonal wavelet approach for solving boundary value problems, Preprint, Inst.f.Geom.Prakt.Math., RWTH Aachen, 1993.

[D]   W.Dahmen, Decomposition of refinable spaces and applications to operator equations, to appear in: Algorithms for Approximation III, M.G.Cox, J.Mason (eds.), Oxford, 1993.

[DK2]   W.Dahmen, and A.Kunoth, Multilevel preconditioning, Numer. Math. **63**, 315-344, 1992.

[DM]   W.Dahmen, and C.A.Micchelli, Using the refinement equation for evaluating integrals of wavelets, Siam J. Numer. Anal. **30**(2), 507-537, 1993.

[DPS]   W.Dahmen, S.Prössdorf, and R.Schneider, Wavelet approximation methods for pseudodifferential equations II: Matrix compression and fast solution, Preprint No.82, Inst.f.Geom. Prakt.Math., RWTH Aachen, 1993

[Dau]   I.Daubechies, Orthonormal bases of compactly supported wavelets, Comm. Pur. Appl. Math. **41**, 909-996, 1988.

[DH]   P.Deuflhard, and A.Hohmann, Numerische Mathematik, de Gruyter, 1991.

[Ha]   W.Hackbusch, Multigrid Methods and Applications, Springer Series on Computational Mathematics 4, Springer, 1985.

[Ma]   S.Mallat, Multiresolution approximations and wavelet orthonormal bases of $L^2(I\!R)$, Trans. Amer. Math. Soc. **315**(1), 6907, 1989.

[Mey]   Y.Meyer, Ondelettes et Opérateurs I, Hermann, Paris 1990.

[R]   A.Rieder, Semi-algebraic multilevel methods based on wavelet decompositions, Manuscript, Universität des Saarlandes, 1992.

[RS]   J.W.Ruge, and K.Stüben, Algebraic multigrid, in: Multigrid Methods, S.McCormick (ed.), Frontiers in Applied Mathematics, Siam, 1987.

[V]   J.Varga, Matrix Iterative Analysis, Prentice Hall, 1962.

Adaptive Multilevel – Methods

for Obstacle Problems in Three Space Dimensions

B. Erdmann
Konrad–Zuse–Zentrum für Informationstechnik Berlin
Heilbronner Str. 10, 10711 Berlin, Germany
R.H.W. Hoppe
Mathematisches Institut der Technischen Universität München
Arcisstr. 21, D–80333 München, Germany
R. Kornhuber
Konrad–Zuse–Zentrum für Informationstechnik Berlin
Heilbronner Str. 10, 10711 Berlin, Germany

Summary

We consider the discretization of obstacle problems for second order elliptic differential operators in three space dimensions by piecewise linear finite elements. Linearizing the discrete problems by suitable active set strategies, the resulting linear sub–problems are solved iteratively by preconditioned cg–iterations. We propose a variant of the BPX preconditioner and prove an $O(j)$ estimate for the resulting condition number. To allow for local mesh refinement we derive semi–local and local a posteriori error estimates. The theoretical results are illustrated by numerical computations.

1. Introduction

Given a closed subspace $V \subset H^1(\Omega)$, Ω being a bounded polyhedral domain in the Euclidean space $\mathbb{R}^3$, we consider obstacle problems of the form

$$\text{Find } u \in K \text{ such that } \quad \mathcal{J}(u) \leq \mathcal{J}(v), \quad v \in K, \tag{1.1}$$

for the energy functional $\mathcal{J}$,

$$\mathcal{J}(v) = \tfrac{1}{2}a(v,v) - \ell(v), \quad v \in V,$$

and a closed, convex set $K \subset V$,

$$K = \{v \in V \mid v(\mathbf{x}) \leq \varphi(\mathbf{x}) \text{ a.e. in } \Omega\}.$$

Assuming that $\mathcal{J}$ is induced by a symmetric V–elliptic bilinear form $a(\cdot, \cdot)$,

$$a(v,w) = \int_\Omega \sum_{i,j=1}^3 a_{ij}\, \partial_i v\, \partial_j w\, d\mathbf{x},$$

and some functional $\ell \in V'$, it is well–known that (1.1) is equivalent to the variational inequality

$$\text{Find } u \in K \text{ such that } \quad a(u, u - v) \leq \ell(u - v), \quad v \in K. \tag{1.2}$$

For simplicity we restrict our considerations to the case $V = H_0^1(\Omega)$. To ensure existence and uniqueness of the solution u of (1.1) and (1.2), respectively, we assume $\varphi \in H^1(\Omega)$, $\varphi \geq 0$ a.e. on the boundary $\partial\Omega$, and $a_{ij} \in L^\infty(\Omega)$ satisfying

$$a) \quad a_{ij}(\mathbf{x}) \;=\; a_{ji}(\mathbf{x}), \quad 1 \leq i,\, j \leq 3,$$

$$b) \quad \alpha_0 |\xi|^2 \;\leq\; \sum_{i,j=1}^{3} a_{ij}(\mathbf{x}) \xi_i \xi_j \leq \alpha_1 |\xi|^2, \; \xi \in \mathbb{R}^3,\, 0 < \alpha_0 \leq \alpha_1, \tag{1.3}$$

for almost all $\mathbf{x} \in \Omega$.

Discretizing (1.2) in space by continuous, piecewise linear finite elements with respect to a partition $\mathcal{T}$ of Ω in tetrahedra, we consider the efficient solution of the resulting finite dimensional variational inequality together with the appropriate choice of $\mathcal{T}$.

As the convergence rate of standard projected relaxation methods (e.g. [17]) is well–known to deteriorate with increasing refinement, a variety of multigrid methods has been developed in the last decade. For an overview we refer to [23]. In the present paper we consider a linearization based on active set strategies (e.g. [18, 19, 20]). This is a class of iterative schemes where in each iteration step a set of active constraints is prespecified and then a reduced linear sub–problem has to be solved for the computation of the new iterate. In the following section we will briefly recall the algorithm proposed in [19, 20].

The reduced linear problems arising in each step of the active set strategy can be regarded as usual Dirichlet problems on some reduced computational domain and hence may be solved iteratively by appropriate multigrid methods [18, 19, 20] or multilevel precondi-tioned cg–iterations [22, 23]. The construction and analysis of a suitable variant of the BPX preconditioner [10] will be subject of Section 3. Using the well–known interpreta-tion as additive Schwarz methods [7, 34, 37, 38], it is shown that the condition number is bounded in terms of $O(j)$ provided that the free boundary is sufficiently regular. Note that the regularity condition is resulting from the non–local character of the L^2–projection compared to the interpolation arising in the analysis of the hierarchical basis precondi-tioner [23]. Probably, this non–optimal bound can be improved by more sophisticated techniques [9, 12, 31].

For the adaptive construction of a suitable hierarchy of triangulations, efficient and reliable a posteriori error estimates are required. In [23, 26] the basic concept of [13, 28] relying on suitable localizations of the discretized defect problem has been extended to variational inequalities. Related results in three space dimensions are presented in Section 4. Note that a posteriori estimates for a penalty method have been proposed in [25], while a recent generalization of the well–known Bank/Weiser estimator [4] can be found in [1].

The final section is devoted to some numerical experiments supporting the theoretical findings.

2. Active Set Strategies

A partition $\mathcal{T}$ of the computational domain $\Omega \subset \mathbb{R}^3$ in tetrahedra is called triangulation. Henceforth we only consider triangulations which are conforming in the sense that the intersection of two different tetrahedra $t, t' \in \mathcal{T}$ either consists of a common triangular face, a common edge, a common vertex or is empty. The sets of vertices p and edges e which are not part of the boundary $\partial\Omega$ are called $\mathcal{N}$ and $\mathcal{E}$, respectively. We approximate $H_0^1(\Omega)$ by the subspace $\mathcal{S}$ of continuous, piecewise linear finite elements vanishing on the boundary $\partial\Omega$. The nodal basis functions $\lambda_p \in \mathcal{S}$, $p \in \mathcal{N}$ are defined by $\lambda_p(q) = \delta_{pq}$, $p, q \in \mathcal{N}$, (Kronecker delta).

Further, let $\varphi_{\mathcal{T}} \in \mathcal{S}$ be a discrete obstacle approximating the given obstacle φ in an appropriate sense. For example, $\varphi_{\mathcal{T}}$ may be chosen as the L^2–projection of φ onto $\mathcal{S}$ or, if $\varphi \in C(\bar{\Omega})$, as the $\mathcal{S}$–interpolate. Correspondingly, we denote by $K_{\mathcal{T}} = \{v \in \mathcal{S} | v \leq \varphi_{\mathcal{T}}\}$ the set of discrete constraints. Then the finite element approximation of (1.1) amounts to the computation of an element $u_{\mathcal{T}} \in K_{\mathcal{T}}$ satisfying

$$a(u_{\mathcal{T}}, u_{\mathcal{T}} - v) \leq \ell(u_{\mathcal{T}} - v), \ v \in K_{\mathcal{T}}. \tag{2.1}$$

The finite dimensional variational inequality (2.1) will be solved iteratively by the active set strategy proposed in [19, 20]:

Active set strategy

Step 1: Choose an initial iterate $u^{(0)} \in \mathcal{S}$.

Step 2: Given $u^{(\nu)} \in \mathcal{S}$, $\nu \geq 0$, determine the subset of active nodes $\mathcal{N}^{\bullet} \subset \mathcal{N}$ as the set of points $p \in \mathcal{N}$ such that

$$\varphi_{\mathcal{T}}(p) - u^{(\nu)}(p) \leq \ell(\lambda_p) - a(u^{(\nu)}, \lambda_p),$$

while the remaining nodes $\mathcal{N}^\circ = \mathcal{N} \setminus \mathcal{N}^{\bullet}$ are called inactive. Introducing a direct splitting of the finite element space $\mathcal{S}$ into the linear subspaces $\mathcal{S}^\circ, \mathcal{S}^{\bullet} \subset \mathcal{S}$ defined by

$$\mathcal{S}^\circ = \{v \in \mathcal{S} | \ v(p) = 0, p \in \mathcal{N}^{\bullet}\}, \quad \mathcal{S}^{\bullet} = \{v \in \mathcal{S} | \ v(p) = 0, p \in \mathcal{N}^\circ\}$$

the new iterate $u^{(\nu+1)} \in \mathcal{S}$ is computed from

$$u^{(\nu+1)} = u^{\bullet} + u^\circ, \tag{2.2}$$

where $u^{\bullet} \in \mathcal{S}^{\bullet}$ is given by

$$u^{\bullet}(p) = \varphi_{\mathcal{T}}(p), \quad p \in \mathcal{N}^{\bullet}, \tag{2.3}$$

and $u^\circ \in \mathcal{S}^\circ$ is the solution of the reduced linear system

$$a(u^\circ, v) = \ell(v) - a(u^{\bullet}, v), \quad v \in \mathcal{S}^\circ. \tag{2.4}$$

Note that the reduced system (2.4) may be regarded as a Dirichlet problem on the reduced computational domain Ω°,

$$\Omega^\circ = \bigcup_{p \in \mathcal{N}^\circ} \operatorname{supp} \lambda_p. \tag{2.5}$$

Remark 2.1 It is well–known (c.f. [19, 20]) that for arbitrarily given initial iterate $u^{(0)}$ the iterates $u^{(\nu)}$, $\nu \geq 1$, converge monotonically decreasingly to the solution u of (2.1) provided that the finite element discretization is monotone. Note that in three dimensions this condition may be violated even if the underlying triangulation satisfies the Delaunay condition that the circumsphere of the four vertices of any tetrahedron in the triangulation contains no vertices in its interior [29]. However, there is numerical evidence that even the approximate solution of the reduced subproblems (2.4) up to some accuracy κ leads to satisfying results if κ is chosen small enough. The actual choice of κ which seems to depend on regularity properties of the problem is still an open question.

In the inexact case, the convergence of a related most constrained strategy has been proved in [18], providing a stopping criterion for the inner iteration. However, this strategy turns out to be much too pessimistic in actual computations leading to a prohibitive large number of outer iteration steps.

Following [23] the reduced linear subproblems will be solved approximately by preconditioned conjugate gradient iteration. For basic information about the preconditioned *cg*–method we refer to [2] while the construction of appropriate multilevel preconditioners will be considered in the next section.

3. Reduced BPX Preconditioning

Let $\mathcal{T}_0$ be an intentionally coarse conforming triangulation of Ω. Based on some refinement criterion the partition $\mathcal{T}_0$ is refined several times providing a sequence of triangulations and a corresponding sequence of finite element spaces. The underlying refinement algorithm is a straightforward extension of the well–established red/green refinement technique proposed in [5] to three space dimensions. To provide a regular (red) refinement of some tetrahedron t, first the edges of t are bisected as shown in Figure 3.1 leading to a partition in four similar sub–tetrahedra and an octahedron.

To preserve the shape regularity of the elements, the remaining octahedron is subdivided according to the strategy of Bey [6] (see Figure 3.2). After local regular refinement we use three different types of irregular (green) closures depicted in the Figures 3.3–3.5 to obtain a conforming triangulation. For a detailed description of the refinement algorithm we refer to [8].

As usual, a refined tetrahedron is said to be the father of its sub–tetrahedra, which are in turn called sons. The depth of a tetrahedron is given by the number of its ancestors. Finally, the tetrahedra of the initial triangulation $\mathcal{T}_0$ together with all tetrahedra resulting from regular refinement are called regular, while irregular refinement leads to irregular tetrahedra. Using the hierarchical data structures described for instance in the 3-D ELLKASK programmer's manual [16] the triangulations produced by this dynamic refinement process are stored as a sequence $\mathcal{T}_0, \mathcal{T}_1, \cdots, \mathcal{T}_j$ with the following properties.

(T1) Each vertex of $\mathcal{T}_{k+1}$ which does not belong to $\mathcal{T}_k$, is a vertex of a regular tetrahedron, $0 \leq k < j$.

(T2) Irregular tetrahedra have no sons.

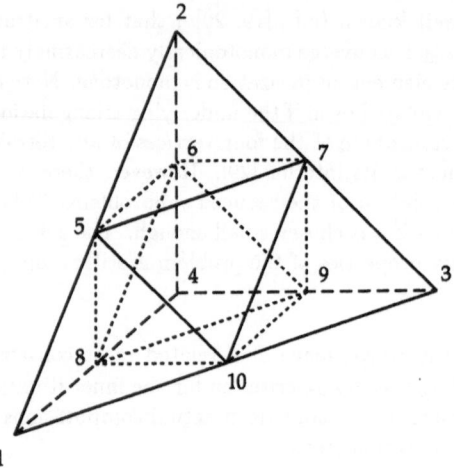

Figure 3.1 Regular Refinement of the Triangular Faces

(T3) The father of each tetrahedron $t \in \mathcal{T}_{k+1} \backslash \mathcal{T}_k$ has depth k, $0 \leq k < j$.

The rule (T3) allows for the reconstruction of the whole sequence $\mathcal{T}_0, \mathcal{T}_1, \cdots, \mathcal{T}_j$ with the properties (T1)–(T3) from the initial triangulation $\mathcal{T}_0$ and the actual triangulation $\mathcal{T}_j$ alone. We emphasize that in general this sequence does not reflect the dynamic refinement process. For further information we refer to [8, 23] and the literature cited therein.

Due to the rules (T1 – T3) the sequence $\mathcal{S}_0, \ldots, \mathcal{S}_j$ of finite element spaces corresponding to $\mathcal{T}_0, \ldots, \mathcal{T}_j$ is nested in the sense that

$$\mathcal{S}_0 \subset \mathcal{S}_1 \subset \ldots \subset \mathcal{S}_j. \tag{3.1}$$

Now assume that we have a disjoint splitting $\mathcal{N}_j = \mathcal{N}_j^\bullet \cup \mathcal{N}_j^\circ$ which may result from an active set strategy applied to (2.1). The remainder of this section will be devoted to the construction of an efficient multilevel preconditioner for the corresponding reduced system

$$\text{Find } u_j^\circ \in \mathcal{S}_j^\circ \text{ such that } a(u_j^\circ, v) = \ell(v) - a(u_j^\bullet, v), \quad v \in \mathcal{S}_j^\circ. \tag{3.2}$$

Following [38, 7, 23] the analysis will be carried out in the framework of additive Schwarz methods. Apparently, this approach was initiated in [14] and meanwhile became standard in the theory of multilevel methods. For excellent overviews we refer to [34, 37].

To provide an appropriate multilevel splitting of the reduced finite element space $\mathcal{S}_j^\circ$, we first derive a sequence of finite element spaces $\mathcal{S}_k^\circ, 0 \leq k < j$, which is nested in the sense of (3.1). For $0 < k \leq j$, let $\mathcal{N}_{k-1}^\circ$ be the set of all nodes $p \in \mathcal{N}_{k-1} \cap \mathcal{N}_k^\circ$ whose k–neighbors $q \in \mathcal{N}_k \backslash \mathcal{N}_{k-1}$ are also contained in $\mathcal{N}_k^\circ$ and let $\mathcal{N}_{k-1}^\bullet = \mathcal{N}_{k-1} \backslash \mathcal{N}_{k-1}^\circ$. As usual, $p, q \in \mathcal{N}_k$ are called k–neighbors if there is an edge $e = (p, q) \in \mathcal{E}_k$. Now the reduced coarse–grid spaces $\mathcal{S}_k^\circ$ are defined as follows

$$\mathcal{S}_k^\circ = \{v \in \mathcal{S}_k \mid v(p) = 0, \ p \in \mathcal{N}_k^\bullet\}, \quad 0 \leq k \leq j. \tag{3.3}$$

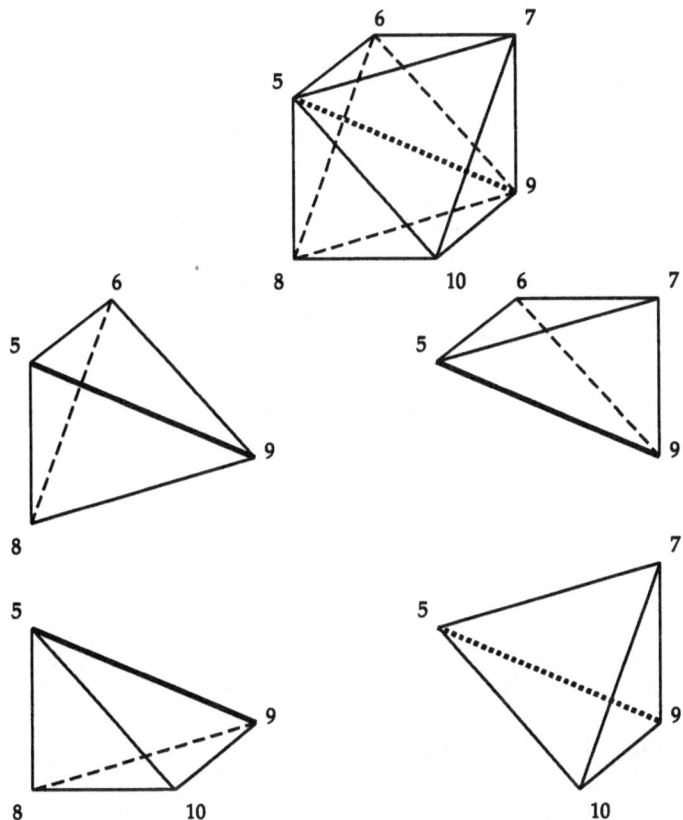

Figure 3.2 Splitting of the Remaining Octahedron in Four Tetrahedra w.r.t. the Diagonal (5,9)

Note that the definition (3.3) may be replaced by

$$v \in \mathcal{S}_k^\circ \Leftrightarrow v \in \mathcal{S}_k \text{ and } \operatorname{supp} v \subset \Omega_j^\circ, \tag{3.4}$$

using the reduced computational domain $\Omega_j^\circ = \bigcup_{p \in \mathcal{N}_j^\circ} \operatorname{supp} \lambda_p^{(j)}$. From (3.4) it is obvious that

$$\mathcal{S}_0^\circ \subset \mathcal{S}_1^\circ \subset \ldots \subset \mathcal{S}_j^\circ \tag{3.5}$$

so that a multilevel splitting of $\mathcal{S}_j^\circ$ can be performed in a straightforward way. In particular we chose the following sets of nodal basis functions,

$$\Lambda_0 = \{\lambda_p^{(0)} | \lambda_p^{(0)} \in \mathcal{S}_0^\circ\},$$

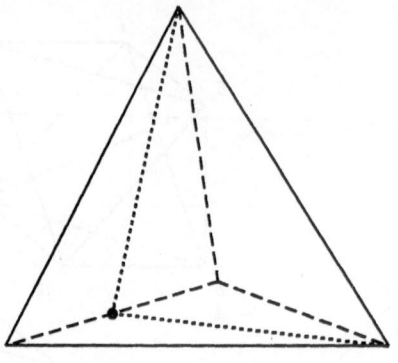

Figure 3.3 Green-I Closure

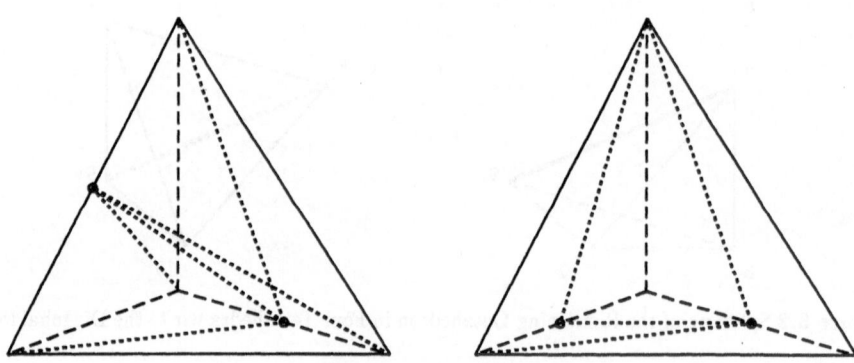

Figure 3.4 Green-II Closure

$$\Lambda_H = \bigcup_{k=1}^{j} \Lambda_k \,, \quad \Lambda_k = \{\lambda_p^{(k)} |\, \lambda_p^{(k)} \in \mathcal{S}_k^\circ \setminus \mathcal{S}_{k-1}^\circ\} \,, \ 1 \le k \le j,$$

and define a multilevel splitting

$$\mathcal{S}_j^\circ = V_0 + \sum_{\lambda \in \Lambda_H} V_\lambda \tag{3.6}$$

into the subspaces

$$V_0 = \mathrm{span}\{\lambda \mid \lambda \in \Lambda_0\}, \quad V_\lambda = \mathrm{span}\{\lambda\}, \ \lambda \in \Lambda_H. \tag{3.7}$$

Applying the well–known machinery of additive Schwarz methods (see for example [15])

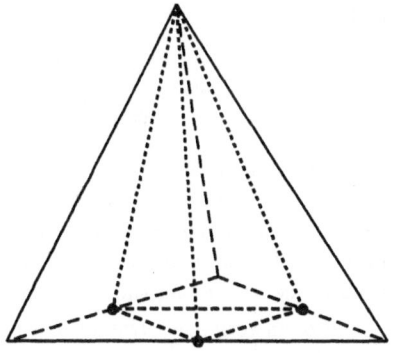

Figure 3.5 Green-III Closure

to the multilevel splitting (3.6) we obtain a reformulation

$$Pu_j^\circ = \ell'$$

of the original problem (3.2). Here

$$P = P_0 + \sum_{\lambda \in \Lambda_H} P_\lambda$$

denotes the sum of the Ritz projections $P_\nu : S_j^\circ \to V_\nu$, $\nu \in \{0, \lambda \in \Lambda_H\}$, defined by

$$a(P_\nu w, v) = a(w, v) , \quad v \in V_\nu$$

for each $w \in S_j^\circ$ and $\ell' \in (S_j^\circ)'$ is chosen appropriately. Denoting by $(\cdot, \cdot)$ the standard L^2–inner product we introduce the L^2–projections $Q_\nu : S_j^\circ \to V_\nu$ and the representation operators $A_\nu : V_\nu \to V_\nu$, $\nu \in \{0, \lambda \in \Lambda_H\}$ according to

$$(Q_\nu w, v) = (w, v) , \quad v \in V_\nu$$

for each $w \in S_j^\circ$ and

$$(A_\nu w, v) = a(w, v) , \quad v \in V_\nu ,$$

for each $w \in V_\nu$, $\nu \in \{0, \lambda \in \Lambda_H\}$. Using the L^2–representation $A_j : S_j^\circ \to S_j^\circ$ of $a(\cdot, \cdot)$ defined by

$$(A_j w, v) = a(w, v) , \quad v \in S_j^\circ ,$$

it is easily verified that $A_\nu P_\nu = Q_\nu A_j$. Hence, the operator P may be rewritten as

$$P = H_j A_j$$

where H_j stands for the preconditioner

$$H_j = A_0^{-1} Q_0 + \sum_{\lambda \in \Lambda_H} A_\lambda^{-1} Q_\lambda .$$

Evaluation of $A_\lambda^{-1}Q_\lambda$ leads to

$$H_j = A_0^{-1}Q_0 + \sum_{\lambda \in \Lambda_H} \frac{(\cdot, \lambda)}{a(\lambda, \lambda)}\lambda. \tag{3.8}$$

Note that H_j may be regarded as multilevel nodal basis preconditioner (c.f. [34]) based on symmetrically truncated basis functions. A possible unsymmetric truncation has been considered in [23]. Note that in the unconstrained case the preconditioner H_j is reducing to a special formulation of the well–known BPX preconditioner [10]. An efficient implementation of H_j is easily derived along the lines indicated in [8].

If H_j is applied in the context of an active set strategy, the coarse grid space V_0 may change in each outer iteration step. For this reason, it may be useful to replace the evaluation of $A_0^{-1}Q_0$ by simple diagonal scaling. For a further discussion we refer to [34, 37] and the literature cited therein.

The subsequent analysis of the condition number of $P = H_j A_j$ will be guided by the following lemma on additive Schwarz methods.

Lemma 3.1   i) Assume that for all $v \in S_j^\circ$ there is a splitting $v = v_0 + \sum_{\lambda \in \Lambda_H} v_\lambda$ such that

$$c\{a(v_0, v_0) + \sum_{\lambda \in \Lambda_H} a(v_\lambda, v_\lambda)\} \le a(v, v) \tag{3.9}$$

holds for some fixed positive constant c. Then we have the estimate

$$ca(v, v) \le a(Pv, v), \quad v \in S_j^\circ.$$

ii) Assume that for all splittings $v = v_0 + \sum_{\lambda \in \Lambda_H} v_\lambda$ of $v \in S_j^\circ$ the estimate

$$a(v, v) \le C\{a(v_0, v_0) + \sum_{\lambda \in \Lambda_H} a(v_\lambda, v_\lambda)\} \tag{3.10}$$

holds for some fixed positive constant C. Then we have the estimate

$$a(Pv, v) \le Ca(v, v), \quad v \in S_j^\circ.$$

Proof.   The assertion i) is the well–known lemma of P.L. Lions [30]. The simple proof is sketched for completeness. Let $v \in S_j^\circ$ and assume that the splitting $v = v_0 + \sum_{\lambda \in \Lambda_H} v_\lambda$ satisfies the condition (3.9). Then it follows from the definition of the orthogonal projections P_ν, $\nu \in \{0, \lambda \in \Lambda_H\}$, and the Cauchy–Schwarz inequality that

$$a(v, v) \le (a(P_0 v, v) + \sum_{\lambda \in \Lambda_H} a(P_\lambda v, v))^{\frac{1}{2}}(a(v_0, v_0) + \sum_{\lambda \in \Lambda_H} a(v_\lambda, v_\lambda))^{\frac{1}{2}}.$$

Application of (3.9) and the definition of P gives the assertion.

To prove ii) we apply (3.10) to the splitting $Pv = P_0 v + \sum_{\lambda \in \Lambda_H} P_\lambda v$ for some fixed $v \in S_j^\circ$ to obtain

$$a(Pv, Pv) \le C\{a(P_0 v, P_0 v) + \sum_{\lambda \in \Lambda_H} a(P_\lambda v, P_\lambda v)\} = Ca(Pv, v)$$

which completes the proof.   ∎

The assumptions (3.9) and (3.10) can be regarded as an asymptotic orthogonality of the subspaces V_ν, $\nu \in \{0, \lambda \in \Lambda_H\}$. Note that (3.10) is frequently established by strengthened Cauchy–Schwarz inequalities measuring the angles between V_ν with respect to $a(\cdot, \cdot)$ or any other symmetric bilinear form which is generating a uniformly equivalent norm on S_j.

In addition to the usual (semi) norms $\| \cdot \|_0$ and $| \cdot |_1$ of $L^2(\Omega)$ and $H^1(\Omega)$ we will make use of the local (semi) norms $\| \cdot \|_{0, \Omega_0}$ and $| \cdot |_{1, \Omega_0}$ induced by

$$(v, w)_{\Omega_0} = \int_{\Omega_0} v(x) \, w(x) \, dx \,, \quad v, w \in L^2(\Omega_0)$$

and the semi–inner product

$$(v, w)_{1, \Omega_0} = \sum_{i=1}^{3} (\partial_i v, \partial_i w) \,, \quad v, w \in H^1(\Omega_0)$$

for measurable $\Omega_0 \subset \Omega$. We introduce the L^2–projection $Q_k : S_j^\circ \to S_k^\circ$ by

$$(Q_k w, v) = (w, v) \,, \quad v \in S_k^\circ \,, \quad 0 \le k \le j,$$

and denote the diameter of a tetrahedron t by $h(t)$. For every node $p \in \mathcal{N}_k$ and every tetrahedron $t \in \mathcal{T}_k$ we define $U(p, k) = \operatorname{supp} \lambda_p^{(k)}$ and

$$U(t, k) = \{t' \in \mathcal{T}_k \mid t \cap t' \ne \emptyset\}$$

as the union of all tetrahedra in $\mathcal{T}_k$ intersecting t. Finally, constants depending only on the ellipticity (1.3) and the shape regularity of $\mathcal{T}_0$ will be denoted by c or C. Other parameters will be indicated explicitly.

We take up the analysis of the preconditioners with the following technical lemma

Lemma 3.2 *For some fixed k, $0 < k \le j$, let $\Lambda_{k-1} = \emptyset$ and $\Lambda_k \ne \emptyset$. Then we have the estimate*

$$\sum_{p \in \mathcal{N}_k^\circ} |v_k(p) \lambda_p^{(k)}|_1^2 \le C |v_k|_1^2, \quad v_k \in S_k^\circ.$$

Proof. Let $p \in \mathcal{N}_{k-1}$. As $\Lambda_{k-1} = \emptyset$, p is either contained in $\mathcal{N}_k^\bullet$ or has at least one k-neighbor $q \in \mathcal{N}_k \setminus \mathcal{N}_{k-1}$ which is contained in $\mathcal{N}_k^\bullet$. Hence, the semi–norm $| \cdot |_1$ is a norm on the restriction of $v_k \in S_k^\circ$ to $U(p, k-1)$. Now it follows from the uniform shape regularity of $\mathcal{T}_k$ and the equivalence of norms on finite dimensional spaces that

$$\sum_{q \in \mathcal{N}_k \cap U(p, k-1)} |v_k(q) \lambda_q^{(k)}|_1^2 \le c |v_k|_{1, U(p, k-1)}^2, \quad v_k \in S_k^\circ,$$

Summing up over all $p \in \mathcal{N}_{k-1}$ gives the assertion. $\blacksquare$

The following assumption is crucial for the stability of the L^2–projections Q_k.

(Q) There is a constant $c_0 > 0$ independent of j such that for $0 \le k \le j$ and all $t \in \mathcal{T}_k$ with the property $t \cap \mathcal{N}_j^\circ \ne \emptyset$ we have the estimate

$$\|v\|_{0, U(t, k)} \le c_0 h(t) |v|_{1, U(t, k)} \,, \quad v \in S_j^\circ. \tag{3.11}$$

Remark 3.1 Recall that the reduced problem (3.2) may be regarded as a Dirichlet problem on the reduced computational domain Ω_j°. It is the basic source of trouble that in general the boundary of Ω_j° is not represented exactly on lower levels. In particular, we cannot control the shape regularity of $U(t,k)$ intersecting the free boundary so that we cannot derive (3.11) from Poincaré's inequality via local transformations to a finite number of reference configurations as in the neighborhood of $\partial\Omega$ (compare the proof of Lemma 4.1 in [36]). Recall that the boundary $\partial\Omega$ of Ω is known to consist of the faces of the coarse tetrahedra $t \in \mathcal{T}_0$. Hence, the assumption (Q) may regarded as an asymptotic regularity assumption on the discrete free boundary $\partial\Omega_0^\circ \setminus \partial\Omega$.

Lemma 3.3 *Assume that (Q) holds. Then the L^2-projections Q_k, $0 \le k \le j$, satisfy the error estimate*

$$\|v - Q_k v\|_0^2 \le c 4^{-k} |v|_1^2, \quad v \in S_k^\circ. \tag{3.12}$$

Moreover, the Q_k are H^1-stable in the sense that

$$|Q_k v|_1^2 \le C |v|_1^2, \quad v \in S_j^\circ. \tag{3.13}$$

The constants c, C depend only on the local geometry of $\mathcal{T}_0$ and the constant c_0 from (Q).

Proof. The proof follows almost literally the arguments of Yserentant [36] in his proofs of Theorem 4.3 and Theorem 4.5. However, the application of Poincaré's inequality in the proof of Lemma 4.1 has to be replaced by (3.11) if $U(t,k)$ intersects the free boundary. ∎

Now we are ready to state the main result of this section.

Theorem 3.1 *Assume that the condition (Q) holds. Then there exist constants K_0, K_1 depending only on α_0, α_1 in (1.3), the shape regularity of $\mathcal{T}_0$ and the constant c_0 in (Q) such that the estimate*

$$K_0(j+1)^{-1} a(v,v) \le a(H_j A_j v, v) \le K_1 a(v,v)$$

holds for all $v \in S_j^\circ$.

Proof. Let us first consider the lower eigenvalue assuming for the moment that $\Lambda_0 \ne \emptyset$. To verify the assumption of Lemma 3.1 i) we consider the splitting

$$v = Q_0 v_0 + \sum_{k=1}^{j} (Q_k v - Q_{k-1} v) \tag{3.14}$$

of some fixed $v \in S_j^\circ$. It is easily seen that (3.14) gives rise to the representation $v = v_0 + \sum_{\lambda \in \Lambda_H} v_\lambda$ where $v_0 \in V_0$ and $v_\lambda \in V_\lambda$, $\lambda \in \Lambda_H$ are uniquely defined by

$$v_0 = Q_0 v, \quad Q_k v - Q_{k-1} v = \sum_{\lambda \in \Lambda_k} v_\lambda, \quad k = 1, \cdots, j. \tag{3.15}$$

Using the continuity of $a(\cdot, \cdot)$ and the the 3–D counterpart of the inverse inequality in Lemma 3.3 of [36], we have

$$\sum_{\lambda \in \Lambda_H} a(v_\lambda, v_\lambda) \le \alpha_1 \sum_{\lambda \in \Lambda_H} |v_\lambda|_1^2 \le c \sum_{k=1}^{j} 4^k \sum_{\lambda \in \Lambda_k} \|v_\lambda\|_0^2. \tag{3.16}$$

A simple computation gives

$$\sum_{\lambda \in \Lambda_k} \|v_\lambda\|_0^2 = \frac{1}{10} \sum_{t \in \mathcal{T}_k \cap \Omega_j^\circ} |t| \sum_{p \in t} |(Q_k v - Q_{k-1}v)(p)|^2$$
$$\leq c \|Q_k v - Q_{k-1}v\|_0^2 \tag{3.17}$$

and finally we have from (3.16), (3.17) and Lemma 3.3 that

$$\sum_{\lambda \in \Lambda_H} a(v_\lambda, v_\lambda) \leq c \sum_{k=1}^{j} 4^k \|Q_k v - Q_{k-1}v\|_0^2 \leq C(j+1)|v|_1^2. \tag{3.18}$$

The proof is completed by the H^1–stability of Q_0, i.e.

$$a(v_0, v_0) \leq \alpha_1 |Q_0 v|_1^2 \leq C|v|_1^2. \tag{3.19}$$

As by definition $V_0 = \text{span}\,\Lambda_0$, we still have to consider the case

$$\Lambda_{k^*} \neq \emptyset, \quad \Lambda_{k^*-1} = \ldots = \Lambda_0 = \emptyset \tag{3.20}$$

for some $k^* > 0$. Thus changing the initial level from 0 to k^* we obtain

$$Q_{k^*} v = \sum_{\lambda \in \Lambda_{k^*}} v_\lambda$$

so that (3.19) has to be replaced by

$$\sum_{\lambda \in \Lambda_{k^*}} a(v_\lambda, v_\lambda) \leq \alpha_1 \sum_{p \in \mathcal{N}_{k^*}^\circ} |v_{\lambda_p^{(k^*)}}|_1^2 \leq c|Q_{k^*} v|_1^2 \tag{3.21}$$

which is an immediate consequence of Lemma 3.2 applied to $v_k = Q_k v$. This completes the proof of the lower bound of $a(H_j A_j v, v)$.

To prove an upper bound we can use the same arguments as in [7] and [38] which rely on a suitable coloring of the nodes and a strengthened Cauchy–Schwarz inequality with respect to $(\cdot, \cdot)_1$. In particular, we decompose Λ_k according to

$$\Lambda_k = \bigcup_{i=1}^{I} \Lambda_{k,i} \tag{3.22}$$

where the $\Lambda_{k,i}$ are chosen such that

$$\text{supp}\,\lambda \cap \text{supp}\,\bar{\lambda} = \emptyset, \quad \lambda, \bar{\lambda} \in \Lambda_{k,i}.$$

Note that due to the refinement rules (T1) – (T3), this can be achieved by a uniformly bounded number I of subsets $\Lambda_{k,i}$ for each level k, $0 < k \leq j$. Based on the partition (3.22) we introduce the spaces

$$V_{k,i} = \text{span}\,\Lambda_{k,i}.$$

Observe that by construction the functions in $V_{k,i}$ are mutually orthogonal.

Now chose an arbitrary splitting of some fixed $v \in \mathcal{S}_j^\circ$ in contributions from the subspaces V_0 and V_λ, $\lambda \in \Lambda_H$

$$v = v_0 + \sum_{\lambda \in \Lambda_H} v_\lambda. \tag{3.23}$$

131

Based on the partition (3.22) the splitting (3.23) can be rewritten in the form

$$v = v_0 + \sum_{k=1}^{j} \sum_{i=1}^{I} v_{k,i} \, , \quad v_{k,i} = \sum_{\lambda \in \Lambda_{k,i}} v_\lambda \in V_{k,i}$$

so that the assertion is an immediate consequence of

$$|v|_1^2 \le C \left(|v_0|_1^2 + \sum_{k=1}^{j} \sum_{i=1}^{I} |v_{k,i}|_1^2 \right) \tag{3.24}$$

exploiting the orthogonality of $v_\lambda \in \Lambda_{k,i}$. However, (3.24) follows from the strengthened Cauchy–Schwarz inequality

$$(v_k, w_l)_1 \le c \left(\frac{1}{\sqrt{2}} \right)^{l-k} |v_k|_1 |w_l|_1 \tag{3.25}$$

for $v_k \in \mathcal{S}_k^\circ$, $w_l \in V_{l,i}$ and $l > k$ which can be derived by standard arguments used for example in [7, 35]. This completes the proof of the theorem. ∎

The estimate derived in Theorem 3.1 is suboptimal compared to the $O(1)$ – results due to [31], [12] and [9]. The corresponding investigation of minimal regularity assumptions on the free boundary, which still allow for optimal condition number estimates will be subject of further research.

However, even in the unconstrained case it frequently happens that the refinement process is stopped for accuracy reasons before the saturation of the condition number occurs. This observation is supported by the numerical results presented in the final section.

4. A Posteriori Error Estimates

Let $u \in H_0^1(\Omega)$ denote the exact solution of (1.2), $u_j \in \mathcal{S}_j$ the exact solution of the approximate problem (2.1) and $\tilde{u}_j \in \mathcal{S}_j$ an approximate solution of (2.1). In particular, $\tilde{u}_j$ may result from a certain number of steps of some iterative solver applied to (2.1). As only $\tilde{u}_j$ is known in actual computations, we are interested in local a posteriori error estimates for the total error $\varepsilon := \|u - \tilde{u}_j\|$ using the energy norm $\|\cdot\| = a(\cdot, \cdot)^{1/2}$ induced by the actual bilinear form. The local contributions to the total error will be utilized as local error indicators in the adaptive refinement process.

For an overview on the variety of well–established concepts in the unconstrained case we refer to [8, 13, 24, 32, 33] and the literature cited therein. Meanwhile there are some generalizations to variational inequalities of obstacle type. See for example [1, 25, 23, 26, 27].

In the present paper we will follow the basic approach of [13] which has been already successfully applied to obstacle problems (see e.g. [23, 26, 27]) to derive a posteriori estimates $\tilde{\varepsilon}$ which are reliable and efficient in the sense that

$$\gamma_0 \tilde{\varepsilon} \le \|u - \tilde{u}_j\| \le \gamma_1 \tilde{\varepsilon} \tag{4.1}$$

holds with positive constants γ_0, γ_1 independent of j. In particular, we will proceed in two main steps:

- Step 1: Approximate the defect problem by piecewise quadratic finite elements.

- Step 2: Approximate the resulting discrete defect problem by a semi–local or local simplification.

We introduce the subspace $\mathcal{Q}_j \subset H_0^1(\Omega)$ of continuous, piecewise quadratic functions vanishing at the boundary and the corresponding approximation

$$K_j^{\mathcal{Q}} = \left\{ v \in \mathcal{Q}_j \middle|\, v(p) \leq \varphi^L(p), p \in \mathcal{N}_j,\ v(e) \leq \varphi^{\mathcal{Q}}(e), e \in \mathcal{E}_j \right\}$$

of the constraints K. Here we used $v(e) := v(\text{midpoint of } e)$, $e \in \mathcal{E}_j$, for functions $v : \Omega \to$ IR and suitable restrictions φ^L, $\varphi^{\mathcal{Q}}$ of the obstacle φ to $\mathcal{N}_j$ and $\mathcal{E}_j$, respectively. Recall that $\mathcal{E}_j$ is denoting the set of interior edges of $\mathcal{T}_j$. The piecewise quadratic approximation $U_j \in K_j^{\mathcal{Q}}$ of u is obtained from

$$\text{Find } U_j \in K_j^{\mathcal{Q}} \text{ such that } \quad a(U_j, U_j - v) \leq \ell(U_j - v), \quad v \in K_j^{\mathcal{Q}}. \tag{4.2}$$

For notational convenience the index j will be suppressed in the following notations. Now the approximate error $d = U_j - \tilde{u}_j \in \mathcal{Q}_j$ may be computed from (4.2) or directly from the following defect problem

$$\text{Find } d \in D \text{ such that } \quad a(d, d - v) \leq r(d - v), \quad v \in D. \tag{4.3}$$

The constraints are given by

$$D = D(\tilde{u}_j) := \{ v \in \mathcal{Q}_j |\, v + \tilde{u}_j \in K_j^{\mathcal{Q}} \}$$

and the right–hand side is the residual $r := \ell - a(\tilde{u}_j, \cdot)$.

In Step 2 we concentrate on the simplification of (4.3) replacing $a(\cdot, \cdot)$ by a suitable quadratic form $\tilde{a}(\cdot, \cdot)$. For this reason, we introduce the two–level splitting

$$\mathcal{Q}_j = \mathcal{S}^L \oplus \mathcal{S}^{\mathcal{Q}} \tag{4.4}$$

consisting of the linear part $\mathcal{S}^L = \mathcal{S}_j$ and the remaining quadratic part $\mathcal{S}^{\mathcal{Q}} = \text{span}\{\mu_e | e \in \mathcal{E}_j\}$, where the quadratic bubbles $\mu_e \in \mathcal{Q}_j$ are defined by $\mu_e(p) = 0$, $p \in \mathcal{N}_j$, and $\mu_e(\bar{e}) = \delta_{e,\bar{e}}$, $\bar{e} \in \mathcal{E}_j$ (Kronecker delta). This splitting is independent of the space dimension. Utilizing the representation $v = v_L + \sum_{e \in \mathcal{E}_j} v_e \mu_e$, $v \in \mathcal{Q}_j$, the quadratic form $\tilde{a}(\cdot, \cdot)$ is defined by

$$\tilde{a}(v, w) = a(v^L, w^L) + \sum_{e \in \mathcal{E}_j} v_e w_e a(\mu_e, \mu_e), \quad v, w \in \mathcal{Q}_j. \tag{4.5}$$

It is well–known from [8, 13] that $a(\cdot, \cdot)$ and $\tilde{a}(\cdot, \cdot)$ are spectrally equivalent in the sense that

$$c\tilde{a}(v, v) \leq a(v, v) \leq C\tilde{a}(v, v), \quad v \in \mathcal{S}_j. \tag{4.6}$$

Now we can state the main result of this section.

Theorem 4.1 *Assume that the piecewise quadratic approximation $U_j \in \mathcal{Q}_j$ is of higher accuracy than the piecewise linear approximation $u_j \in \mathcal{S}_j$ in the sense that*

$$\|u - U_j\| \leq q\|u - u_j\|, \quad 0 \leq q < 1, \tag{4.7}$$

133

and that $\tilde{u}_j \in S_j$ satisfies

$$\|u - u_j\| \le \sigma \|u - \tilde{u}_j\| \tag{4.8}$$

with $q\sigma < 1$ and q, σ not depending on j. Let $\tilde{d}$ be the solution of the semi–local problem

$$\text{Find } \tilde{d} \in D \text{ such that } \tilde{a}(\tilde{d}, \tilde{d} - v) \le r(\tilde{d} - v), \quad v \in D. \tag{4.9}$$

Then (4.1) holds for $\tilde{\varepsilon}$ defined by

$$\tilde{\varepsilon}^2 = \tilde{a}(\tilde{d}, \tilde{d}) \tag{4.10}$$

and constants γ_0, γ_1 depending only on $q\sigma$, the ellipticity of $a(\cdot, \cdot)$ and the shape regularity of $\mathcal{T}_0$.

Proof. Theorem 4.1 is an immediate consequence of the Lemmas 4.1 and 4.2 in [23]. ∎

Note that it is a sufficient condition for (4.8) that

$$\|u_j - \tilde{u}_j\| \le (1 - 1/\sigma)\|u - u_j\|$$

holds with $\sigma < q^{-1}$. This may be regarded as an accuracy assumption on $\tilde{u}_j$. For a further discussion of (4.7) and (4.8) we refer to [23].

The error estimate (4.9) is called semi–local, because $\tilde{d}^L$ and $\tilde{d}^Q$ are decoupled with respect to the quadratic form but coupled by the set of constraints.

In our numerical experiments we will use the local contributions

$$\eta_e = (\tilde{d}_e^Q)^2 a(\mu_e, \mu_e), \quad e \in \mathcal{E}_j \tag{4.11}$$

of $a^Q(\tilde{d}^Q, \tilde{d}^Q)$ as local error indicators in the adaptive refinement process. As we cannot expect the active region of the continuous defect problem to coincide with the active region of the simplified discretization (4.9) there are no local variants of the inclusion (4.1). Indeed, consider a linear obstacle function φ and let $u = \varphi$ on some tetrahedron t. Then it is not clear that the corresponding indicators η_e vanish though it is known from Theorem 4.1 that asymptotically they cannot be too large. This explains why the semi–local estimate sometimes tends to be too pessimistic.

In practical computations (4.9) may be solved approximately using the active set strategy described above. To provide a good initial iterate the linear and the quadratic part in (4.9) are decoupled by one Gauss–Seidel step applied to the initial iterate zero. More precisely, we compute an estimate $\delta = \delta^L + \delta^Q$ from

$$\text{Find } \delta^L \in D^L \text{ such that } a(\delta^L, \delta^L - v) \le r^L(\delta^L - v), \quad v \in D^L \tag{4.12}$$

and

$$\text{Find } \delta^Q \in D^Q(\delta^L) \text{ such that } \\ a^Q(\delta^Q, \delta^Q - v) \le r^Q(\delta^Q - v), \quad v \in D^Q(\delta^L) \tag{4.13}$$

where r^L, r^Q denote the restriction of r to S^L, S^Q and D^L, $D^Q(\delta^L)$ are defined by

$$D^L = S^L \cap D, \quad D^Q(w^L) = \{v^Q \in S^Q | v^Q + w^L \in D\}, \quad w^L \in S^L.$$

Note that in the case of

$$K_j = \{v \in S_j | v(p) \le \varphi^L(p), p \in \mathcal{N}_j\} \subset K_j^Q$$

the linear defect problem is recovered by (4.12) with the consequence

$$\delta^L = u_j - \tilde{u}_j.$$

Moreover, each component δ_e^Q of δ^Q can be computed separately, giving

$$\delta_e^Q = \min\{r^Q(\mu_e)/a(\mu_e, \mu_e), (\varphi^Q - \delta^L - \tilde{u}_j)(e)\}, \quad e \in \mathcal{E}_j. \tag{4.14}$$

Remark 4.1 Assuming that $K_j \subset K_j^\Omega$ and that the iterative error δ^L is known, we obtain the local error estimate $\bar{\varepsilon}^2 := \tilde{a}(\delta, \delta)$ introduced in [26]. As a consequence of the theoretical and numerical considerations in [23] this estimate is likely to underestimate the error, but works very satisfactory as soon as the reduced domain $\Omega^\circ = \{x \in \Omega | u(x) < \varphi(x)\}$ is resolved properly by the discretization Ω_j°.

5. Numerical Results

We consider the elasto–plastic torsion of a cylindrical bar $\Omega = (0,1)^3$ which is twisted at its upper end around the longitudinal axis in such a way that the lateral surface remains stress free. Modelling the plastic region according to the von Mises yield criterion and normalizing physical constants, it is well–known (e.g. [17]) that for positive twist angle C per unit length the stress potential u is the solution of the variational inequality (1.2) with $a(\cdot, \cdot)$, $\ell(\cdot)$ given by

$$a(v, w) = (v, w)_1 , \quad \ell(v) = 2C \int_\Omega v \, d(x, y)$$

and the constraints K,

$$K = \{v \in V | \, v(x) \le \text{dist}(x, \Gamma_0), \ \text{a.e. in } \Omega\},$$

with $\Gamma_0 = \{x \in \partial\Omega \mid x = (x, y, z), 0 < z < 1\}$ denoting the vertical faces of the bar. The solution space V consists of all functions $v \in H^1(\Omega)$ satisfying homogeneous Dirichlet conditions at Γ_0. The inactive part $\Omega^\circ = \{x | u(x) < \text{dist}(x, \Gamma_0)\}$ of Ω characterizes the elastic region, while the material is considered plastic in the active points. Note that the elastic region becomes arbitraryly small for increasing C providing a challenging test example both for the preconditioner and the adaptive algorithm.

Of course, there is an equivalent 2–D formulation of this problem which has been already considered in the context of multilevel methods [23]. In the present paper the adaptive multilevel algorithm described in the sequel is applied to the 3–D formulation to allow for a comparison with these former results.

Table 5.1 Iteration History

Level	Depth	Nodes	Iterations	
			Solution	Error Estimate
0	0	27	1/0.0	3/1.7
1	1	125	3/2.3	3/2.0
2	2	223	5/3.3	3/2.7
3	3	665	4/6.3	2/0.0
4	4	2715	4/7.8	2/0.0
5	4	5651	4/8.8	3/0.7
6	5	29773	4/8.3	2/0.0
7	6	44075	5/7.6	2/0.0

On each refinement level j we apply the active–set strategy described in Section 2 until the active set remains invariant. The iteration is started with the interpolated approximation

135

from the previous level where the value at each node having at least one active neighbor is projected to the obstacle. On the first level the obstacle function is used as initial iterate. Each step of the outer iteration requires the solution of the linear subproblem (2.4) which is performed iteratively by cg–iterations preconditioned by the reduced BPX preconditioner introduced in Section 3. This inner iteration is stopped as soon as the estimated linear iteration error κ satisfies $\kappa \leq \kappa_0$. Here estimate κ is computed as described in [8]. Recall that the threshold κ_0 has to be chosen small enough to ensure the convergence of the outer iteration (c.f. Remark 2.1). In the following computations $\kappa_0 = 10^{-4}$ is used.

The same algorithm with $\kappa_0 = 10^{-2}$ is applied to the semi–local defect problem (4.9) providing local error indicators $\eta_e, e \in \mathcal{E}_j$ and an inexact semi–local error estimate $\tilde{\varepsilon}$ according to (4.11) and (4.10), respectively. As initial iterate we use $\delta^Q \in \mathcal{Q}_j$ which is computed from the local obstacle problems (4.14). Recall from Remark 4.1 that δ^Q gives rise to a local error estimate. Now a tetrahedron $t \in \mathcal{T}_j$ is marked for refinement if for at least one edge e of t the contribution η_e exceeds a certain threshold $\sigma\bar{\eta}$. We determine $\bar{\eta}$ by extrapolation as proposed in [3] (see [26] for details) and choose $\sigma = 0.5$.

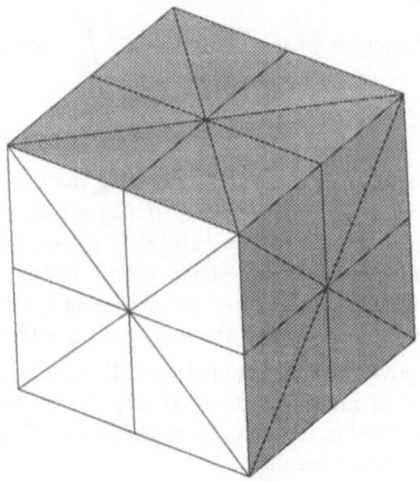

Figure 5.1 Initial Triangulation $\mathcal{T}_0$

Starting with the initial triangulation $\mathcal{T}_0$ depicted in Figure 5.1 and choosing $C = 5$, no elastic region is detected on the initial level. Note that in this case the local error estimates (4.14) provide the start iterate zero which obviously is a too optimistic guess. Compare the corresponding theoretical results and numerical observations in [23].

The algorithm is producing the final triangulation $\mathcal{T}_7$ with maximal depth $j = 6$ and the subscript now indicating the number of 7 refinement steps. The Figures 5.2 and 5.3 show the triangular faces and the level curves of the solution at the cutting plane $z = 0.5$. Note that the free boundary is emphasized by shading the faces of all tetrahedra which are fully contained in the plastic region. Obviously the refinement concentrates on the elastic part where the solution cannot be represented by piecewise linear functions. Though the refinement concentrates on the elastic region the semi–local error indicators seem to be too pessimistic introducing points in the plastic part of Ω here and there. Recall the discussion in the previous section.

A picture of the complete free boundary of the elastic region Ω_7° is shown in Figure 5.4.

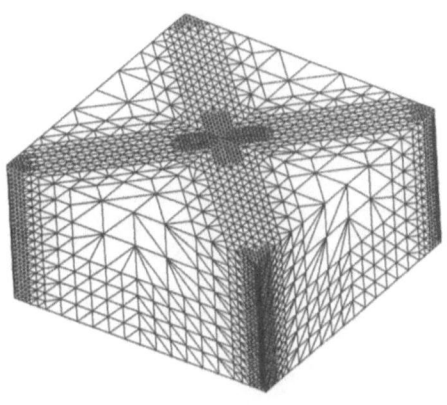

Figure 5.2 Final Triangulation $\mathcal{T}_7$ on the Cutting Plane $z = 0.5$

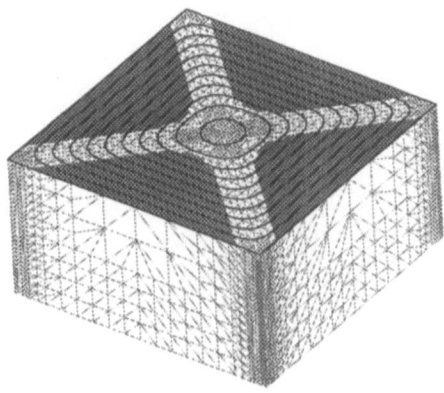

Figure 5.3 Final Solution and Free Boundary on the Cutting Plane $z = 0.5$

The behavior of the semi–local and local a posteriori error estimates up to refinement level 5 is illustrated in Figure 5.5 in comparison with the "exact" error resulting from a uniform refinement of $\mathcal{T}_5$. As mentioned above the local estimate fails on the initial level but works quite satisfactory later on.

For a detailed history of the solution process we refer to Table 5.1 . The data are presented in the form "number of outer iterations / average number of inner iterations" both needed for the solution and the semi–local error estimate, respectively. Observe that the semi–local error estimate reduces to the local error estimate with increasing refinement. Indeed, the outer iterations do not change the initial guess and may be skipped.

To illustrate the behavior of the reduced BPX preconditioner in more detail, we choose κ_0 unreasonable small, i.e., $\kappa_0 = 10^{-8}$ and the initial iterate is fixed to zero for all inner

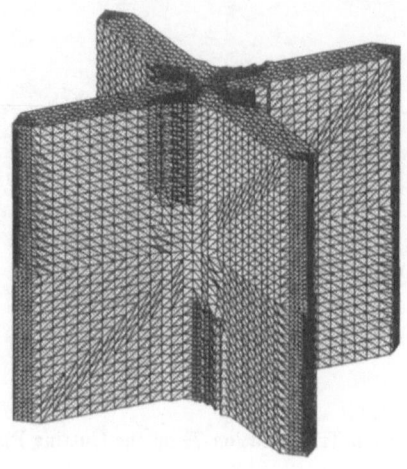

Figure 5.4 Final Approximation of the Elastic Region

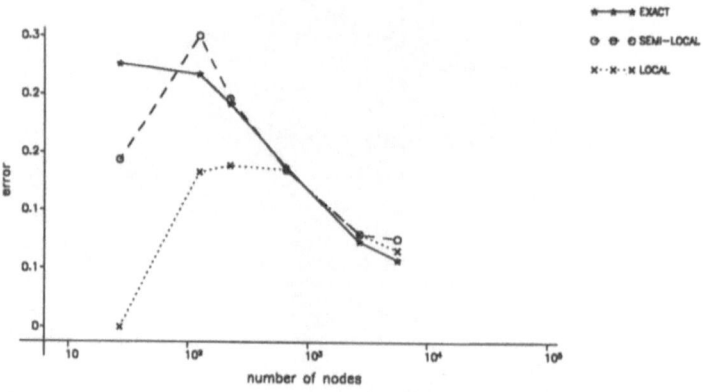

Figure 5.5 Behavior of the Error Estimate

iterations. In this case the number of (preconditioned) iterations may be used as a measure of the condition number of the corresponding linear system. For each refinement level we choose the linear sub–problem requiring the maximal number of (preconditioned) cg–iteration steps which are reported as a function of the number of unknowns in Figure 5.6. As expected from the theoretical considerations we observe a linear increase of the number of multilevel preconditioned iterations while without preconditioning the number of iterations grows exponentially with increasing refinement. Obviously, this behavior occurs as soon as the resolution of the elastic region allows for an adequate representation on coarser triangulations. We emphasize that due to Lemma 3.2 the condition number cannot be too bad before this situation is reached.

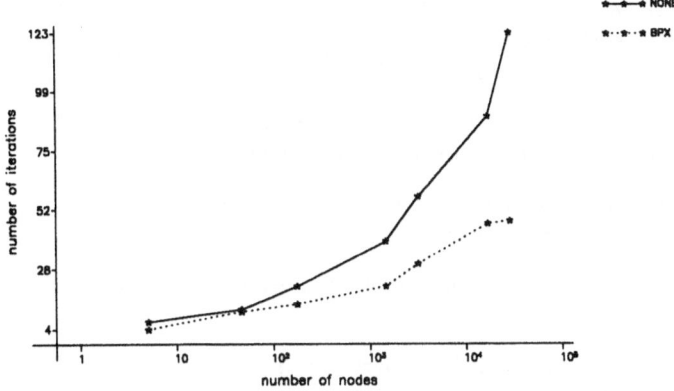

Figure 5.6 Behavior of the Reduced BPX Preconditioner

Acknowledgements. The authors would like to thank H. Yserentant for pointing out the difficulties concerning the stability of the L^2-projections resulting from the presence of a free boundary, F. Bornemann and R. Roitzsch for their assistance in implementation and P. Deuflhard for his continuous support. The figures have been generated with the help of the graphical environment GRAPE.

References

[1] M. Ainsworth, J.T. Oden, C.Y. Lee: *Local A Posteriori Error Estimators for Variational Inequalities.* Numer. Meth. Part. Diff. Eq. **9**, p. 23–33 (1993).

[2] O. Axelsson, V.A. Barker: *Finite Element Solution of Boundary Value Problems.* Academic Press, New York (1984).

[3] I. Babuška, W.C. Rheinboldt: *Estimates for Adaptive Finite Element Computations.* SIAM J. Numer. Anal. **15**, p. 736–754 (1978).

[4] R. E. Bank, A. Weiser: *Some a posteriori Error Estimators for Elliptic Partial Differential Equations.* Math. Comp. **44**, p. 283–301 (1985).

[5] R.E. Bank, A.H. Sherman, H. Weiser: *Refinement Algorithm and Data Structures for Regular Local Mesh Refinement.* In: Scientific Computing, R. Stepleman et al. (eds.), Amsterdam: IMACS North–Holland, p. 3–17 (1983).

[6] J. Bey: *Analyse und Simulation eines Konjugierte–Gradienten–Verfahrens mit einem Multilevel–Präkonditionierer zur Lösung dreidimensionaler, elliptischer Randwertprobleme für massiv parallele Rechner.* Diplomarbeit, RWTH Aachen (1991).

[7] F. Bornemann: *A sharpened Condition Number Estimate for the BPX Preconditioner of Elliptic Finite Element Problems on Highly Nonuniform Triangulations.* Konrad–Zuse–Zentrum Berlin, Preprint SC 91-9 (1991).

[8] F. Bornemann, B. Erdmann, R. Kornhuber: *Adaptive Multilevel–Methods in Three Space Dimensions.* To appear in Int. J. Numer. Methods Eng. (1993).

[9] F. Bornemann, H. Yserentant: *A Basic Norm Equivalence for the Theory of Multilevel Methods.* To appear in Numer. Math. (1993).

[10] J.H. Bramble, J.E. Pasciak, J. Xu: *Parallel Multilevel Preconditioners.* Math. Comp., **55**, p. 1–22 (1990).

[11] F. Brezzi, W.W. Hager and P.A. Raviart: *Error Estimates for the Finite Element Solution of Variational Inequalities I.* Numer. Math. **28**, p. 431–443 (1977).

[12] W. Dahmen, A. Kunoth: *Multilevel Preconditioning.* Numer. Math. **63**, p. 315– 344 (1992).

[13] P. Deuflhard, P. Leinen, H. Yserentant: *Concepts of an Adaptive Hierarchical Finite Element Code.* IMPACT 1, p. 3–35 (1989).

[14] M. Dryja, O.B. Widlund: *Multilevel Additive Methods for Elliptic Finite Element Problems.* Tech. Rep. 507, Dept. of Comp. Science, Courant Inst. New York (1990).

[15] M. Dryja, O.B. Widlund: *Towards a Unified Theory of Domain Decomposition Algorithms for Elliptic Problems.* In: Domain Decomposition Methods for Partial Differential Equations. T.F. Chan et al. (eds.), SIAM, Philadelphia, p. 3–21 (1989).

[16] B. Erdmann, R. Roitzsch: *3-D ELLKASK Programmers Manual 1.0 .* To appear as Technical Report, Konrad–Zuse–Zentrum Berlin (ZIB).

[17] R. Glowinski, J.L. Lions, R. Trémolières: *Numerical Analysis of Variational Inequalities.* North–Holland, Amsterdam (1981).

[18] W. Hackbusch, H.D. Mittelmann: *On Multigrid Methods for Variational Inequalities.* Numer. Math., **42**, p. 65–76 (1983).

[19] R.H.W. Hoppe: *Multigrid Algorithms for Variational Inequalities.* SIAM J. Numer. Anal., **24**, p. 1046–1065 (1987).

[20] R.H.W. Hoppe: *Two-sided Approximations for Unilateral Variational Inequalities by Multigrid Methods.* Optimization, **18**, p. 867–881 (1987).

[21] R.H.W. Hoppe: *Une méthode multigrille pour la solution des problèmes d'obstacle.* M^2 AN, **24**, p. 711–736 (1990).

[22] R.H.W. Hoppe, R. Kornhuber: *Multilevel Preconditioned CG–Iterations for Variational Inequalities.* In: Preliminary Proceedings of the 5th Copper Mountain Conference on Multigrid Methods, St. McCormick et al.(eds.), Copper Mountain Colorado (1991).

[23] R.H.W. Hoppe, R. Kornhuber: *Adaptive Multilevel–Methods for Obstacle Problems.* to appear in SIAM J. Num. Anal. (1993).

[24] C. Johnson: *Numerical Solutions of Partial Differential Equations by the Finite Element Method.* Cambridge University Press, Cambridge (1987).

[25] C. Johnson: *Adaptive Finite Element Methods for the Obstacle Problem.* Preprint NO 1991-25, Chalmers University of Technology, Göteborg (1991).

[26] R. Kornhuber, R. Roitzsch: *Self Adaptive Finite Element Simulation of Bipolar, Strongly Reverse Biased pn-Junctions.* To appear in Comm. Num. Meth. Engrg. (1993).

[27] R. Kornhuber, R. Roitzsch: *Self Adaptive Computation of the Breakdown Voltage of Planar pn-Junctions with Multistep Field Plates.* In: Proceedings of the 4th International Conference of Simulation of Semiconductor Devices and Processes. (Fichtner et al., eds.), Zurich, p. 535-543 (1991).

[28] P. Leinen: *Ein schneller adaptiver Löser für elliptische Randwertprobleme.* Dissertation, Dortmund (1990).

[29] F. W. Letniowski: *Three-Dimensional Delaunay Triangulations for Finite Element Approximations to a Second-Order Diffusion Problem.* SIAM J. Sci. Stat. Comput., **13**, p.765-770 (1992).

[30] P.L. Lions: *On the Schwarz Alternating Method I.* In: Domain Decomposition Methods for Partial Differential Equations. R.Glowinski et al. (eds.), SIAM, Philadelphia (1988).

[31] P. Oswald: *On Discrete Norm Estimates Related to Multilevel Preconditioners in the Finite Element Method.* In: Proceedings of the International Conference on the Constructive Theory of Functions, Varna 1991 (to appear).

[32] B. Szabó, I. Babuška: *Finite Element Analysis.* J. Wiley & Sons, New York (1991)

[33] R. Verfürth: *A Posteriori Error estimation and Adaptive Mesh-Refinement Techniques.* Preprint Universität Zürich (1992).

[34] J. Xu: *Iterative Methods by Space Decomposition and Subspace Correction.* SIAM Review **34** p. 581-613 (1992).

[35] H. Yserentant: *On the Multilevel Splitting of Finite Element Spaces.* Numer. Math. **49**, p. 379-412 (1986).

[36] H. Yserentant: *Two Preconditioners Based on the Multilevel Splitting of Finite Element Spaces.* Numer. Math. **58**, p. 163-184 (1990).

[37] H. Yserentant: *Old and New Convergence Proofs of Multigrid Methods.* to appear in Acta Numerica (1993).

[38] X. Zhang: *Multilevel Schwarz Methods.* Numer. Math. **63**, p. 521-539 (1992).

Adaptive Point Block Methods

M. Griebel and S. Zimmer
Institut für Informatik
Technische Universität München
Arcisstraße 21, D-W 8000 München 2, Germany

Summary

We present adaptive algorithms for the solution of elliptic PDEs that are based on the so-called sparse grid discretization technique and hierarchical error indicators. We discuss both the approximation qualities of the adaptively generated sparse grids and the efficient iterative solution of the arising linear systems by means of certain grid- and point-oriented multilevel methods.

The results of numerical experiments regarding both topics are presented.

1 Introduction

For the discretization of a function u on the unit square $\Omega := (0,1)^2$, we use a tableau of grids

$$\Omega_{m,n} = \{(i \cdot h_m, j \cdot h_n) : i, j \in \mathbb{N}\} \cap \Omega, \quad m, n \in \mathbb{N}, \tag{1}$$

that are independently refined in x- and y-direction. Here, $h_m := 2^{-m}$ and $h_n := 2^{-n}$ denote the mesh sizes with respect to the x- and y-direction, respectively.

On each grid, we consider the space $V_{m,n}$ of piecewise bilinear functions with the usual nodal basis

$$B_{m,n} := \{\phi_x^{(m,n)}, x \in \Omega_{m,n}\} \tag{2}$$

with basis functions $\phi_x^{(m,n)} \in V_{m,n}$ that satisfy $\phi_x^{(m,n)}(y) = \delta_{x,y}$, $y \in \Omega_{m,n}$.

Our goal is to select a finite-dimensional subspace V^A from the infinite dimensional-space

$$V := \bigcup_{m=1}^{\infty} \bigcup_{n=1}^{\infty} V_{m,n}, \tag{3}$$

e.g. by adaptive refinement strategies, that allows an approximation of u within a prescribed accuracy and possesses only as few degrees of freedom as necessary, i.e. a possibly minimal $\dim(V^A)$. We describe such adaptively generated subspaces V^A as the span of certain subsets H^A of a hierarchical basis H for V.

In order to construct H, we define for each m, n the set of hierarchical grid points

$$\tilde{\Omega}_{m,n} := \begin{cases} \Omega_{1,1} & \text{for } m = n = 1 \\ \Omega_{m,1} \setminus \Omega_{m-1,1} & \text{for } m > 1, n = 1 \\ \Omega_{1,n} \setminus \Omega_{1,n-1} & \text{for } m = 1, n > 1 \\ \Omega_{m,n} \setminus (\Omega_{n-1,m} \cup \Omega_{m,n-1}) & \text{for } m > 1, n > 1 \end{cases} \tag{4}$$

and the associated set of nodal basis functions

$$\tilde{B}_{m,n} := \{\phi_x^{(m,n)}, x \in \tilde{\Omega}_{m,n}\}. \tag{5}$$

Then, we obtain by

$$H := \bigcup_{m=1}^{\infty} \bigcup_{n=1}^{\infty} \tilde{B}_{m,n} \tag{6}$$

a basis for V. Note that for all m, n each $\phi_x^{(m,n)} \in H$ corresponds to exactly one $x \in \tilde{\Omega}_{m,n}$. Note further that in contrast to the hierarchical basis introduced by Yserentant [11, 12], H can be viewed as the product of two one-dimensional hierarchical bases with respect to the x- and y-direction, respectively, and contains basis functions with extremely distorted support, since we use grids $\tilde{\Omega}_{m,n}$ with $n \ll m$ or $m \ll n$.

We will describe certain types of adaptively refined grids and spaces by selection of arbitrary finite subsets $H^A \subset H$ (specified equivalently by the sets of corresponding grid points Ω^A) that only have to satisfy the following condition which guarantees the hierarchical structure of H^A:

$$\forall \phi_1, \phi_2 \in H : (\phi_1 \in H^A \wedge \text{supp}(\phi_1) \subset \text{supp}(\phi_2) \Rightarrow \phi_2 \in H^A). \tag{7}$$

In this paper we will focus on two major questions arising with adaptive refinement strategies for the solution of PDEs. We will consider both the proper selection of H^A and the efficient solution of the linear systems arising from the discretization of an elliptic linear PDE using such adaptive grids.

For reasons of simplicity, we restrict our study to the case of the Poisson equation with Dirichlet boundary conditions, denoted in the weak formulation as

$$u \in V : \forall v \in V : a(u,v) = f(v) \tag{8}$$

with $a(u,v) := \int_\Omega \nabla u \cdot \nabla v \, d\Omega$ and the linear functional f expressing the boundary conditions and the right hand side terms.

The remainder of the paper is organized as follows. In section 2, we give a brief outline of the sparse grid discretization method introduced in [13] that can be viewed as a special a priori choice of V^A with significantly reduced storage requirements compared to standard full grids, provided that u is sufficiently smooth. In section 3, we discuss the efficient solution of the linear systems arising from the sparse grid discretization, using multigrid-type and so called point-block techniques as presented in [5]. By using a hierarchical error indicator [3, 14], we generalize the sparse grid discretization method in section 4 to the case of adaptive refinement and demonstrate approximation qualities of adaptive sparse grids. These grids turn out to preserve the advantages of sparse grids over locally refined full grids even in the case of non-smooth solutions. In section 5, we turn again to the solution of the arising linear systems and show that the point-block techniques are effective in the adaptive case as well. Finally, we demonstrate in section 6 that the point-block principle is useful not only with respect to the solution process, but also for the construction of subspaces. It might allow an a priori generation of a suitable Ω^A driven only by the input data (e.g. the boundary conditions) of the problem.

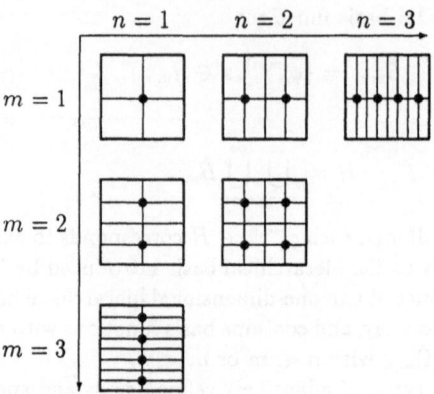

Figure 1: The triangular scheme of grids $\tilde{\Omega}_{m,n}$, $m + n \leq k + 1$, that are contained in the sparse grid Ω_k^S, $k = 3$.

2 Regular sparse grids

We assume in this section that the solution u is smooth in the sense

$$\frac{\partial^4 u}{\partial^2 x \partial^2 y} \in \mathcal{C}(\Omega) \tag{9}$$

and denote by $u_{M,N}^I \subset V$ the interpolation of u on some $\Omega_{M,N}$, represented in the hierarchical basis as

$$u_{M,N}^I = \sum_{\phi \in H} u_\phi^H \cdot \phi. \tag{10}$$

Then, following [13], it can be shown that the coefficients u_ϕ^H are of the order $O(h_m^2 h_n^2)$ for all $\phi \in \tilde{B}_{m,n}$, where the bounding constant only depends on u and not on M and N. This motivates a choice of H^A that only uses basis functions from the grids $\tilde{\Omega}_{m,n}$ where $h_m^2 h_n^2$ is large, and omits the basis functions that are associated to the grids $\tilde{\Omega}_{m,n}$ where $h_m^2 h_n^2$ is small and therefore have no significant contribution to the interpolant of u. So, we define for $k \in \mathbb{N}$ the finite hierarchical basis

$$H_k := \bigcup_{m=1}^{k} \bigcup_{n=1}^{k+1-m} \tilde{B}_{m,n} \tag{11}$$

that spans the sparse grid space

$$V_k := \operatorname{span} H_k. \tag{12}$$

We call the set of corresponding grid points

$$\Omega_k^S := \bigcup_{m=1}^{k} \bigcup_{n=1}^{k+1-m} \tilde{\Omega}_{m,n} \tag{13}$$

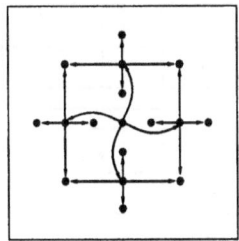

Figure 2: Representation of a sparse grid ($k = 3$) in a graph-like data structure.

a (regular) *sparse grid*, see Figure 1.

It can be shown that under the smoothness assumption (9), the interpolation error for the sparse grid space is $O(h_k^2 \log h_k^{-1})$ with respect to the L_2-norm, which is only slightly worse than $O(h_k^2)$ resulting for the full grid space $V_{k,k}$. With respect to the H^1-norm, the interpolation error is $O(h)$ which is even of the same order as for the corresponding full grid [3]. However, the number of grid points used, is substantially reduced from $O(h_k^{-2})$ for the full grid $\Omega_{k,k}$ to $O(h_k^{-1} \log h_k^{-1})$, since the grids used in (13) contain comparatively few grid points. Therefore, the choice of $V^A := V_k$ gives an efficient representation of sufficiently smooth smooth functions that fulfill condition (9).

Suitable data structures for the implementation of sparse grid algorithms are graph-like structures like those discussed in [1, 3] or that shown in Figure 2. There, the pointer structure reflects the hierarchical structure of the basis. Each node represents one grid point and has at most two 'son' nodes per coordinate direction, which are grid points from the next finer grid in that direction.

3   Efficient iterative methods for sparse grid systems

For the iterative solution of the linear systems that arise from the Galerkin discretization of the Poisson equation, we implemented several types of subspace correction schemes [10]. First, we simply use the one-dimensional subspaces spanned by the basis functions $\phi \in H_k$. This is the Gauss-Seidel relaxation in the hierarchical basis H_k, similar to the HB-MG method [2], see also Figure 3.

Hierarchical basis method
for $n = 1...k$
   for $m = 1...k + 1 - n$
      for $\phi \in \tilde{B}_{m,n}$
$$u \leftarrow u - \frac{a(u,\phi) - f(\phi)}{a(\phi,\phi)} \cdot \phi$$

Note that the corresponding hierarchical stiffness matrix is not as sparse as in the case of the standard full grid discretization with the nodal basis $B_{k,k}$. Due to its special

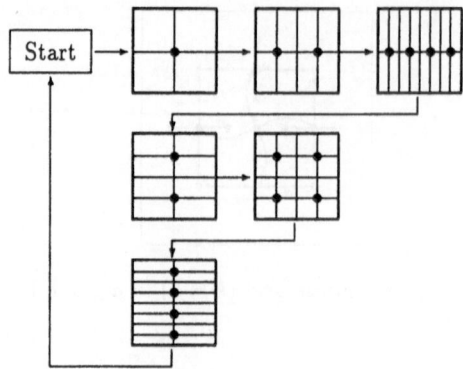

Figure 3: Sequence of relaxations for the hierarchical basis method.

Table 1: Reduction rates and numbers of iterations for the hierarchical basis method.

k	3	4	5	6	7	8	9	10
ρ	0.56	0.68	0.77	0.83	0.88	0.913	0.938	0.955
it	40	60	88	120	180	250	360	500

structure, however, the matrix never needs to be assembled explicitly and the update of the residuals $a(u, \phi) - f(\phi)$ and therefore the relaxation can be performed with a constant amount of work per grid point [3]. Note further that the sequence of relaxations inside each grid is arbitrary, since the basis functions involved have non-overlapping supports.

Table 1 shows for $k = 3, \ldots, 10$ the reduction rate ρ per iteration step and the number of iterations

$$it := \frac{-10}{\log \rho} \tag{14}$$

that are needed to reduce an arbitrary initial error by the factor 10^{-10}. It can clearly be seen that (in contrast to the conventional HB-MG method) the iteration count it grows exponentially with k.

Therefore, we have to search for more efficient solvers. In [5], such algorithms are constructed using a generating system

$$E_k := \bigcup_{m=1}^{k} \bigcup_{n=1}^{k+1-m} B_{m,n} \tag{15}$$

for V_k instead of the basis H_k. With the subspaces spanned by the generating functions $\phi \in E_k$, various multigrid-like methods can be constructed by means of Gauss-Seidel relaxations for the generating system E_k, depending on the ordering of E_k. For example, a grid-oriented algorithm similar to the methods in [6, 7, 8, 9] is given for the sparse grid case.

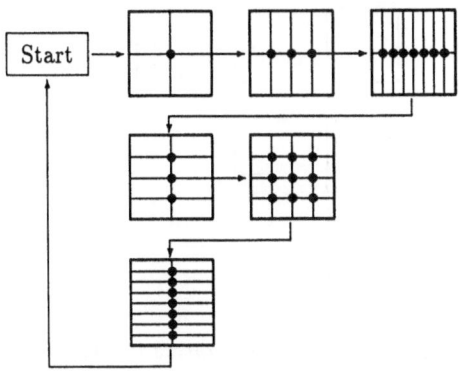

Figure 4: Sequence of relaxations for the multigrid method (outer loop).

Table 2: Reduction rates and numbers of iterations for the multigrid method.

k	3	4	5	6	7	8	9	10
ρ	0.066	0.10	0.18	0.22	0.27	0.32	0.33	0.34
it	8.5	10	13	15	18	20	21	21

Multigrid method
for $n = 1...k$
  for $m = 1...k + 1 - n$
    for $\phi \in B_{m,n}$
      $u \leftarrow u - \frac{a(u,\phi) - f(\phi)}{a(\phi,\phi)} \cdot \phi$

Inside each level, the sequence of relaxation is no longer arbitrary. We use a lexicographical ordering on each grid as an inner loop, while the outer loop switches from grid to grid in the sequence shown in Figure 4.

Table 2 shows that the reduction rates and iteration counts for this method are fairly small and bounded independently of k. Further experiments with other traversal orderings through the different grids $\Omega_{m,n}$, $m + n \leq k + 1$, revealed similar reduction rates.

Beside modeling multigrid-type methods, the generating system approach also leads in a natural way towards quite different types of relaxation schemes. As an example, we will discuss the so-called point-block techniques.

The key idea behind these schemes is a subspace correction where all generating functions from different levels belonging to one grid point are grouped together to span a *point-block* subspace. Thus, we set

$$B_x := \{\phi_x^{(m,n)} \in E_k : x \in \Omega_{m,n}\} \tag{16}$$

and obtain the corresponding point-space

$$V_x := \text{span } B_x, \tag{17}$$

147

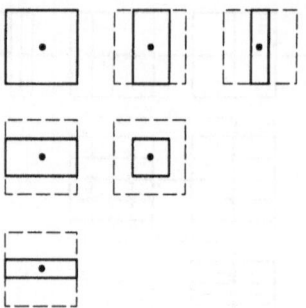

Figure 5: Supports of the generating functions $B_{(\frac{1}{2}, \frac{1}{2})}$.

Table 3: Reduction rates and numbers of iterations for the point-block relaxation.

k	3	4	5	6	7	8	9	10	11	12	13
ρ	0.0022	0.0042	0.0086	0.015	0.018	0.019	0.021	0.023	0.024	0.024	0.024
it	3.8	4.2	4.8	5.5	5.7	5.8	6.0	6.1	6.2	6.2	6.2

see Figure 5 for an example. Now, we perform the Gauss-Seidel relaxation block-wise in the same sequence of grid points as in the hierarchical basis method.

Point-block method
    for $n = 1...k$
        for $m = 1...k + 1 - n$
            for $x \in \tilde{\Omega}_{m,n}$
                find $\delta u \in V_x$ with $\forall \phi \in B_x : a(u - \delta u, \phi) = f(\phi)$ (18)
                $u \leftarrow u - \delta u$

To compute δu from equation (18), the comparatively small subsystem belonging to the point-block space V_x has to be solved, e.g. by a direct solver for the corresponding submatrix of the stiffness matrix.

The rapid convergence of this method, shown in Table 3, is not surprising, since the space V_x allows a good approximation to the Green functions of the Laplacian.

The point-block method has other favorable properties. Note first, that after the relaxations for the points on the center line $(y = 1/2)$ are finished, no further communication between the two halfs of Ω above and below this line is needed. So, this method has parallelization properties like a domain decomposition method, without bothering about the Schur complements. This applies of course recursively to both half squares as well as on the relaxation within one line.

Additionally, for an anisotropic model problem, the point-block method behaved much more robust than the multigrid method above [5].

4   Adaptive sparse grids

While regular sparse grids are efficient in the case of smooth solutions u, their favorable approximation properties break down in the presence of singularities. However, investigating the data structure used for the grid representation (Figure 2), we see that, from an implementational point of view, the criterion $m + n \leq k + 1$ used in equation (11) for the selection of the grid points in Ω_k^S seems to be artificial in some sense. Every other set H^A of basis functions could be stored as well, as long as condition (7) is satisfied. The algorithm that computes the update of the residual can also be performed as long as condition (7) holds. Therefore, the concept of sparse grids can be easily extended to the case of adaptively generated grids [3].

The construction of a grid is now a successive process that is controlled by a local error indicator applied to some approximate solution $\hat{u}$. Since a hierarchical basis is used, we apply a hierarchical error indicator [14] for driving the refinement process. It measures the contribution of a basis function ϕ:

$$|\hat{u}_\phi^H \cdot \phi| > \epsilon, \tag{19}$$

where $\phi \in H^A$, $\hat{u}_\phi^H$ is the coefficient in the representation $\hat{u} = \sum \hat{u}_\phi^H \cdot \phi$, ϵ is a positive threshold value, either user specified and fixed or computed by some refinement strategy, and $|.|$ denotes some appropriate norm (the L_2-norm in our examples). Now, starting with a coarse grid $\Omega^{A,old}$, the adaption process reads as follows:

1. Refine the grid *globally*, i.e. add 'son' nodes to all nodes in the grid graph, if not already present. Denote the resulting grid by $\Omega^{A,new1}$.

2. Compute an approximate solution $\hat{u}$ on $\Omega^{A,new1}$ (see section 5).

3. Delete from $\Omega^{A,new1}$ all nodes that do not satisfy condition (19) and that are not needed to satisfy condition (7). Denote the resulting grid by $\Omega^{A,new2}$.

4. Set $\Omega^{A,old} := \Omega^{A,new2}$ and repeat these steps until a termination criterion (see below) is reached.

In this paper, we do not focus on strategies to determine appropriate values of ϵ, see [14] for a discussion of this topic. Instead, our experiments are carried out with a priori fixed values of ϵ. The refinement process stops, if further refinement steps generate no new grid points. This strategy is in practice not particularly efficient, since the last refinement steps that are performed turn out to generate only a few new grid points with no substantial gain in accuracy, so a weaker termination criterion or a parameter driven relative threshold strategy [14] could save much work. However, in order to demonstrate the approximation quality of the resulting grids and the performance of the solvers in the adaptive case, this simple criterion is sufficient.

We applied this algorithm first to the smooth model problem with the solution

$$u(x,y) = \frac{\sinh(\pi(1-x)) \cdot \sin(\pi y)}{\sinh(\pi)} \tag{20}$$

and secondly, to the non-smooth model problem with the solution

$$u(x,y) = \mathrm{Re}\sqrt{x + i(y - \frac{1}{2})} \tag{21}$$

that possesses a singularity of type $r^{1/2}$ at the point $(0, 1/2)$. The results are presented in Figure 6. For the case of *regular*, i.e. non-adapted, sparse grids Ω_k^S, $k = 2, \ldots, 10$, the first diagram shows the difference of the computed solution u_k to the interpolation u_k^I of the exact solution in the L_2-norm. Note the poor performance of the second model problem.

The second diagram shows the case of *adaptively* generated grids with different, but fixed, values of $\epsilon = 4^{-i}$, $i = 2, \ldots, 13$. While there is no significant improvement for the first model problem, the results for the second model problem exhibit an approximation quality as in the smooth case. This shows that adaptive refinement can overcome the smoothness restriction of regular sparse grid discretizations.

Figure 7 shows an example for an adaptive sparse grid for the second model problem. Note that in regions where the solution is smooth, structures similar to regular sparse grids develop, so these regions are represented by only a few grid points. Thus, the reduction in storage requirements that was observed for regular sparse grids is preserved in the adaptive case.

5 Efficient relaxation methods in the adaptive case

Now we address the question of the efficient computation of an approximate solution on the adaptively refined grids arising in the refinement procedure above.

To transfer the multigrid and point-block solvers from the regular to the adaptive case, we have to construct the generating system E^A corresponding to a given basis H^A. For that purpose, we decompose the grid level-wise into components $\tilde{\Omega}_{m,n}^A := \tilde{\Omega}_{m,n} \cap \Omega^A$. Then, we "fill the holes" in $\tilde{\Omega}_{m,n}^A$ by adding all points from $\Omega_{m,n}$ that have in $\tilde{\Omega}_{m,n}^A$

- two neighbors in the same coordinate direction or

- all four neighbors in diagonal direction,

and denote the result by $\Omega_{m,n}^A$, cf. equations (1) and (4) and, for an example, Figure 8. Now, set $B_{m,n}^A := \{\phi_x^{(m,n)}, x \in \Omega_{m,n}^A\}$ and

$$E^A := \bigcup_{m=1}^{\infty} \bigcup_{n=1}^{\infty} B_{m,n}^A. \tag{22}$$

The hierarchical basis method and the multigrid method can be directly applied to the case of an adaptive grid: in the inner loops of the algorithms, running over (in the hierarchical basis case) $\tilde{B}_{m,n}$, respectively (in the multigrid case) $B_{m,n}$, the generating functions ϕ that are not contained in the basis H^A or the generating system E^A, respectively, are skipped in the relaxation process. For the point-block method, the subsystem belonging to a grid point x is

$$B_x^A := \{\phi_x^{(m,n)} \in E^A : x \in \Omega_{m,n}\} \tag{23}$$

(cf. (16)). Except for these obvious modifications, the algorithms, especially the update of the residual, work unchanged in the adaptive case.

Now, we apply the different relaxation schemes to the grids created by adaptive refinement for our second model problem (21). In Table 4, we give for each grid

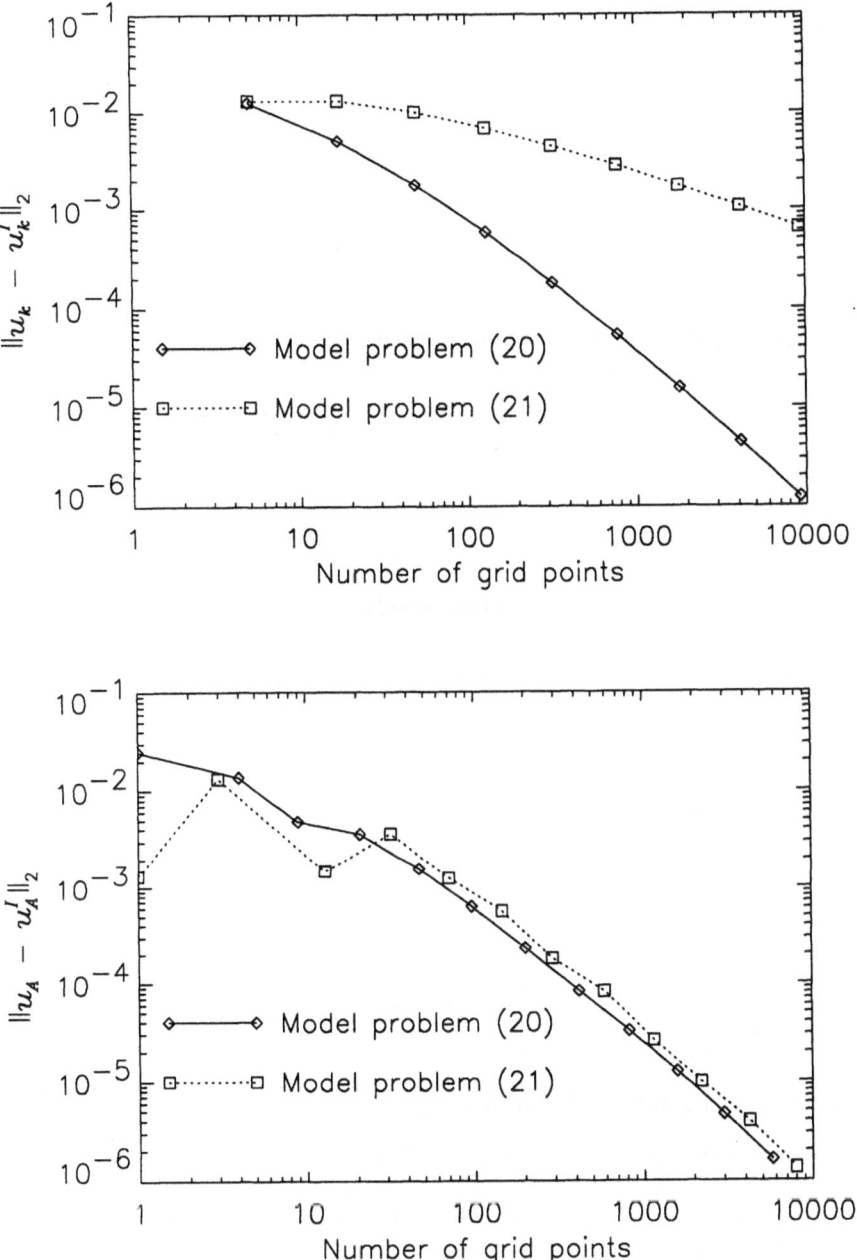

Figure 6: Error for regular (top) and adaptive sparse grids.

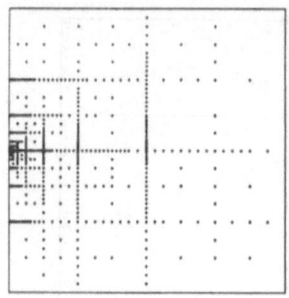 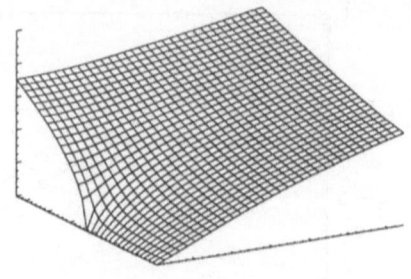

Figure 7: Example for an adaptive grid: problem (21), $\epsilon = 4^{-9}$ and the solution u.

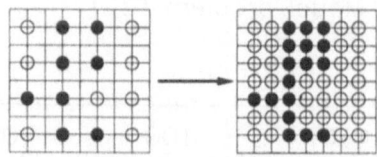

Figure 8: Constructing an $\Omega_{3,3}^A$ from an $\tilde{\Omega}_{3,3}^A$.

Table 4: Results from the relaxation methods for some adaptively generated grids.

ϵ	$|H^A|$	$|E^A|$	$\max|B_x^A|$	ρ^{HB}	ρ^{MG}	ρ^{PB}
4^{-9}	575	1188	20	0.9921	0.57	0.56
4^{-10}	1134	2537	26	0.9956	0.58	0.57
4^{-11}	2203	5261	32	0.9981	0.65	0.57

- the accuracy bound ϵ for the error estimator,

- the number of grid points $|H^A|$,

- the number of generating functions $|E^A|$,

- the size of the largest point block $\max\limits_{x \in \Omega^A} |B_x^A|$ and

- the reduction rates ρ^{HB}, ρ^{MG} and ρ^{PB} for the hierarchical basis, the multigrid and the point block relaxation, respectively.

While the reduction rates ρ^{HB} for the hierarchical basis method are tending to one, the multigrid and the point-block methods exhibit relatively good reduction rates in the adaptive case, bounded independently of the number of grid points. The rate for the point-block technique is now only slightly better than the rate for for the multigrid

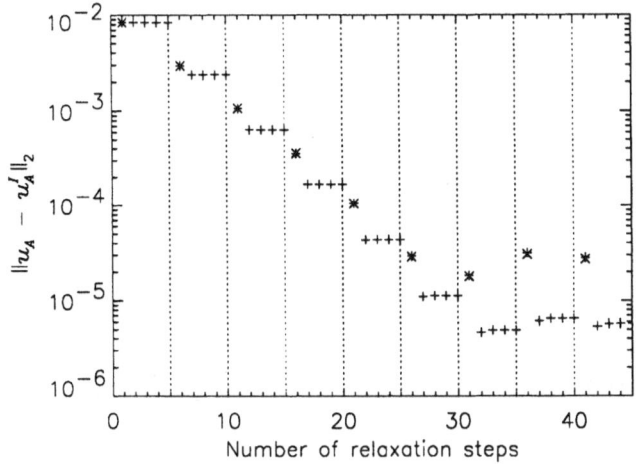

Figure 9: Convergence history of the adaptive algorithm for the model problem (20) with one ($\times$) and five ($+$) point-block relaxation steps on each grid, respectively.

method, but of course the advantage of the domain-decomposition-like structure for the point-block technique remains.

For an adaptive algorithm, the asymptotic reduction rate ρ is much less important than the actual error reduction in the case of a grid that results from the refinement of some other grid, where a good approximation of the solution already exists. Figure 9 shows the performance of the point-block algorithm in this situation. There, we see the convergence history of the complete algorithm for our first model problem (20) with $\epsilon = 4^{-9}$. The solution is evaluated on nine successively refined grids (as explained in section 4) with five relaxation steps on each grid. The $+$-marks show the error after each relaxation step (i.e. the difference to the interpolant of the exact solution, measured in the L_2-norm). It can be seen clearly, that already the first step on each grid reduces the error to the same magnitude as the discretization error on this grid. The results from the same algorithm with only one point-block relaxation step per level are also shown in Figure 9 by the $\times$-marks that almost coincide with the first $+$-marks on each level. Therefore, only one step on each grid is already sufficient to compute an intermediate approximation for the next refinement step. Note, that, due to the fixed value of ϵ, there is little change in the solution on the last two refinement levels shown. Although still some grid points are generated or deleted, the magnitude of the error remains the same.

Since the smooth solution of the first model problem causes a large gain in accuracy by the global refinement step, we can even expect better results for the less smooth second model problem (21), where the difference between the coarse and the fine grid is much smaller. Figure 10, that gives the results analogous to Figure 9 for the second model problem, shows that this is true. Here, on each level, all relaxation steps except the first one reveal no significant improvement.

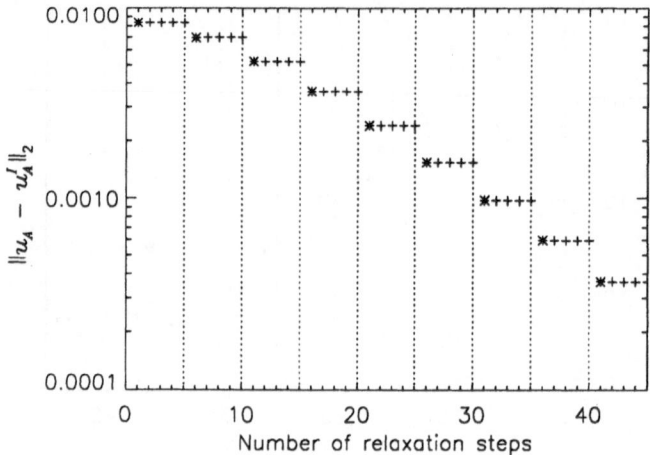

Figure 10: Convergence history of the adaptive algorithm for the model problem (21) with one ($\times$) and five ($+$) point-block relaxation steps on each grid, respectively.

For comparison, Figure 11 shows the results for five hierarchical basis relaxation steps on each level and the first model problem. Note the much slower error reduction compared to the point-block case (Figure 9). The five relaxation steps are not sufficient to reduce the error on each grid to the order of the discretization error.

6 Point-block adaptivity

So far, our concept of adaptivity used the basis H only, whereas the point-blocks were independently used for the solution of the linear systems. We conclude this report of our experiments with an observation indicating that it might be useful to overcome this distinction.

Until now, we only considered grid points inside the domain Ω. Now we will extend the point-block principle to the grid points on the boundary $\partial\Omega$. Since this is the location of the Dirichlet data, which causes the singularity in the second model problem, we want to study, whether it is possible to absorb most of the pollution effect of the singularity by properly chosen point blocks on the boundary.

To make this idea clear, we first consider the one-dimensional case. Figure 12 shows a typical one-dimensional example of a pattern of grid points generated for a singularity at the left side of the interval, together with the corresponding basis functions and one additional basis function located in the left boundary point. Now, note that the spanned finite element space is equivalently generated by one point block located in the left boundary point, see Figure 13.

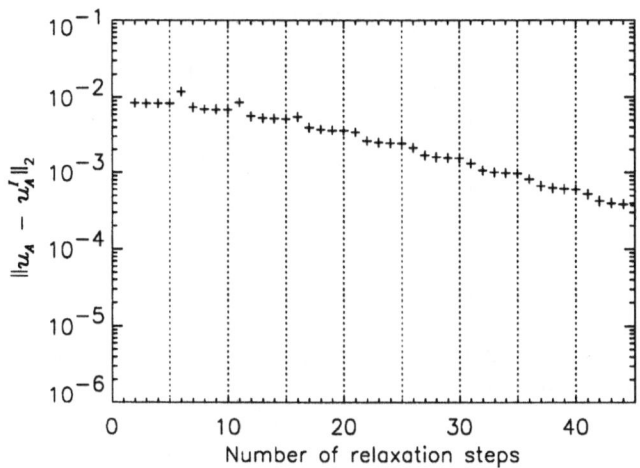

Figure 11: Convergence history of the adaptive algorithm for the model problem (20) with five hierarchical basis relaxation steps on each grid.

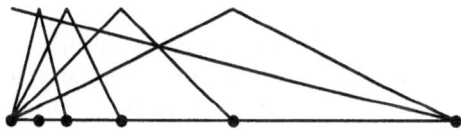

Figure 12: 1-D example for a grid pattern near a singularity.

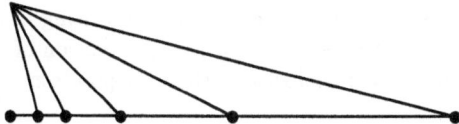

Figure 13: 1-D example for a point-block in a boundary point.

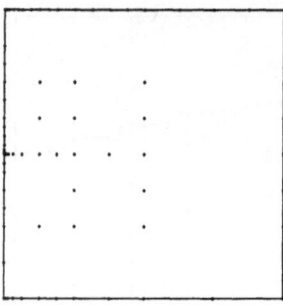

Figure 14: The grid from Figure 7 represented by boundary point-blocks and the usual basis functions in the interior.

In the two-dimensional case, we extend the grids $\Omega_{m,n}$ and the corresponding bases $B_{m,n}$ to the boundary and define the point-block systems B_x for boundary points x analogously to the case of interior points as the collection of all basis functions from different levels belonging to x and again denote the corresponding point-block spaces by $V_x := \operatorname{span} B_x$.

For an example we take the adaptive grid from Figure 7 and remove all these grid points whose corresponding basis functions $\phi \in H^A$ are contained in the point-block space V_x for a boundary point x. The result, shown in Figure 14, consists now of boundary point-blocks and only 20 remaining interior grid points, that show a much more regular structure compared to Figure 7, since the effect of the singularity is mostly absorbed by the boundary point-blocks.

This is of course no loss of information but just a shifting. But since the Dirichlet boundary values are the input data to our algorithm, one can think of a grid of *a priori* generated boundary point-blocks and e.g. a small *regular* sparse grid in the interior. This would eliminate the need for the successive grid generation process and would allow the computation on a static data structure that can be constructed in a setup phase by means of a sufficient resolution of the Dirichlet data.

In the case of the Poisson equation, the singularities arising from the right hand side terms might analogously be absorbed by an a priori extension of a fixed regular sparse grid *in depth*, i.e. associating point-blocks to the grid points near the singularity in a setup phase.

Although the smoothness of the solution is fully determined by the input data (boundary values and right hand side function) and therefore, in principle, there is no need for an adaption strategy driven by an approximate solution on some intermediate test grids, we have not yet been able to derive a sound strategy for the a priori generation of grids.

Therefore, the relation between the resolution accuracy of the Dirichlet and the right hand side data on one hand and the accuracy of the solution on the other hand will be subject of further studies.

References

[1] R. BALDER AND C. ZENGER, *The d-dimensional Helmholtz equation on sparse grids*, SFB-Report 342/21/92 A; TUM-I9232, TU München, Institut für Informatik, 1992.

[2] R. BANK, T. DUPONT, AND H. YSERENTANT, *The hierarchical basis multigrid method*, Num. Math., 52 (1988), pp. 427–458.

[3] H.-J. BUNGARTZ, *Dünne Gitter und deren Anwendung bei der adaptiven Lösung der dreidimensionalen Poisson-Gleichung*, Dissertation, Institut für Informatik, TU München, 1992.

[4] M. GRIEBEL, *Grid- and point-oriented multilevel algorithms*, in Incomplete Decomposition (ILU): Theory, Technique and Application, Proceedings of the Eigth GAMM-Seminar, Kiel January 24-26, 1992, W. Hackbusch, ed., Vieweg-Verlag, 1992.

[5] M. GRIEBEL AND S. ZIMMER, *Multilevel Gauss-Seidel-algorithms for full and sparse grid problems*. To appear in Computing.

[6] W. HACKBUSCH, *A new appoach to robust multi-grid solvers*, Report, Institut für Informatik und praktische Mathematik, CAU Kiel, 1987.

[7] ——, *The frequency decomposition multi-grid method, part I: Applications to anisotropic equations*, Numerische Mathematik, 56 (1989), pp. 229–245.

[8] W. MULDER, *A new multigrid approach to convection problems*, Report 88-04, CAM, 1988.

[9] N. H. NAIK AND J. VAN ROSENDALE, *The improved robustness of multigrid elliptic solvers based on multiple semicoarsened grids*, Report 55-70, ICASE, 1991.

[10] J. XU, *Iterative methods by space decomposition and subspace correction: A unifying approach*, SIAM Review, 34 (1992), pp. 581–613.

[11] H. YSERENTANT, *On the multilevel splitting of finite element spaces*, Numerische Mathematik, 49 (1986), pp. 379–412.

[12] ——, *Hierarchical bases*, in Proceedings of the Second International Conference on Industrial and Applied Mathematics, J. R. E. O'Malley, ed., Philadelphia, 1992, SIAM.

[13] C. ZENGER, *Sparse grids*, in Parallel Algorithms for Partial Differential Equations, Proceedings of the Sixth GAMM-Seminar, Kiel, January 19-21, 1990, W. Hackbusch, ed., Vieweg-Verlag, 1991.

[14] O. C. ZIENKIEWICZ AND A. CRAIG, *Adaptive refinement, error estimates, multigrid soultion and hierarchic finite element concepts*, in Accuracy Estimates and Adaptive refinements in Finite Element Computations, I. Babuška, O. C. Zienkiewicz, J. Gago, and E. R. de A. Oliveira, eds., John Wiley & Sons, 1986.

ADAPTIVE COMPUTATION OF COMPRESSIBLE FLUID FLOW

VOLKER HANNEMANN, DANIEL HEMPEL, THOMAS SONAR
Institut für Theoretische Strömungsmechanik
DLR Göttingen
Bunsenstraße 10
3400 Göttingen

ABSTRACT. We describe an adaptive algorithm for the numerical computation of
two-dimensional, inviscid, compressible flow of an ideal gas. The computations are
done with an upwind finite volume method on triangulations. The code uses MUSCL
extrapolation and is based on the Riemann solver of Osher and Solomon. Adaptive techniques are outlined and the use of the finite element residual as refinement
indicator is described.

1. GOVERNING EQUATIONS

The system of hyperbolic conservation laws

$$\partial_t u + \sum_{i=1}^{2} \partial_{x_i} f_i(u) = 0 \tag{1.1}$$

where $\mathbb{R}_0^+ \times \mathbb{R}^2 \ni (t,x) \xmapsto{u} u(t,x) = (\rho(t,x), \rho v_1(t,x), \rho v_2(t,x), \rho E(t,x))^T \in \mathbb{R}^4$,
$f_i(u) = (\rho v_i, \rho v_1 v_i + p\delta_{1i}, \rho v_2 v_i + p\delta_{2i}, \rho H v_i)^T$. $\rho, v_1, v_2, E, p, H = E + \frac{p}{\rho}$ denote density,
velocity in x_1-direction, velocity in x_2-direction, total energy, pressure and enthalpy of
the fluid is known as Euler's equations and governs inviscid compressible fluid flow in
the plane. The equation of state for a perfect gas reads as

$$p = (\kappa - 1)\rho \left[E - \frac{1}{2}|v|^2 \right]. \tag{1.2}$$

Here κ denotes the ratio of specific heats and $v := (v_1, v_2)^T$. The system is to be solved
under initial as well as under boundary conditions in a bounded domain $\Omega \subset \mathbb{R}^2$.
Due to the nonlinearity of the system (1.1) discontinuities like shocks and contact
discontinuities are substantial parts of the solution. Therefore a weak solution $u \in L^\infty \cap L^1(\mathbb{R}^+ \times \mathbb{R}^2; \mathbb{R}^4)$ satisfying

$$\frac{d}{dt} \int_C u \, dx = - \oint_{\partial C} \{f_1(u)n_1 + f_2(u)n_2\} \, ds \tag{1.3}$$

for all control volumes $C \subset \Omega$ with unit outer normal vector $n := (n_1, n_2)^T$ is sought.
In this context a control volume is a subset of $\mathbb{R}^2$ on which the Gauss integral theorem
is applicable and we remark that (1.3) is essentially equivalent to the usual definition
of weak solutions using test functions in a space-time setting.

The domain Ω is discretised by means of a conforming triangulation T^h. Let $K_j \in T^h$ denote the triangles in the mesh and let i denote a node of the triangulation. We define $\mathcal{K}_i := \{K_j \in T^h \mid \text{node } i \text{ is vertex of } K_j\}$ and consider a box C_i around each node i in T^h. Let the boundary of each box consist of polygons which are constructed in each $K_j \in \mathcal{K}_i$ by connecting the midpoints of the two edges having node i in common with the barycenter of K_j. We assume the numerical solution u^h to be piecewise constant on the boxes. Replacing the exact weak solution u in (1.3) by u^h leads to trouble since the line integral on the right hand side in (1.3) is not defined for step functions with jumps on the box boundaries. We therefore introduce a numerical flux function $\mathbb{R}^4 \times \mathbb{R}^4 \ni (u_i, u_j) \overset{H}{\longmapsto} h(u_i, u_j) \in \mathbb{R}^4$ satisfying the fundamental consistency condition $H(s,s) = f_1(s)n_1 + f_2(s)n_2$ for all $s \in \mathbb{R}^4$. Without specifying the numerical flux function we formulate our basic finite volume method in the form

$$\frac{du_i(t)}{dt} = -\frac{1}{|C_i|} \sum_{j \in N(i)} H(u_i, u_j) \int_{\partial C_i \cap \partial C_j} ds \qquad (2.1)$$

where $N(i) := \{j \in \mathbb{N} \mid \overline{C}_i \cap \overline{C}_j \neq \emptyset\}$ defines the neighbourhood of box C_i and $u_i := u^h|_{C_i}$.

As a numerical flux function H any of the well known methods developed in a finite difference framework may be chosen [5]. We use the approximate Riemann solver of Osher and Solomon [7].

If the approximate Riemann solver is used in (2.1) and a time integration procedure is applied the resulting scheme shows an order of accuracy of one. To achieve a higher order of accuracy which is necessary for practical problems we use the generalized MUSCL approach as described by Osher in [6]. On each box C_i a linear recovery function

$$w_i^h(t) = u_i^h(t) + \Theta_i \nabla_{(x_1,x_2)} w^h(t) \cdot (x_1 - x_1^0, x_2 - x_2^0) \qquad (2.2)$$

is sought for density, pressure and velocities, i.e. $w \in \{\rho, p, v_1, v_2\}$. The point (x_1^0, x_2^0) is the barycenter of C_i, $\nabla_{(x_1,x_2)} w^h(t)$ denotes a gradient to be computed and Θ_i denotes a slope limiter to supress pre- and postshock oscillations. To assign a gradient of the numerical solution to each box C_i we use a dual representation of u^h. While u^h was assumed to be piecewise constant on the boxes we can view the numerical solution as being piecewise linear on the triangles and continuous in Ω, thanks to the dual meshes of the basic box method. This linear continuous function is denoted by $\tilde{u}^h$ and allows the computation of a projected gradient by using the gradients on the triangles. We compute the gradient of the linear recovery function in box C_i by means of

$$\nabla_{(x_1,x_2)} w^h(t)|_{C_i} := \frac{1}{|C_i|} \sum_{j \in \mathcal{K}_i} \nabla_{(x_1,x_2)} \tilde{u}^h(t)|_{K_j} |C_i \cap K_j|$$

giving the symbol $\nabla_{(x_1,x_2)} w^h(t)$ a precise meaning. The functions $w^h(t)$ are evaluated for density, pressure and velocities at the midpoints of $\partial C_i \cap \partial C_j$, a new vector of conservative variables $\hat{u}^h$ is computed at these points and the values inserted in the Osher-Solomon numerical flux function to give the resulting scheme

$$\frac{du_i(t)}{dt} = -\frac{1}{|C_i|} \sum_{j \in N(i)} H(\hat{u}_i(t), \hat{u}_j(t)) \int_{\partial C_i \cap \partial C_j} ds \qquad (2.3)$$

which is formally of second order in the space discretisation. To avoid over- and undershoots at shocks the slope limiter Θ_i is introduced in each box. Any of the several limiters used in practice can be chosen here [5]. We use one given by Barth and Jespersen in [1] which is one of the less diffusive limiters.

To discretize in time direction we use a three-stage TVD-Runge-Kutta method as constructed by Shu and Osher [8]. This kind of time stepping preserves the monotonicity properties of the space discretization but as a drawback the CFL number is bounded by one. A detailed description of the scheme is given in [10].

3. ADAPTIVE TECHNIQUES

Let us assume that we have already developed a refinement indicator which, after some time steps, gives a list of triangles to be defined as well as a list of triangles which may be too small. Let us define

$$\Sigma_I := \{K \in T^h \mid K \text{ has to be refined}\} \tag{3.1}$$

$$\Sigma_R := \{K \in T^h \mid K \text{ may be removed}\}. \tag{3.2}$$

Let us first discuss the possible point insertion techniques. One quite popular appproach is the isotropic red/green refinement as described by Bank et. al. [3]. The two kinds of refinement are shown in figures 1 and 2. All triangles $K \in \Sigma_I$ are red refined. Since

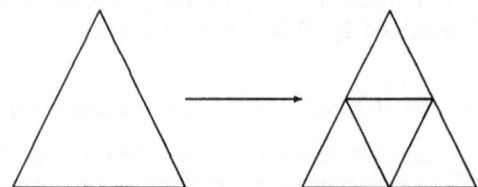

FIGURE 1. Red refinement

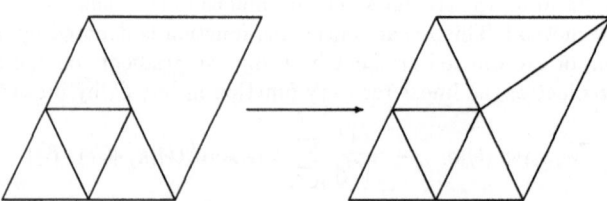

FIGURE 2. Green refinement

this step results in a non-conforming triangulation with hanging nodes all those nodes are removed by a final green refinement. Since the green refinement deteriorates the angles within the triangulation care has to be taken to avoid zero limit angles. Hence, at the beginning of each refinement cycle all previous green refinements are removed from the grid.

Another possible strategie for a point insertion technique is a strategy developed by Rivara and generalised by Bänsch [2] which we will call insertion by halfing. In each triangel $K \in \mathcal{T}^h$ one edge is marked to be the refinement edge. All $K \in \Sigma_I$ are then treated by the rule shown in 3, i.e. they are halfed at the middle of the refinement edge and the two new triangles get their refinement edges according to the rule in 3. After the process of refining all $K \in \Sigma_I$ the remaining triangulation, again called $\mathcal{T}^h$

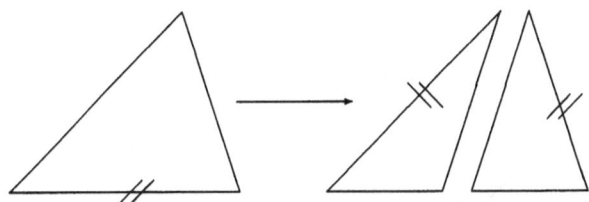

FIGURE 3. Refinement by halfing

for simplicity, contains triangles with hanging nodes. We redefine

$$\Sigma_I := \{K \in \mathcal{T}^h \mid K \text{ has a hanging node}\}$$

and start the algorithm again. This cycle is repeated until a conforming triangulation is achieved.

What we found unsatisfactory in practical experiments with both of the refinement techniques was the spreading of refinement zones. Therefore a local point insertion technique was developed. In each $K \in \Sigma_I$ a point is introduced at the barycenter of K and connected with the nodes of K. Thus three new triangles are created for each triangle to be refined. The angles of the new triangles are computed and an edge swapping algorithm is called if some of the angles are below a certain bound to avoid stability problems. Both steps of the algorithm are shown in figures 4 and 5. This technique is also easily extendable into 3-d, the only sophisticated part being the face swapping in this case. If a point has to be removed an algorithm tries to swap edges around this point in order to arrive at a situation where there are only three edges emanating from this point left. The point is then removed by removing the three edges. It may occur that there are four edges left where two of them are co-linear. Then the four edges left are removed. For details see [4].

4. RESIDUAL CONTROLLED REFINEMENT INDICATORS

The goal of a reliable refinement indicator should be the refinement of the grid such that the error

$$e^h := u - u^h, \tag{4.1}$$

i.e. the difference between the true and the numerical solution, is below some given bound TOL if measured in some norm. Thus, true error control means

$$\|e^h\|_X \leq \text{TOL} \tag{4.2}$$

as the ultimate result of an adaptive algorithm. Since the true solution is not available in problems of practical interest some other quantity has to be detected which bounds

161

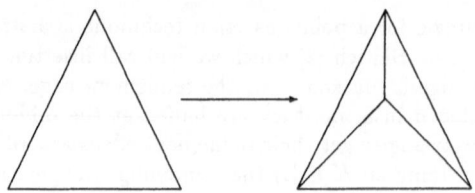

FIGURE 4. Local refinement

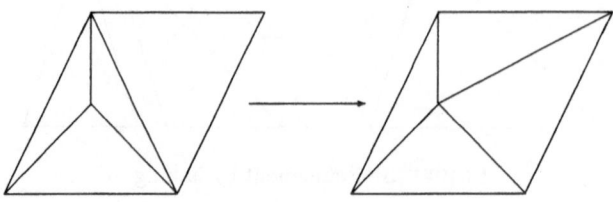

FIGURE 5. Edge swapping

the error. It is well known from linear elliptic equations that the residual is a bound for the error. Since the equations of gas dynamics are nonlinear there is no evidence other than numerical experiments that this is also the case for fluid flow. The residual is defined by

$$r^h := \partial_t \tilde{u}^h + \sum_{i=1}^{2} \partial_{x_i} f_i(\tilde{u}^h). \tag{4.3}$$

Since $\tilde{u}^h$ is continuous in Ω the space derivatives can be easily computed. A more sophisticated problem is the computation of the time derivative since we are using finite differences (Runge-Kutta) in time. The best we can do is to consider the jump of $\tilde{u}^h$ in time, i.e. we replace the time derivative in (4.3) by the forward difference $(u^{n+1} - u^n)/\Delta t$ where the superscripts refer to time slabs.

Since the residual is a vector valued quantity a norm is needed to give a refinement indicator. The indicator is to be evaluated locally on each triangle K of the triangulation such that it can be decided which triangle has to be refined (or removed). In the elliptic finite element context the L^2-Norm is used but it can be shown that $\|r^h\|_{L^2(K)}$ blows up like $1/\sqrt{\Delta x}$ at discontinuities in 1-d scalar equations [9]. To get proper scaling we therefore use

$$R_2 := h\|r^h\|_{L^2(K)} \tag{4.4}$$

as the refinement indicator. Here, h denotes the longest edge of the triangle. As was suggested by Süli [11] we also use the weak norm indicator

$$R_{-1} := \|r^h\|_{H^{-1}(K)} \tag{4.5}$$

which leads to nearly the same results as the indicator above. The performance of both indicators is examined in [9].

5. Numerical Examples

To show the ability of the algorithms to resolve the flow features in compressible fluid flow the forward facing step example of Woodward and Colella [12] was considered. The Mach number of the incoming flow is $Ma = 3$. Due to the step a bow shock will develop with time and a Mach stem will occur on the upper wall. A contact discontinuity is one of the constituting parts of the stem. If we non-dimensionalize the system of Euler equations using the velocity and density of the flow outside the channel the most interesting shock configuration shows up at $t = 8$, which corresponds to $t = 4$ in [12]. The grid at time $t = 8$ is shown in figure 6 while the Mach number distribution can be seen in figure 7. The adaptive algorithm used the weighted L^2-norm as described above. As can be seen the refinement indicator was able to detect all relevant flow features. The shocks are sharply captured and the contact discontinuity is very well resolved. Even the corner singularity at the step was detected and refined.

As a second example we consider steady flow in a channel with a wedge. The onflow Mach number is $Ma = 2$. Since this configuration is part of a combustion chamber the computation was done using a slightly changed isentropic exponent of $\kappa = 1.36$.

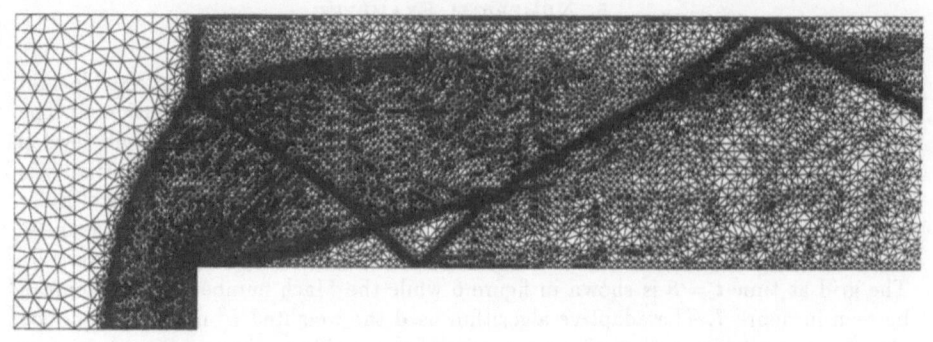

FIGURE 6. Adapted grid at time $t = 8$

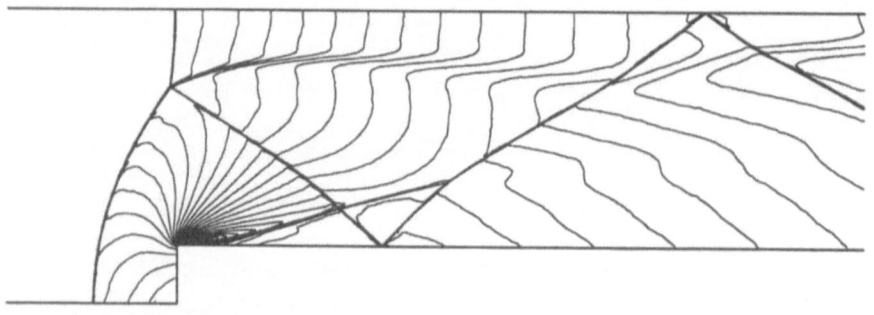

FIGURE 7. Mach number distribution

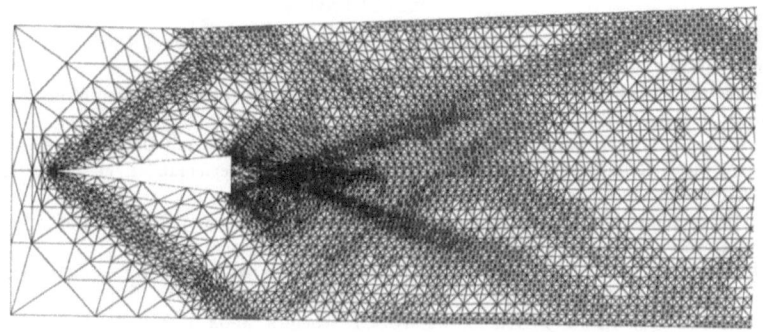

FIGURE 8. Adapted grid

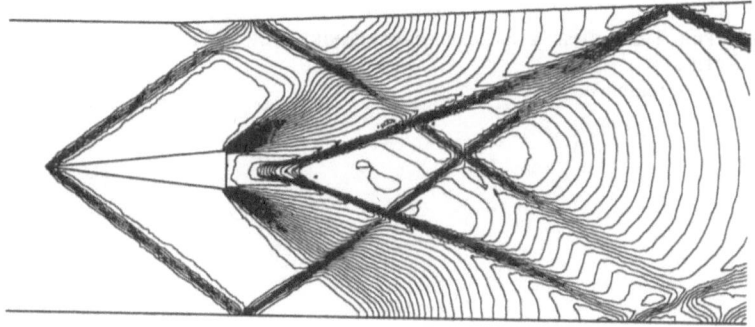

FIGURE 9. Pressure distribution

References

1. T.J. Barth, D.C. Jespersen - The Design and Application of Upwind Schemes on Unstructured Meshes. *AIAA-89-0366, (unpublished), (1989)*.
2. E. Bänsch - Local Mesh Refinement in 2 and 3 Dimensions. *Impact of Computing in Science and Engineering, Vol.3, No.3, (1991)*.
3. R.E. Bank, A.H. Sherman, A. Weiser. - Refinement Algorithms and Data Structures for Regular Local Mesh Refinements. *(in: Scientific Computing; R. Stepleman et. al. (eds.), Amsterdam: IMACS North-Holland, (1983))*.
4. V. Hannemann, D. Hempel, Th. Sonar - Dynamic Adaptivity and Residual Control in Unsteady Compressible Flow Computation. *Manuscript, DLR Göttingen, (1993)*.
5. C. Hirsch - Numerical Computation of Internal and External Flows, Vol. 2. *John Wiley & Sons, Chichester, New York, Brisbane, Toronto, Singapore, (1990)*.
6. S. Osher - Convergence of Generalised MUSCL Schemes. *Math. Comp. 38, 339-374, (1982)*
7. S. Osher, F. Solomon - Upwind Difference Schemes for Hyperbolic Conservation Laws. *SIAM. J. Num. Anal. 22, 947-961, (1985)*.
8. C.-W. Shu, S. Osher - Efficient Implementation of Essentially Non-Oscillatory Shock-Capturing Schemes. *J. Comp. Phys. 77, 439-471, (1988)*.
9. Th. Sonar - Strong and Weak Norm Error Indicators Based on the Finite Element Residual for Compressible Flow Computation, I. The Steady Case. *to be published in Impact of Computing in Science and Engineering (1993)*.
10. Th. Sonar - On the Design of an Upwind Scheme for Compressible Flow on General Triangulations. *Numerical Algorithms (1993)*.
11. E. Süli - private communication .
12. P. Woodward, P. Colella - The Numerical Simulation of Two-Dimensional Fluid Flow with Strong Shocks. *(J. Comp. Phys. 54, 115-173, (1984))*.

INSTITUT FÜR THEORETISCHE STRÖMUNGSMECHANIK, DLR GÖTTINGEN, BUNSENSTRASSE 10, 3400 GÖTTINGEN

E-mail address: tommi@ts.go.dlr.de

ON NUMERICAL EXPERIMENTS WITH CENTRAL DIFFERENCE OPERATORS ON SPECIAL PIECEWISE UNIFORM MESHES FOR PROBLEMS WITH BOUNDARY LAYERS

ALAN F. HEGARTY

Department of Mathematics and Statistics, University of Limerick, Limerick, Ireland.

JOHN J.H. MILLER

Department of Mathematics, Trinity College, Dublin 2, Ireland.

EUGENE O'RIORDAN

Department of Mathematics, Regional Technical College, Tallaght, Dublin 24, Ireland.

AND

G.I. SHISHKIN

Institute of Mathematics and Mechanics, Russian Academy of Sciences, Ekaterinburg, Russia.

SUMMARY

Singularly perturbed second order elliptic equations with boundary layers are considered. Numerical methods composed of central difference operators on special piecewise uniform meshes are constructed for the above problems. Numerical results are obtained which show that these methods give approximate solutions with error estimates that are independent of the singular perturbation parameter.

INTRODUCTION

Partial differential equations with a small parameter (denoted here by ε) multiplying the highest derivatives arise in many areas of engineering (e.g., computational fluid dynamics, heat and mass transfer, oil reservoir simulation, semiconductor device modelling). These equations are usually said to be singularly perturbed. It is well known that classical finite difference or finite element methods give numerical approximations to the solution of such problems, which have spurious oscillations with an amplitude which is either unbounded or comparable to the modulus of the exact solution when $\varepsilon < h$. These oscillations may be eliminated by using upwind difference operators on uniform meshes. By using these upwind difference operators on special piecewise uniform meshes convergent numerical solutions are obtained which have error estimates independent of ε (called ε-uniform henceforth).

FORMULATION OF THE PROBLEM

Consider the following singularly perturbed elliptic problem

$$\varepsilon\Delta u + a_1 u_x + a_0 u = f, \quad \text{in} \quad \Omega = (0,1) \times (0,1), \tag{1a}$$

$$u = g \quad \text{on} \quad \partial\Omega, \tag{1b}$$

$$a_1 \geq \alpha > 0 \quad \text{in} \quad \bar{\Omega}, \tag{1c}$$

$$0 < \varepsilon \leq 1. \tag{1d}$$

Here u may be thought of as a concentration of some quantity (e.g., heat, a pollutant etc.) that is driven by a known advective velocity field $\mathbf{a} = (a_1, 0)$ where ε is the diffusivity. Equation (1) may also be regarded as a linear model for the Navier-Stokes flow equations in the unit square Ω. If $\varepsilon << 1$ the solution has a regular layer on the side $x = 0$ and, because a is parallel to the sides $y = 0$ and $y = 1$, it also has a parabolic layer on each of these. This is shown schematically in Fig. 1.

Fig. 1. Location of boundary layers for problem (1)

DISCRETIZATION OF DOMAIN AND DERIVATIVES

In this paper both uniform and special piecewise-uniform meshes of the form $\Omega_N \equiv \{(x_i, y_j); 0 \leq i, j \leq N\}$ are considered. A uniform mesh is denoted by Ω_N^u and in this case $x_i = i/N, y_j = j/N$. A special piecewise uniform mesh $\Omega_N^* \equiv \{(x_i^*, y_j^*) : 0 \leq i, j \leq N\}$ is now constructed for problem (1). Since a boundary layer of width $O(\varepsilon)$ appears in the solution of (1) near the outflow boundary, it is natural to refine the mesh near this boundary. This is done by defining

$$x_i^* = \begin{cases} ih_1, & \text{for } 0 \le i \le N/2, \\ \sigma_x + (i - N/2)h_2, & \text{for } N/2 \le i \le N; \end{cases}$$

with

$$h_1 = 2\sigma_x/N, \quad \text{and} \quad h_2 = 2(1 - \sigma_x)/N.$$

The transition point σ_x, which depends on both ε and N, is defined by

$$\sigma_x \equiv \min\{1/2, C_1 \varepsilon \ln N\}, \tag{2}$$

where C_1 is a constant. One of the main aims of this paper is to investigate numerically the effect of the choice of the constant C_1 on the numerical solution.

Similarly near the characteristic boundaries, where parabolic boundary layers of width $O(\sqrt{\varepsilon})$ appear, the mesh is refined by defining

$$y_j^* = \begin{cases} jk_1, & \text{for } 0 \le j \le N/4, \\ \sigma_y + (j - N/4)k_2, & \text{for } N/4 \le j \le 3N/4, \\ 1 - \sigma_y + (j - 3N/4)k_1, & \text{for } 3N/4 \le j \le N; \end{cases}$$

with

$$k_1 = 4\sigma_y/N, \quad \text{and} \quad k_2 = 2(1 - 2\sigma_y)/N.$$

The transition point σ_y, which depends on both ε and N, is defined by

$$\sigma_y \equiv \min\{1/4, \sqrt{\varepsilon} \ln N\}.$$

The finite difference methods considered here use the central difference operators D_x^0, δ_x^2 and δ_y^2, defined as follows

$$D_x^0 u_N(x_i, y_j) \equiv (u_N(x_{i+1}, y_j) - u_N(x_{i-1}, y_j))/\bar{h}_i,$$

$$D_x^+ u_N(x_i, y_j) \equiv (u_N(x_{i+1}, y_j) - u_N(x_i, y_j))/h_{i+1},$$

$$D_x^- u_N(x_i, y_j) \equiv (u_N(x_i, y_j) - u_N(x_{i-1}, y_j))/h_i,$$

$$\delta_x^2 u_N(x_i, y_j) \equiv (D_x^+ u_N(x_i, y_j) - D_x^- u_N(x_i, y_j))/\bar{h}_i,$$

where $h_i = x_i - x_{i-1}$ and $\bar{h}_i \equiv (h_{i+1} + h_i)/2$ and δ_y^2 is defined analogously.

Shishkin [4] proved analytically for a wide class of singular perturbation problems that the choice of a special piecewise uniform mesh guarantees ε-uniform convergence as long as the difference operator is monotone. In [1] and [3] this theoretical result was confirmed numerically for several problems using a specific monotone difference operator (upwind).

The key point of the present paper is to report the experimental result that the same choice of mesh leads to numerical methods that appear to be ε-uniform even when the difference operator is non-monotone (in particular, a central difference operator). It is shown that the numerical solution may, for some special meshes described below, contain spurious oscillations, but that appropriate choice of a constant C_1 will eliminate these oscillations.

The following central difference method is used to obtain a numerical approximation u_N to u

$$[\varepsilon(\delta_x^2 + \delta_y^2) + a_1(x_i, y_j)D_x^0 + a_0(x_i, y_j)]u_N(x_i, y_j) = f(x_i, y_j), \tag{3a}$$

$$u_N = u \quad \text{on} \quad \partial\Omega_N. \tag{3b}$$

On a uniform mesh Ω_N^u the solutions of (3) for moderate values of ε are satisfactory. However, for small ε it is well known that the solutions become unstable unless the mesh parameter $h = 1/N$ is of the same order of magnitude as ε.

It is now shown numerically that the solutions of (3) on the special piecewise uniform meshes Ω_N^* constructed above are ε-uniform. The errors in the numerical solutions obtained from (3) on the special piecewise uniform meshes are estimated in the following way. The pointwise nodal errors are approximated, for successive values of ε, on the five special meshes Ω_8^*, Ω_{16}^*, Ω_{32}^*, Ω_{64}^*, Ω_{128}^*, for successive values of ε, by $e_{\varepsilon,N}(i,j) = |u_N(x_i^*, y_j^*) - u_{comp}(x_i^*, y_j^*)|$, where the subscript N indicates the number of mesh elements used and u_{comp} is some accurate ε-uniform numerical approximation to the exact solution u of problem (1). For each ε the maximum nodal error is approximated by

$$E_{\varepsilon,N} = \max_{i,j} e_{\varepsilon,N}(i,j)$$

and the ε-uniform maximum nodal error is approximated by

$$E_N = \max_\varepsilon E_{\varepsilon,N}.$$

Convergence rates for each ε and for $N = 8, 16, 32, 64$, are estimated numerically by $p_{\varepsilon,N}$ where

$$p_{\varepsilon,N} = \log_2\left(\frac{D_{\varepsilon,N}}{D_{\varepsilon,2N}}\right) \quad \text{and} \quad D_{\varepsilon,N} = \max_{\Omega_N}|u_N - u_{2N}^I|. \tag{4}$$

Here u_{2N}^I is the bilinear interpolant of u_{2N} on the mesh Ω_N^*. The ε-uniform convergence rate for each N is then estimated numerically by p_N where for each value of N the ε-uniform quantities

$$p_N = \log_2\left(\frac{D_N}{D_{2N}}\right) \quad \text{and} \quad D_N = \max_\varepsilon D_{\varepsilon,N}. \tag{5}$$

NUMERICAL RESULTS

Numerical solutions are computed for the specific problem

$$a_1 = 1 + x^2 + y^2, \quad f \equiv 0, \ a_0 \equiv 0, \tag{6a}$$

with the boundary conditions

$$u(x,0) = x^3; \quad u(x,1) = x^2; \quad u(0,y) = 0; \quad u(1,y) = 1. \tag{6b}$$

The numerical solution obtained by using (3) on the special mesh Ω_{32}^* with $C_1 = 2$ for $\varepsilon = 0.0001$ is displayed in Fig. 2.

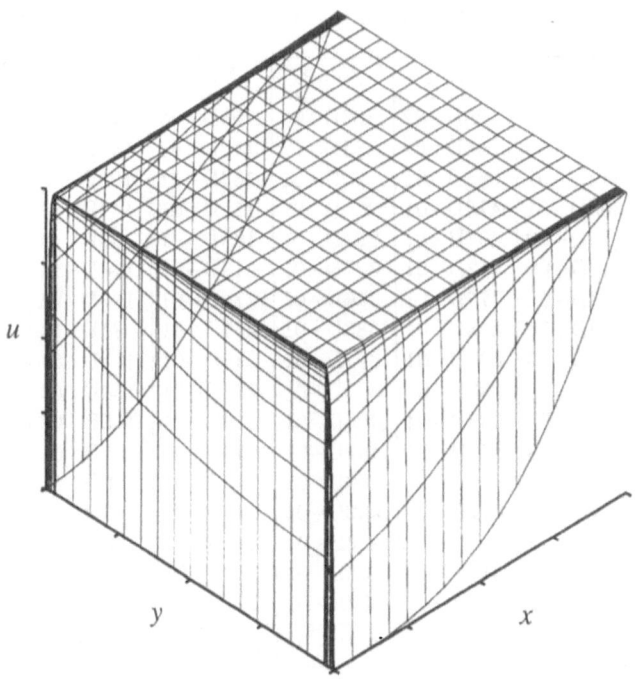

Fig. 2. Numerical solution of (6) for $\varepsilon = 0.0001$ on Ω_{32}^* with $C_1 = 2$

The numerical approximation u_{comp} to the exact solution u of the specific problem (6) having ε-uniform accuracy was obtained in [1]. This uniform accuracy is supported by the theoretical ε-uniform convergence results of Shishkin [4] for upwind finite difference operators on special piecewise uniform meshes Ω_N^*. The nodal errors $E_{\varepsilon,N}$ and E_N, obtained by using (3) on the special meshes Ω_N^* defined above with $C_1 = 2$, are presented in Table 1. This indicates numerically that, for each fixed N, the errors stabilize at a finite value as ε decreases and that these values themselves decrease with increasing N.

The estimated convergence rates $p_{\varepsilon,N}$ for each ε and N and the estimated ε-uniform convergence rates p_N for each N are presented in Table 2. The values of p_N indicate numerically that the method has an ε-uniform convergence rate of at least 1.

The importance of the choice of the constant C_1 is now investigated. The numerical solution obtained by using (3) on the special mesh Ω_{32}^* with $C_1 = 1$ for $\varepsilon = 0.0001$ is displayed in Fig. 3. In contrast with the case when $C_1 = 2$, on this occasion there are notable spurious oscillations in the numerical solution. However, if C_1 is fixed at 1 and N is increased, the amplitude of the oscillations is reduced and tends toward zero as $N \to \infty$. Thus, it is of interest to determine whether the numerical method (3) remains ε-uniformly convergent in this case and, if so, how much the presence of these oscillations affects the rate of ε-uniform convergence.

171

Table 1. Values of $E_{\varepsilon,N}$ for the Scheme (3) on Special Meshes Ω_N^* with $C_1 = 2$

ε	$N=8$	$N=16$	$N=32$	$N=64$	$N=128$
1	.474D$-$02	.125D$-$02	.363D$-$03	.155D$-$03	.112D$-$03
2^{-2}	.865D$-$02	.246D$-$02	.138D$-$02	.142D$-$02	.144D$-$02
2^{-4}	.613D$-$01	.365D$-$01	.190D$-$01	.132D$-$01	.108D$-$01
2^{-6}	.808D$-$01	.476D$-$01	.251D$-$01	.171D$-$01	.141D$-$01
2^{-8}	.118D$+$00	.597D$-$01	.291D$-$01	.192D$-$01	.158D$-$01
2^{-10}	.141D$+$00	.712D$-$01	.337D$-$01	.212D$-$01	.170D$-$01
2^{-12}	.153D$+$00	.775D$-$01	.366D$-$01	.224D$-$01	.177D$-$01
2^{-14}	.159D$+$00	.808D$-$01	.383D$-$01	.231D$-$01	.181D$-$01
2^{-16}	.163D$+$00	.825D$-$01	.392D$-$01	.235D$-$01	.183D$-$01
2^{-18}	.164D$+$00	.833D$-$01	.397D$-$01	.237D$-$01	.184D$-$01
2^{-20}	.165D$+$00	.838D$-$01	.400D$-$01	.238D$-$01	.185D$-$01
2^{-22}	.165D$+$00	.840D$-$01	.401D$-$01	.239D$-$01	.185D$-$01
2^{-24}	.165D$+$00	.841D$-$01	.402D$-$01	.239D$-$01	.185D$-$01
2^{-26}	.165D$+$00	.841D$-$01	.402D$-$01	.239D$-$01	.185D$-$01
2^{-28}	.166D$+$00	.842D$-$01	.402D$-$01	.239D$-$01	.185D$-$01
2^{-30}	.166D$+$00	.842D$-$01	.402D$-$01	.239D$-$01	.185D$-$01
2^{-32}	.166D$+$00	.842D$-$01	.402D$-$01	.239D$-$01	.185D$-$01
E_N	.166D$+$00	.842D$-$01	.402D$-$01	.239D$-$01	.185D$-$01

Table 2. Values of $p_{\varepsilon,N}$ for the Scheme (3) on Special Meshes Ω_N^* with $C_1 = 2$.

ε	$N=8$	$N=16$	$N=32$	$N=64$
1	1.96	1.99	1.98	1.98
2^{-2}	1.77	1.80	1.92	1.95
2^{-4}	0.73	1.34	2.39	2.00
2^{-6}	0.77	1.17	1.49	1.46
2^{-8}	1.00	1.25	1.53	1.46
2^{-10}	1.01	1.28	1.50	1.49
2^{-12}	1.00	1.28	1.51	1.49
2^{-14}	1.00	1.27	1.51	1.50
2^{-16}	1.00	1.26	1.51	1.50
2^{-18}	1.00	1.26	1.51	1.50
2^{-20}	1.00	1.26	1.51	1.50
2^{-22}	1.00	1.25	1.50	1.50
2^{-24}	1.00	1.25	1.50	1.50
2^{-26}	1.00	1.25	1.50	1.50
2^{-28}	1.00	1.25	1.50	1.50
2^{-30}	1.00	1.25	1.50	1.50
2^{-32}	1.00	1.25	1.50	1.50
p_N	1.00	1.25	1.50	1.50

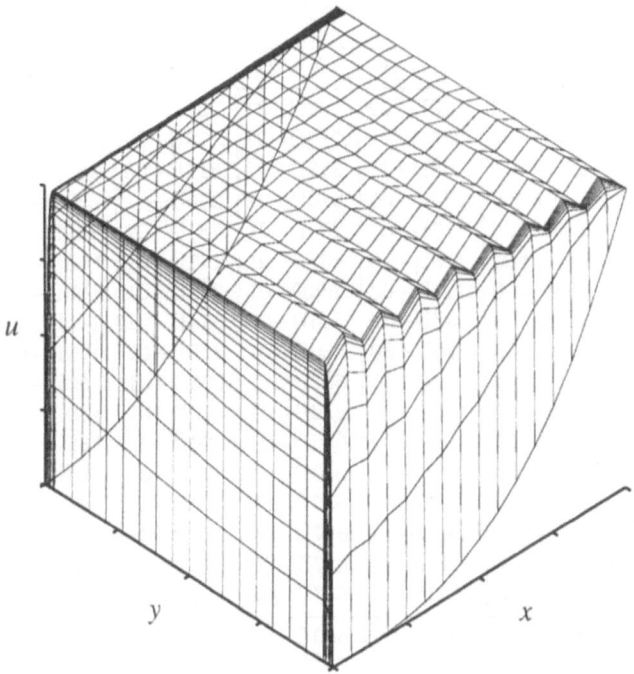

Fig. 3. Numerical solution of (6) for $\varepsilon = 0.0001$ on Ω_{32}^* with $C_1 = 1$

The nodal errors $E_{\varepsilon,N}$ and E_N, obtained by using (3) on the special meshes Ω_N^* defined above with $C = 1$, are presented in Table 3. This indicates numerically that as in Table 1, for each fixed N, the errors stabilize at a finite value as ε decreases and that these values themselves decrease with increasing N.

The values of E_N are larger in Table 3 than in Table 1, except for $N = 128$. No obvious conclusion can be drawn from a comparison of these tables. The estimated convergence rates $p_{\varepsilon,N}$ are presented in Table 4. These indicate numerically that the method may have an ε-uniform convergence rate of approximately 1, but the rate is certainly lower than in the case $C_1 = 2$. Thus the oscillations appear to lower the rate of ε-uniform convergence.

To investigate further the role of C_1, values of p_N are presented in Table 5 for the same problem (6) for various values of C_1.

Table 3. Values of $E_{\varepsilon,N}$ for the Scheme (3) on Special Meshes Ω_N^* with $C_1 = 1$

ε	$N= 8$	$N= 16$	$N= 32$	$N= 64$	$N=128$
1	.470D$-$02	.120D$-$02	.298D$-$03	.721D$-$04	.145D$-$04
2^{-2}	.885D$-$02	.257D$-$02	.710D$-$03	.178D$-$03	.366D$-$04
2^{-4}	.275D$-$01	.106D$-$01	.313D$-$02	.814D$-$03	.453D$-$03
2^{-6}	.100D$+$00	.254D$-$01	.810D$-$02	.270D$-$02	.637D$-$03
2^{-8}	.179D$+$00	.677D$-$01	.209D$-$01	.751D$-$02	.209D$-$02
2^{-10}	.236D$+$00	.105D$+$00	.407D$-$01	.137D$-$01	.668D$-$02
2^{-12}	.266D$+$00	.124D$+$00	.568D$-$01	.238D$-$01	.103D$-$01
2^{-14}	.281D$+$00	.133D$+$00	.645D$-$01	.314D$-$01	.146D$-$01
2^{-16}	.288D$+$00	.136D$+$00	.660D$-$01	.328D$-$01	.163D$-$01
2^{-18}	.291D$+$00	.137D$+$00	.657D$-$01	.323D$-$01	.160D$-$01
2^{-20}	.292D$+$00	.137D$+$00	.656D$-$01	.320D$-$01	.159D$-$01
2^{-22}	.293D$+$00	.137D$+$00	.656D$-$01	.320D$-$01	.158D$-$01
2^{-24}	.294D$+$00	.138D$+$00	.657D$-$01	.320D$-$01	.158D$-$01
2^{-26}	.294D$+$00	.138D$+$00	.657D$-$01	.320D$-$01	.158D$-$01
2^{-28}	.294D$+$00	.138D$+$00	.657D$-$01	.320D$-$01	.158D$-$01
2^{-30}	.294D$+$00	.138D$+$00	.657D$-$01	.320D$-$01	.158D$-$01
2^{-32}	.294D$+$00	.138D$+$00	.657D$-$01	.320D$-$01	.158D$-$01
E_N	.294D$+$00	.138D$+$00	.660D$-$01	.328D$-$01	.163D$-$01

Table 4. Values of $p_{\varepsilon,N}$ for the Scheme (3) on Special Meshes Ω_N^* with $C_1 = 1$.

ε	$N= 8$	$N= 16$	$N= 32$	$N= 64$
1	1.96	1.99	1.98	1.98
2^{-2}	1.77	1.80	1.92	1.95
2^{-4}	1.18	1.32	1.46	1.50
2^{-6}	2.06	2.03	1.67	1.70
2^{-8}	1.28	1.76	1.67	1.91
2^{-10}	1.08	1.25	1.56	1.39
2^{-12}	1.08	1.07	1.19	1.44
2^{-14}	1.09	1.08	1.02	1.15
2^{-16}	1.10	1.09	1.03	0.99
2^{-18}	1.10	1.10	1.05	1.00
2^{-20}	1.10	1.10	1.05	1.02
2^{-22}	1.10	1.10	1.05	1.02
2^{-24}	1.10	1.10	1.06	1.02
2^{-26}	1.10	1.10	1.06	1.02
2^{-28}	1.10	1.10	1.06	1.02
2^{-30}	1.10	1.10	1.06	1.02
2^{-32}	1.10	1.10	1.06	1.02
p_N	1.10	1.10	1.05	0.99

Table 5. Values of p_N for the Scheme (3) on Special Meshes Ω_N^*.

C_1	$N=8$	$N=16$	$N=32$	$N=64$
0.5	0.72	0.66	0.57	0.52
1.0	1.10	1.10	1.05	0.99
2.0	1.00	1.25	1.50	1.50
4.0	0.73	0.96	1.29	1.56
8.0	0.91	0.65	0.99	1.32

It is clear from Table 5 that, for values of C_1 less than 1, the oscillations in the numerical solution, which grow as $C_1 \to 0$, progressively degrade the performance of the numerical method as $C_1 \to 0$. Large values of C_1 appear to have a similar effect on the ε-uniform rate of convergence, although there are no oscillations apparent in the numerical solution for $C_1 \geq 2$. Further experimentation is needed for different problems, but Table 5 suggests that the optimal value of C_1 for this problem is approximately 2. Cross-sections at $y = 0.5$ of the numerical solutions for $C_1 = 0.5, 1$ and 2 are shown in Fig. 4. The increased oscillations for $C_1 = 0.5$ are apparent.

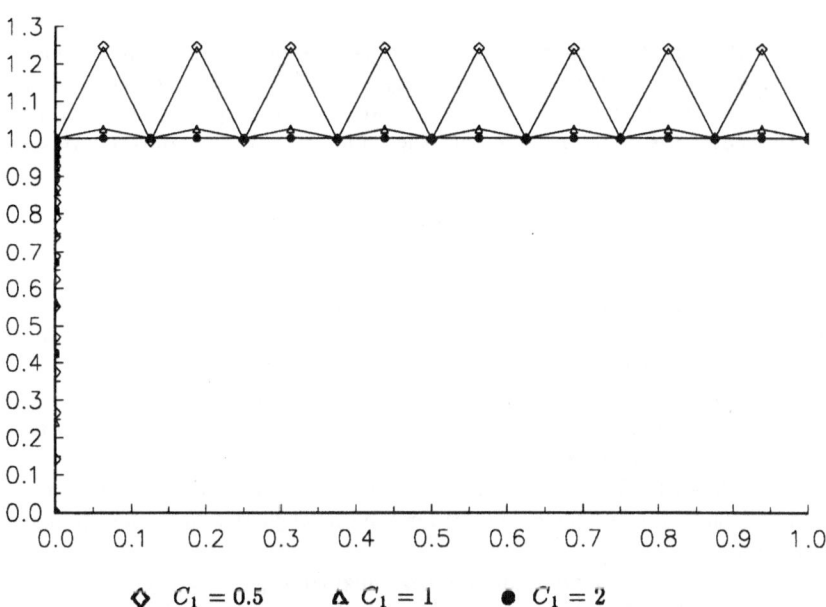

$\diamondsuit \quad C_1 = 0.5 \qquad \triangle \quad C_1 = 1 \qquad \bullet \quad C_1 = 2$

Fig. 4. Cross-sections at $y = 0.5$ of numerical solution of (6) for $\varepsilon = 0.0001$ on Ω_{32}^* with $C_1 = 0.5, 1$ and 2

For problem (6), none of the choices of C_1 had any effect on the performance of the iterative linear solver used (preconditioned CGS). In particular the number of iterations required for convergence is independent of ε. It should be noted however, that for a problem of the form

$$\varepsilon \Delta u + a_1 u_x + a_2 u_y + a_0 u = f, \quad \text{in} \quad \Omega = (0, 1) \times (0, 1), \tag{7a}$$

$$u = g \quad \text{on} \quad \partial\Omega, \tag{7b}$$

$$a_1 \geq \alpha_1 > 0 \quad a_2 \geq \alpha_2 > 0 \quad \text{in} \quad \bar{\Omega}, \tag{7c}$$

$$0 < \varepsilon \leq 1, \tag{7d}$$

the solution of which has a regular layer along the side $x = 0$ and the side $y = 0$, it is not clear that central difference operators can be used with success. The design of a special piecewise uniform mesh for this problem and numerical experiments using an upwind finite difference operator may be found in [2]. When a central difference operator is used, an appropriate choice of constants analogous to C_1 is necessary to suppress spurious oscillations in the numerical solution. However, even with an appropriate choice of these constants, convergence of the linear solver appears to be highly ε-dependent.

CONCLUSIONS

Numerical methods composed of central difference operators on special piecewise uniform meshes were used to solve a singularly perturbed elliptic problem with a boundary layer. The results show experimentally that, with appropriate choice of the constant C_1, the method is ε-uniform and that the ε-uniform order of convergence is at least 1.

REFERENCES

1. A.F. Hegarty, J.J.H. Miller, E. O'Riordan and G.I. Shishkin, 'Special meshes for finite difference approximations to an advection-diffusion equation with parabolic layers', (submitted for publication).
2. A.F. Hegarty, J.J.H. Miller, E. O'Riordan and G.I. Shishkin, 'On a novel mesh for the regular boundary layers arising in advection-dominated transport in two dimensions', (submitted for publication).
3. J.J.H. Miller, E. O'Riordan and G.I. Shishkin, 'On piecewise uniform meshes for upwind and central difference operators for solving singularly perturbed problems', (submitted for publication).
4. G.I. Shishkin, 'Grid approximation of singularly perturbed elliptic and parabolic equations', second Doctoral thesis, Keldysh Institute of Applied Mathematics, U.S.S.R. Academy of Sciences, Moscow, 1990.

The box method for elliptic interface problems on locally refined meshes

Bernd Heinrich

Technische Universität Chemnitz-Zwickau, Fachbereich Mathematik,
PSF 964, D - 09009 Chemnitz, Germany

Abstract

Consider some second order elliptic boundary value problem of the plane, where the solution involves singularities due to interfaces containing corners or intersecting the boundary of the given domain. The box method is applied to these problems and some results concerning the numerical treatment of interface singularities by appropriate local mesh refinement are presented. For triangular meshes with grading, basic inequalities for norms of grid functions on such meshes and some analytic and matrix properties of the box approximation operator are discussed. Finally, a priori error estimates on locally refined triangular meshes are derived, which yield the same rate of convergence as known for regular solutions.

1. Introduction

The box method, sometimes also called finite volume method or balance method, can be understood as a generalized finite difference approach working on general triangulations as well as for boundary value problems (briefly: BVPs) with non-smooth solutions. The first fact is known since the fifties, cf. [14], and many authors applied this discretization method to the numerical solution of field problems in physics and mechanics. Recently the box method has been studied under the aspects of numerical analysis too, see e.g. [1, 6, 7] and the references therein. But, the box method under the assumption of non-smooth solutions caused by so-called interface corner singularities ([11]) seems not to be treated in the literature. It is well known that the accuracy of standard finite element methods applied to such problems is reduced significantly and that a systematic local mesh refinement near the corners may provide a degree of accuracy which is known for regular solutions, cf. [17] for problems with corners on the boundary. Due to the similarity of box schemes to some finite element schemes, it can be conjectured that similar procedures will help to improve the accuracy of box methods.

In this paper, we shall consider an elliptic problem of second order on a plane polygonal domain, where the coefficient k in the operator $\operatorname{div}(k \operatorname{grad} u)$ may have a discontinuity along a polygonal line called the interface. It is admitted that there are corners on the interface or that the interface touches the boundary. For simplicity, we shall neglect curved boundaries and consider Dirichlet boundary conditions only. Nevertheless, the analysis described subsequently can also be extended to more general problems. In Section 2, the formulation of the BVP, some assumptions and a short description of the solution properties are given. In Section 3, meshes of triangles with refinement governed by some

177

parameter $\mu \in (0, 1]$, grid functions and discrete norms are introduced and, furthermore, basic inequalities for grid functions on locally refined meshes are established, such as Friedrichs' inequality. The derivation of box schemes and the characterization of their analytical and matrix properties are carried out in Section 4. Especially, symmetry, positive definiteness, monotonicity and the condition number of the matrix associated with the box scheme are investigated. Finally, in Section 5 error estimates and convergence of the approximate discrete box solution to the solution of the BVP are discussed. Depending on the refinement parameter $\mu \in (0, 1]$, the rate of convergence is estimated. For an appropriate choice of μ, this rate is equal to that one known for regular solutions $u \in H^2$.

2. The BVP and analytical background

We consider the following BVP in the plane, with an elliptic operator of second order containing a discontinuous, piecewise constant coefficient k, and with homogeneous Dirichlet boundary conditions, viz.

$$
\begin{aligned}
(Lu)(x) &:= -\sum_{i=1}^{2} \frac{\partial}{\partial x_i}\left(k\frac{\partial u}{\partial x_i}\right)(x) = f(x), \quad x \in \Omega \subset R^2, \\
(lu)(x) &:= u(x) = 0, \quad x \in \Gamma := \partial\Omega,
\end{aligned}
\tag{1}
$$

briefly: $Au = F$, with $A = (L, l)^T$, $F = (f, 0)^T$. Assume that Ω is a bounded polygon, with $\Gamma \in C^{0,1}$, and that the boundary Γ is composed of straightline segments Γ_j, $\Gamma = \bigcup\limits_{j=1}^{N} \Gamma_j$. The data should satisfy $f \in L_2(\Omega)$ at least and

$$
0 < k_0 \le k(x) \le k_1, \quad k(x) := k^{\pm} \text{ for } x \in \Omega^{\pm},
\tag{2}
$$

where $k^{\pm}$ are constants. Here, Ω^- and Ω^+ denote two polygonal subdomains of Ω, with $\Omega^+ \cap \Omega^- = \varnothing$, $\bar{\Omega}^- \cup \bar{\Omega}^+ = \bar{\Omega}$, and with boundaries $\Gamma^{\pm} := \partial\Omega^{\pm} \in C^{0,1}$ being of the same

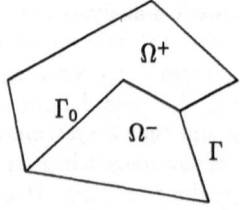

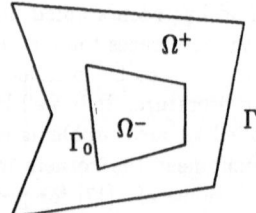

Fig. 1a   Fig. 1b

type as Γ. The line of discontinuity of k, viz. $\Gamma_0 := \bar{\Omega}^- \cap \bar{\Omega}^+$, also called interface of k, is assumed to be a polygonal curve within $\bar{\Omega}$, with $\Gamma \cap \Gamma_0 \ne \emptyset$, see Fig. 1a, or a polygonal and closed curve lying within Ω, i.e. $\Gamma \cap \Gamma_0 = \emptyset$, see Fig. 1b. In the first case, we suppose that $\Gamma \cap \Gamma_0$ consists of two different points lying on Γ. On Γ_0, the solution u of BVP (1) should fulfil the so-called interface conditions

$$
u|_{\Gamma_0-0} = u|_{\Gamma_0+0}, \quad k\frac{\partial u}{\partial n}\Big|_{\Gamma_0-0} = k\frac{\partial u}{\partial n}\Big|_{\Gamma_0+0}.
\tag{3}
$$

Here, n denotes the outward normal to (say) $\partial\Omega^- \cap \Gamma_0$. The symbol $a|_{\Gamma_0\pm 0}$ is taken as indicating the left-and right-hand limits $a(x^\pm) \to a(x\pm 0)$ as $x^\pm \to x$, $x^\pm \in \Omega^\pm, x \in \Gamma_0$, or as a trace operator for functions a on $\Omega^\pm$ or Ω. The generalized formulation of the interface problem $((1), (2), (3))$ is given by

$$\text{find} \quad u \in \overset{\circ}{H}{}^1(\Omega) := \{v \in H^1(\Omega) : v|_\Gamma = 0\} \text{ such that}$$

$$a(u,v) := \int_\Omega (k\nabla u, \nabla v)dx = \int_\Omega fvdx =: f(v) \text{ for any } v \in H^1(\Omega), \tag{4}$$

where H^s (s real), here for $s = 1$, denotes as usual the Sobolev (-Slobodetskii) space and $(.,.)$ the scalar product in R^2. By the Lax–Milgram theorem, the existence of a unique solution $u \in \overset{\circ}{H}{}^1(\Omega)$ of problem (4) is obvious. Moreover, the interface conditions (3) are satisfied, too, in the spaces $H^{1/2}(\Gamma_0)$ and $H^{-1/2}(\Gamma_0)$, respectively. This follows from $u \in \overset{\circ}{H}{}^1(\Omega)$ and the generalized Green's formula, as e.g. written for $\Omega^\pm$:

$$\int_{\Omega^\pm} (Lu)vdx = \int_{\Omega^\pm} (k\nabla u, \nabla v)dx - <t_1u, t_0v>_{\partial\Omega^\pm}, \tag{5}$$

which holds for $u \in H^1_2(\Omega^\pm) := \{v \in H^1(\Omega^\pm) : Lv \in L_2(\Omega^\pm)\}$ and $v \in H^1(\Omega^\pm)$, with the duality pairing $<.,.>_{\partial\Omega^\pm} (H^{-1/2} \times H^{1/2})$ and trace operators t_i of order i, see e.g. [3].

It is well-known that for problems like (4), the shift theorem , here: $Lu = f \in H^0(\Omega)$ implies $u \in H^2(\Omega)$, does not hold. For a discussion of this topic, see e.g. [5, 11, 12, 13]. Especially, due to corners on Γ, Γ_0 and to the intersection $\Gamma \cap \Gamma_0 \neq \emptyset$, so-called singularity functions may occur in the solution u, and even for convex Ω and smooth Γ_0, where $\Gamma_0 \subset \Omega$, the relation $u \notin H^2(\Omega)$ holds, in general. Nevertheless, the solution u of (4) can be split into a locally acting singular part u_s and a remaining part u_r which is regular at least on $\Omega^\pm$.

In order to derive a representation of $u = u_s + u_r$, we define the set C of points which consists of corners on Γ, Γ_0 and of points of intersection $\Gamma \cap \Gamma_0 \neq \emptyset$. Moreover, introduce around each point $P \in C$ local polar coordinates (r_P, φ_P) $(0 < r_P < r'_{0P}, 0 < \varphi_P < \varphi_{0P})$ and, in the usual way, a smooth cut-off function $\eta_P(r_P)$ assigned to P. Then, the solution u of problem (4) can be represented by

$$u = u_s + u_r = \sum_{P\in C_s} \sum_{i=1}^{I_P} c_{P_i} s_{P_i} + u_r, \tag{6}$$

with real numbers c_{P_i} and functions s_{P_i}, u_r satisfying

$$s_{P_i} := \eta_P r_P^{\lambda_{P_i}} \Phi_{P_i}(\varphi_P), \ Ls_{P_i} \in L_2(\Omega^\pm), \ u_r \in H^2(\Omega^\pm), \quad \|u_r\|_{H^2(\Omega^\pm)} \leq C\|f\|_{L_2(\Omega)}.$$

The functions $\Phi_{P_i}(\varphi_P)$ are eigenfunctions of some selfadjoint Sturm–Liouville problem (see e.g. [11]); Φ_{P_i} is continuous on $[0, \varphi_{0P}]$, C^∞-smooth for $P \in C_s \cap (\Gamma \setminus \Gamma_0)$ and piecewise C^∞-smooth for $P \in C_s \cap \Gamma_0$, with jumps of the derivatives at the interface legs. Here, $\frac{1}{4} < \lambda_{P_i} < 1, I_P = 1$ or $I_P = 2$, and $C_s \subset C$ holds, with $C_s \neq \emptyset$ in general, i.e. C_s consists of that points of C where singularity functions occur. For a proof of these relations, we refer to [8]. It should be noted that more than two domains of the type $\Omega^\pm$, e.g. the combination of the cases shown in Figs. 1a, 1b, can be treated analogously. Similar assertions can be found e.g. in [4, 11, 18].

3. Meshes, refinement and basic inequalities

Let T_h denote a family of triangulations of $\bar{\Omega}$ utilizing triangles $K \in T_h$ in the usual way, i.e. $\bar{\Omega} = \bigcup_{K \in T_h} K$, and the intersection of different triangles may be only a common vertex or edge, see e.g. [2]. We demand that the subtriangulations $T_h^{\pm}$ of T_h on $\bar{\Omega}^{\pm}$ are disjoint, $\bar{\Omega}^{\pm} = \bigcup_{K \in T_h^{\pm}} K$, and that $P \in C$ (cf. Section 2) is always a vertex of some $K \in T_h$.

Let $\bar{\omega}$ and $\bar{\omega}'$ denote the sets of the vertices of $K \in T_h$ and of the midpoints of all triangle edges, respectively. Take $x, \xi \in \bar{\omega}$ as neighbouring nodes defining a triangle edge $\mathbf{h}(x, \xi) := [x, \xi]$, with length $h(x, \xi)$, and x'_ξ (or ξ'_x), briefly x', as the midpoint of $\mathbf{h}(x, \xi)$. Thus, using x'_ξ or x' as an edge index, we shall also write $h(x'_\xi)$ or $h(x')$ instead of $h(x, \xi)$, etc. Introduce subsets of $\bar{\omega}$ and $\bar{\omega}'$ by

$$\gamma := \bar{\omega} \cap \Gamma, \quad \omega := \bar{\omega} \cup \Omega, \quad \gamma_0 := \omega \cap \Gamma_0, \quad \gamma' := \bar{\omega}' \cap \Gamma, \quad \omega' := \bar{\omega}' \cap \Omega, \quad \gamma_0' := \bar{\omega}' \cap \Gamma_0. \quad (7)$$

Furthermore, use the symbol $S'(x) := \{\xi \in \bar{\omega} : \mathbf{h}(x, \xi) \text{ is a triangle edge}\}$ for denoting the nearest neighbours ξ of the grid point $x \in \bar{\omega}$, and $S(x) := S'(x) \cup \{x\}$. The global mesh parameter h is given by $h := \varepsilon^{-1} \max_{x' \in \bar{\omega}} h(x')$, with some constant ε, $0 < \varepsilon \leq 1$.

Assumption 3.1. (i) The triangles K are 'shape regular', i.e. the diameter $h_K := \operatorname{diam} K$ of K, the length $h(x')$ of any triangle edge $\mathbf{h}(x') \subset \partial K$ and the interior angles Θ at the vertices of K satisfy the relations (with fixed ε_0, Θ_0)

$$\varepsilon_0 h_K \leq h(x') \leq h_K \quad (0 < \varepsilon_0 < 1), \quad 0 < \Theta_0 \leq \Theta \leq \pi - \Theta_0 \quad (0 < \Theta_0 < \frac{\pi}{2}), \quad (8)$$

(ii) For each pair of interior angles $\Theta^{\pm}$ at the vertices opposite to the common edge $\mathbf{h}(x')$ of any two adjacent triangles, relation

$$\Theta^+ + \Theta^- \leq \pi - \Theta_0 \quad \text{or} \quad \Theta^+ + \Theta^- = \pi \tag{9}$$

is satisfied and, if $\mathbf{h}(x') \subset \Gamma_0$ holds, even $\Theta^{\pm} \leq \frac{\pi}{2} - \Theta_0$. For $\mathbf{h}(x') \subset \Gamma$, relation $\Theta \leq \frac{\pi}{2}$ is required for the angle Θ at the vertex opposite to $\mathbf{h}(x')$. $\equiv$

In Assumption 3.1, (ii) is a sufficient condition for defining correctly box meshes of type PB (perpendicular bisectors) as well as for carrying out error estimates on such boxes. For boxes of the type MD (medians, cf. [7]), (ii) can be omitted.

In a sufficently small neighbourhood of each point $P \in C_s$, we shall define a graded refinement of the mesh, which is determined by some parameter μ_P, with $0 < \mu_P \leq 1$. For simplicity, we shall consider an arbitrary point $P \in C_s$ and neglect temporarily the subscript P. Introduce the distance r_K of a triangle K to $P \in C_s$, viz. $r_K := \operatorname{dist}(K, P) := \inf_{Q \in K} |Q - P|$, $n := n(h)$ as an integer of the order h^{-1} ($n := [ah^{-1}]$ for some real $a > 0$), some grading function r_i (with some real $b > 0$) and local mesh parameters h_i,

$$r_i := b(ih)^{\frac{1}{\mu}} \ (i = 0, 1, \ldots, n), \quad h_i := r_i - r_{i-1} \ (i = 1, 2, \ldots, n). \tag{10}$$

Assumption 3.2. The triangulation T_h satisfies Assumption 3.1 and, moreover, the asymptotic behaviour of $h_K := \operatorname{diam} K$ is determined by

$$\begin{aligned} &\varepsilon_1 h_i \leq h_K \leq \varepsilon_1^{-1} h_i \quad \text{for } K \in T_h : r_{i-1} \leq r_K < r_i \ (i = 1, \ldots, n), \\ &\varepsilon_2 h \leq h_K \leq \varepsilon_2^{-1} h \quad \text{for } K \in T_h : r_n \leq r_K, \end{aligned} \tag{11}$$

with $0 < \varepsilon_j \le 1$, $(j = 1, 2)$ and h_i taken from (10). $\equiv$

Some asymptotic relations of the graded refinement are given by the following lemma (for the proof, cf. [9]).

Lemma 3.3. For h, h_i, r_i and μ $(0 < h \le h_0,\ 0 < \mu < 1)$ the following relations hold,

$$b^\mu h r_i^{1-\mu} \le h_i \le \frac{b^\mu}{\mu} h r_i^{1-\mu}\ (i = 1,\dots,n)\quad h_{i-1} < h_i \le (2^{\frac{1}{\mu}} - 1) h_{i-1}\ (i = 2,\dots,n). \equiv \quad (12)$$

Now we shall cover $\bar\Omega$ by a secondary triangulation $\mathbf{T_H}$ using boxes $\mathbf{H}(x)$ defined e.g. by the perpendicular bisectors (PB), cf. Fig. 2a. Thus, we have $\bar\Omega = \bigcup_{x \in \bar\omega} \mathbf{H}(x)$. Due to Assumption 3.1 (ii), there is always a secondary triangulation of PB–type. For an edge $\mathbf{h}(x, \xi)$, with $\mathbf{h} \not\subset \Gamma$, two circumcentres of the adjacent triangles correspond to $\mathbf{h}(x, \xi)$ and are connected by the segment $\mathbf{s}(x'_\xi) := \mathbf{s}(x, \xi)$. Thus, for $x \in \omega$ relation $\partial\mathbf{H}(x) = \bigcup_{\xi \in S'(x)} \mathbf{s}(x, \xi)$ holds. The boundary $\partial\mathbf{H}(x)$ and its components are oriented counterclockwise around x. The length of the segments is denoted by $s(x') := s(x'_\xi) := |\mathbf{s}(x'_\xi)|$, the area of the box $\mathbf{H}(x)$ by $H(x) := |\mathbf{H}(x)|$. Moreover, define $H(x') := h(x')s(x')$ for $x' \in \omega'$, ω' from (7).

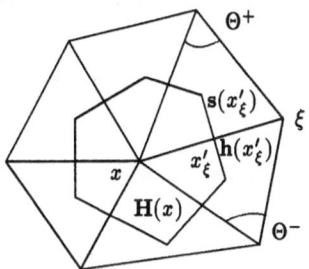

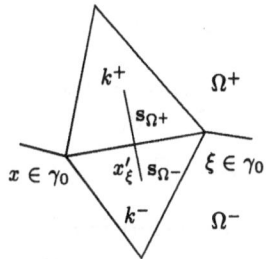

Fig. 2a　　　　　　　　　　　Fig. 2b

The sets of all grid functions defined on $\bar\omega$ and ω' are denoted by D and $\tilde D'$, respectively, furthermore $D_0 := \{y \in D : y(x) = 0 \text{ for } x \in \gamma\}$ (according to [7]). Using $H(x)$ and $H(x')$ introduced previously, we define the discrete scalar products and norms by

$$(y, v) := \sum_{x \in \omega} y(x) v(x) H(x) \text{ for } y, v \in D_0,\quad \|y\|_0 := (y, y)^{\frac{1}{2}} \text{ for } y \in D_0,$$

$$(y, v)' := \sum_{x' \in \omega'} y(x') v(x') H(x') \text{ for } y, v \in \tilde D',\quad \|y\|_0' := (y, y)'^{\frac{1}{2}} \text{ for } y \in \tilde D',$$

$$\|y\|_1 := \{\|y\|_0^2 + |y|_1^2\}^{\frac{1}{2}},\quad |y|_1 := \|y_\mathbf{h}\|_0' \text{ for } y \in D_0,$$

$$\text{with } y_\mathbf{h}(x'_\xi) := \frac{y(\xi) - y(x)}{h(x, \xi)} \text{ for } y \in D,\quad \|y\|_{C(\bar\omega)} := \max_{x \in \bar\omega} |y(x)|.$$

$$(13)$$

Especially, $y_\mathbf{h} \in \tilde D'$ given by (13) defines a first order difference quotient on the mesh segment $\mathbf{h}(x, \xi)$.

We now consider some inequalities for grid functions $y \in D_0$, which have been derived in [7] for several types of primary and secondary triangulations. It turns out that these inequalities are also valid for triangulations with local refinement.

Lemma 3.4. Let the triangulation T_h of $\bar{\Omega}$ be refined locally around each $P \in C_s$, with $0 < \mu_P \leq 1$, according to Assumption 3.2. Then, the inequalities

$$\|y\|_1 \leq M_0 |y|_1, \quad \|y\|_{C(\bar{\omega})} \leq M_1 |ln\,h|^{\frac{1}{2}} \|y\|_1 \quad \text{for } y \in D_0, \tag{14}$$

hold, where the constants $M_i (i = 0, 1)$ are independent of $h, h \leq h_0$. $\equiv$

Clearly, these relations reflect discrete versions of Friedrichs' inequality and of the weak imbedding of H^1 in C. Their proofs follow from [7, pp. 78–80, 80–90] and [9, pp. 7–10], where discrete Friedrichs–Poincaré type inequalities of more general form (e. g., for mixed boundary conditions) are derived, too.

4. A box scheme and properties

In order to derive box schemes and to get a tool for error estimates, Gauss' formula and the so-called balance equations on boxes $\mathbf{H}(x)$ are of interest. For solutions u satisfying $u \in H^2(\mathbf{H}(x))$, this topic has been discussed e.g. in [7]. But for $x \in \gamma_0$, the relation $u \notin H^2(\mathbf{H}^\pm(x))$ holds, in general.

Lemma 4.1. For the solution u of the interface problem (4) and any $x \in \omega$, the balance equations (15) holds, i.e.

$$- \int_{\partial\mathbf{H}(x)} k\frac{\partial u}{\partial n} ds = \int_{\mathbf{H}(x)} f(\bar{x}) d\bar{x}, \quad x \in \omega. \equiv \tag{15}$$

Here, $\bar{x} \in \text{int } \mathbf{H}(x)$. The proof is essentially based on Gauss' formula

$$\int_{\mathbf{H}(x)} \text{div} \, (k \, \text{grad} \, u) \, d\bar{x} = \int_{\partial\mathbf{H}(x)} k\frac{\partial u}{\partial n} ds, \quad x \in \omega, \tag{16}$$

which is valid for the solution u of problem (4), too, cf. [8]. For $x \in \omega\backslash\gamma_0$ (see (7)), we get $u \in H^2(\mathbf{H}(x))$, and relations (15), (16) are evident. Utilizing (5) and (6), we obtain (16) and, consequently, (15) for $x \in \gamma_0$, too.

It is well-known that the approximation of (15), together with essential boundary conditions $u(x) = 0$, $x \in \gamma$, yields box finite difference schemes assigned to BVP (1), cf. [1, 6, 7]. For $x \in \omega\backslash\gamma_0$, i.e. $u \in H^2(\mathbf{H}(x))$, various approximations are given e.g. in [7], where general elliptic second order operators, different types of boundary conditions and of secondary triangulations are treated, too. It remains to specify the approximations at $x \in \gamma_0$. Since $\partial\mathbf{H}(x) = \bigcup_{\xi \in S'(x)} \mathbf{s}(x, \xi)$ holds, it is sufficient to consider two adjacent triangles with common edge $\mathbf{h}(x, \xi)$, and partial segments $\mathbf{s}_{\Omega\pm}(x, \xi) := \mathbf{s}(x, \xi) \cap \bar{\Omega}^\pm$ of the perpendicular bisectors, cf. Fig. 2b, with length $s_{\Omega\pm}(x'_\xi)$. Approximate as follows,

$$\int_{\partial\mathbf{H}(x)} k\frac{\partial u}{\partial n} ds = \sum_{\xi \in S'(x)} \int_{\mathbf{s}(x'_\xi)} k\frac{\partial u}{\partial n} ds \approx \sum_{\xi \in S'(x)} k_h(x'_\xi) s(x'_\xi) u_\mathbf{h}(x'_\xi) \quad \text{for } x \in \omega,$$

$$\text{with } k_h(x'_\xi) := \frac{1}{s(x'_\xi)}\{k(x'_\xi - 0)s_{\Omega-}(x'_\xi) + k(x'_\xi + 0)s_{\Omega+}(x'_\xi)\}, \tag{17}$$

$$\text{especially } k_h(x'_\xi) := k(x'_\xi), \text{ if } \xi \in \omega \setminus \gamma_0, \text{ and } u_\mathbf{h}(x'_\xi) := \frac{u(\xi) - u(x)}{h(x'_\xi)}.$$

It should be noted that approximation (17) is also suitable for nonconstant $k^\pm$, where $k(x'_\xi \pm 0)$ could be substituted approximately by $\frac{1}{2}\{k(x \pm 0) + k(\xi \pm 0)\}$. The right-hand side of (15) can be approximated e.g. by

$$\int_{\mathbf{H}(x)} f(\bar{x})d\bar{x} \approx f_h(x) := f(x-0)H^-(x) + f(x+0)H^+(x) \quad \text{for } f \in L_2(\Omega) \cap C(\bar{\Omega}^\pm \backslash \partial\Omega), (18)$$

where $H^\pm(x)$ denotes the area of $\mathbf{H}(x) \cap \bar{\Omega}^\pm$, or by $f(x)H(x)$, if $f \in L_2(\Omega) \cap C(\Omega)$.

Inserting (17), (18) into (15) and replacing $u(x)$, $x \in \bar\omega$, by an unknown grid function $y \in D_0$ which restores the equality in (15), then scaling (15) with $H^{-1}(x)$, we get a box finite difference scheme (FDS) $A_h y = F_h$ being a discrete analogue of the BVP $Au = F$, viz.

$$\begin{aligned}
(L_h y)(x) &:= H^{-1}(x) \sum_{\xi \in S'(x)} k_h(x'_\xi)s(x'_\xi)y_\mathbf{h}(x'_\xi) = f_h(x), \quad x \in \omega, \\
(l_h y)(x) &:= y(x) = 0, \quad x \in \gamma,
\end{aligned} \tag{19}$$

briefly $A_h y = F_h$, with $A_h = (L_h, l_h)^T$, $F_h = (f_h, 0)^T$. Clearly, the FDS $A_h y = F_h$ (19) is associated with a system (using the same denotation) $A_h y = F_h$ of $n := |\bar\omega|$ linear equations. Inserting the boundary values $y = 0$ on γ and rescaling each row of the system by $H(x)$, we get the system

$$B_h y = G_h, \quad B_h = (b_{ij})_{\tilde{n} \times \tilde{n}}, \quad G_h = (G_i)_{\tilde{n}}, \quad \tilde{n} := |\omega| = |\bar\omega| - |\gamma|, \tag{20}$$

with $\tilde{n}$ linear equations, matrix entries $b_{ij} = O(1)$ and vector components $G_i = O(\mathbf{H}(x))$, where x is in correspondence with i. For details, cf. [7, pp. 55–57].

For quasiuniform triangulations and continuous k, the properties of A_h and B_h have been studied extensively in [7, Section 4]. It becomes apparent that symmetry, energy and monotonicity properties of the finite difference operators and matrices are preserved, if triangulations with local refinement and discontinuous coefficients k are admitted.

Lemma 4.2. Let Assumption 3.2 be fulfilled and $h \leq h_0$, h_0 sufficiently small. Then, the FDS $A_h y = F_h$ (19) is correctly defined and there is a unique solution y. Moreover,

(i) $(A_h y, v) = (y, A_h v)$ for $y, v \in D_0$, i.e. A_h is symmetric on D_0;

(ii) A_h satisfies the estimates

$$\|y\|_1 \leq M_2(A_h y, y), \quad \|y\|_1 \leq M_3\|A_h y\|_{-1} := M_3 \sup_{v \in D_0, v \neq 0} \frac{|(A_h y, v)|}{\|v\|_1} \quad \text{for } y \in D_0, \tag{21}$$

i.e. A_h is coercive, positive definite and stable in pairs of discrete spaces $H^{-1}\text{-}H^1$, $H^0\text{-}H^0$;

(iii) A_h is 'of monotonic type' on D, i.e. $A_h y \geq 0$ on $\bar\omega$ implies $y \geq 0$ on $\bar\omega$, which holds in the strict sense, too. $\equiv$

The proof of Lemma 4.2 is given in [8] and based upon the corresponding proof for quasiuniform triangulations in [7, pp. 58–59, 85–88]. To get the approximating grid function $y(x) \approx u(x)$, $x \in \omega$, the system $B_h y = G_h$ of (20) has to be solved and, therefore, the properties of the matrix B_h are of particular interest. Some of them are consequences of Lemma 4.2.

Theorem 4.3. Let T_h be a family of triangulations with graded local refinement according to Assumption 3.2, and $h \leq h_0$, h_0 sufficiently small. Then, the $\tilde{n} \times \tilde{n}$ − matrix B_h given by (20) and associated with the operator A_h from (19) is provided with the following properties,

(i) $B_h = B_h^T > 0$, i.e. B_h is symmetric and positive definite; the relations

$$M_4 h^2 (y,y)_{R^{\tilde{n}}} \leq (B_h y, y)_{R^{\tilde{n}}}, \quad \beta(B_h) := \frac{\lambda_{max}}{\lambda_{min}} \leq M_5 h^{-2}, \tag{22}$$

hold, where $(.,.)_{R^{\tilde{n}}}$ is the usual scalar product in $R^{\tilde{n}}$, λ_{max} and λ_{min} are the greatest and smallest eigenvalues of B_h, respectively, and $\beta(B_h)$ denotes the condition number of B_h;

(ii) B_h is an irreducibly diagonally dominant matrix, with positive diagonal and nonpositive off-diagonal entries; B_h is an M-matrix and all entries of B_h^{-1} are positive. $\equiv$

Essential parts of the proof of Theorem 4.3 are already given in [7] (quasiuniform triangulations), for extension to meshes with graded local refinement, cf. [9] and [8]. Some part of the proof of (22) is based on a discrete analogue of a Hardy–like inequality. The condition number $\beta(B_h)$ on locally refined meshes of type (11) does not grow faster than on quasiuniform triangulation without refinement. We note that simple estimates like $\|y\|_1^2 \geq \|y\|_0^2 \geq \min_{x \in \omega} H(x)(y,y)_{R^{\tilde{n}}} = M_6 h^{2/\mu}(y,y)_{R^{\tilde{n}}}$ would not yield the optimal result given at (22).

5. Error estimates and convergence

Now we shall study the convergence of the finite difference solution y of the FDS $A_h y = F_h$ to the exact solution u of the BVP $Au = F$, i.e. $y \to u$ in the sense of $\|y - u\| \to 0$ as $h \to 0$, where discrete norms $\| \cdot \| = \| \cdot \|_1$ and $\| \cdot \| = \| \cdot \|_C$ will be employed. Moreover, the choice of the refinement parameter $\mu \in (0,1]$ and the rate of convergence $y \to u$ will be discussed in terms of h, μ and λ.

Introduce the local approximation error z of y, with respect to u, as well as the 'flux' and 'source' approximation errors $\kappa(x'_\xi)$ and $\psi_f(x)$ by

$$z(x) := y(x) - u(x) \text{ for } x \in \bar{\omega}, \quad \psi_f(x) := f_h(x) - H^{-1}(x) \int\limits_{H(x)} f(\bar{x}) \, d\bar{x}$$

$$\kappa(x'_\xi) := \frac{1}{s(x'_\xi)} \int\limits_{s(x'_\xi)} k \frac{\partial u}{\partial n} \, ds - k_h(x'_\xi) u_h(x'_\xi), \quad \xi \in S'(x), \ x \in \omega. \tag{23}$$

Lemma 5.1. Let Assumption 3.2 be fulfilled. Then, the errors z, ψ_f and κ given by (23) are defined correctly, and the discrete H^1-norm $\|z\|_1$ can be estimated by

$$\|z\|_1 \leq M_7 \|A_h z\|_{-1} \leq M_7 \{ \|\kappa\|_0' + \|\psi_f\|_0 \}. \quad \equiv \tag{24}$$

Relation (24) yields an upper bound of $\|A_h z\|_{-1}$, which is appropriate for further estimates. The proof of (24) is based on the coerciveness of A_h and on discrete analogues of Green's formula and Cauchy's inequality. For meshes with $\mu = 1$, this is shown in [7], and for $0 < \mu < 1$, (24) is also true, cf. [8]. Clearly, bounds of $\|z\|_1$ in terms of powers of h will be get by evaluating the norms of the error functionals κ and ψ_f given by (23), (24).

Theorem 5.2. Let u be the solution of the BVP $Au = F$ (1), e.g. with $f \in H^s(\Omega)$, $s > 1$, or $f \in C^s(\bar{\Omega})$, $s \geq 1$ (or PC^1, W^s_∞), and y the solution of the FDS $A_h y = F_h$ (19). Suppose that Assumption 3.2 is satisfied. Then, error estimates of the type

$$
\|y - u\|_1 \leq M_8 \chi(h), \text{ with } \chi(h) := \begin{cases} h^\sigma & \lambda_P < \mu_P \leq 1 \\ h|lnh|^{\frac{1}{2}} & \text{for} \quad \lambda_P = \mu_P \quad \text{and } P \in C_s, (25) \\ h & 0 < \mu_P < \lambda_P \end{cases}
$$

$\lambda_P := \min\limits_{i \in I_P} \lambda_{P_i}$, $\sigma := \min\limits_{P \in C_s} \frac{\lambda_P}{\mu_P}$, and $\|y - u\|_{C(\bar{\omega})} \leq M_9 \chi(h)|lnh|^{\frac{1}{2}}$ hold. The constants M_8 and M_9 do not depend on h, $h \leq h_0$. $\equiv$

This means, if the triangulation T_h is refined locally according to Assumption 3.2, where $0 < \mu_P < \lambda_P$ holds for each $P \in C_s$, we shall get $\|y - u\|_1 \leq M_8 h$ and $\|y - u\|_{C(\bar{\omega})} \leq M_9 h|lnh|^{\frac{1}{2}}$, i.e. the same rate of convergence as it is proved for regular solutions $u \in H^2(\Omega)$ in [7]. The proof of Theorem 5.2 is given in [8] and employs Lemma 5.1 and estimates of the norms of $\kappa(u_s + u_r)$, ψ_f, cf. (24). In comparison with [7], estimates of $\|\kappa(u_s)\|'_0$ are to be added. Moreover, relations (8), (9), (11) and (12) are used essentially. The estimate of the C–norm follows from (25) and the second inequality of (14).

Concluding Remarks. Instead of BVP (1), more general problems with mixed boundary conditions and variable coefficients k_{ij} $(i, j = 1, 2)$ in the principal part of the operator L can also be approximated by box finite difference schemes. Moreover, general 'shape regular' triangulations without the restrictive Assumption 3.1 (ii) can be admitted. But in such cases, we should change to secondary triangulations of type MD, where the medians of the triangles define the box boundaries, and the centres of gravity of K can be chosen as appropriate quadrature points for approximating the 'fluxes' through $\partial\mathbf{H}(x) \cap K$, $K \in T_h$. This approach has been studied in [7]. Obviously, the FDSs $A_h y = F_h$ derived in [7, pp.48-54] can also be defined for discontinuous coefficients k_{ij} $(i, j = 1, 2; k_{ij} \in C^1(\bar{\Omega}^\pm))$, if the triangulation T_h is consistent with Γ, Γ_0 and with the singularity points due to corners and collision points of boundary conditions of different type. Moreover, if the triangulation is refined locally (cf. Assumption 3.2, $0 < \mu \leq 1$), the matrix B_h assigned to the box difference operator A_h is symmetric, positive definite, and the condition number $\beta(B_h)$ does not grow faster than with h^{-2}. Unfortunately, for general coefficients k_{ij} and triangulations T_h, the monotonicity properties contained in Lemma 4.2 (iii) and Theorem 4.3 (ii) can not be preserved, in general.

References

[1] Bank, R., and D. J. Rose, *Some error estimates for the box method*, SIAM J. Num. Anal. 24 (1987), 777-787.

[2] Ciarlet, Ph. G., *The finite element method for elliptic problems*, North–Holland, Amsterdam 1978.

[3] Costabel, M., *Starke Elliptizität von Randintegraloperatoren erster Art*, Habilitationsschrift, TH Darmstadt, Fachbereich Mathematik, Preprint 868, 1984.

[4] Dobrowolski, M., *Numerical approximation of elliptic interface and corner problems*, Habilitationsschrift, Rheinische Friedrich–Wilhelms–Universität, Bonn 1981.

[5] Grisvard, P., *Elliptic problems in nonsmooth domains*, Pitman, London 1985.

[6] Hackbusch, W., *On first and second order box schemes*, Computing 41 (1989), 277-296.

[7] Heinrich, B., *Finite difference methods on irregular networks*, Int. Series of Num. Math., vol. 82, Birkhäuser Verlag, Basel 1987.

[8] Heinrich, B., *The box method for elliptic interface problems*, TU Chemnitz, 1993, in preparation.

[9] Heinrich, B. und M. Meisel, *Integralbilanzmethode für elliptische Probleme auf lokal verfeinerten Netzen*, Preprint Nr. 115, TU Chemnitz (formerly Karl-Marx-Stadt), Juli 1989.

[10] Heinrich, B., and M. Meisel, *The box method for elliptic problems on plane polygons*, TU Chemnitz, 1993, in preparation.

[11] Kellogg, R. B., *Singularities in interface problems*, pp. 351-400, in 'Numerical solution of Partial Differential Equations–II', B. Hubbard ed., Academic Press, New York 1971.

[12] Kondrat'ev, V. A., *Boundary value problems for elliptic equations in domains with conical or angular points*, (Russian) Trudy Mosc. Mat. Obsc. 16 (1967), 209-292.

[13] Ladyshenskaya, O. A., *Boundary value problems of mathematical physics*, (Russian) Izdatel'stvo Nauka, Moskva 1973.

[14] McNeal, R. H., *An asymmetrical finite difference network*, Quart. Appl. Math. 11 (1953), 295-310.

[15] Meisel, M., *Konvergenz der Integralbilanzmethode für elliptische Probleme mit Randsingularitäten auf lokal verfeinerten Netzen*, Preprint Nr. 89, TU Chemnitz (formerly Karl-Marx-Stadt), Januar 1989.

[16] Oganesyan, L. A., and L. A. Rukhovets, *Variational-difference methods for solving elliptic equations*, (Russian) Izdatel'stvo Akad. Nauk Arm. SSR, Jerevan 1979.

[17] Schatz, A. H., and L. B. Wahlbin, *Maximum norm estimates in the finite element method on plane polygonal domains*, part 1, Math. Comp. 32 (1978), No. 141, 73-109, part 2, Math. Comp. 33 (1979), No. 146, 465-492.

[18] Strang, G., and G. J. Fix, *An analysis of the finite element method*, Prentice–Hall Inc., New Jersey 1973.

PARALLEL STEADY EULER CALCULATIONS USING MULTIGRID METHODS AND ADAPTIVE IRREGULAR MESHES

J. De Keyser [1] and D. Roose

Department of Computer Science, K. U. Leuven,
Celestijnenlaan 200A, B-3001 Leuven, Belgium

SUMMARY

The hydrodynamic Euler equations typically have solutions with local features. A steady Euler solver is described which uses three techniques that facilitate the computation of spatially accurate solutions : irregular mesh discretizations and adaptive refinement, multigrid methods and distributed memory parallelism.

1. THE EULER EQUATIONS

The Euler equations are a special case of the Navier-Stokes equations, describing inviscid fluid flow for an ideal gas with constant specific heats. Flow fields often have localized features, e.g. shocks and boundary layers. In order to limit the computation time needed to achieve a high spatial resolution, simulations may benefit from
  - *irregular mesh discretizations* and *adaptive mesh refinement*, in order to reach maximum accuracy with a fixed-size discrete problem
  - *multigrid methods*, because of their convergence properties
  - *distributed memory parallelism*, in order to achieve a high computation rate
A 2-D steady Euler solver is described that combines these acceleration techniques.

The state of a compressible fluid in a 2-dimensional flow field is defined by the *conservative* variables q : density ρ, momentum in both coordinate directions $m = \rho u$ and $n = \rho v$, with u and v the velocity components, and energy $E = \rho e = p/(\gamma - 1) + \rho(u^2 + v^2)/2$, with p the pressure and γ the ratio of specific heats. Other state variables are the enthalpy $h = e + p/\rho$ and the entropy difference with respect to a reference state (p_0 , T_0, s_0) : $s - s_0 = C_p \ln T/T_0 - R \ln p/p_0$. The fluid is an ideal gas with constant C_p and C_v : $p = \rho R T$, $R = C_p - C_v$, $\gamma = C_p/C_v$ (R is the gas constant and T the absolute temperature). The equations of motion (conservation of mass, momentum, and energy) relative to a fixed inertial frame for an inviscid flow field $q(x, t)$, $x \in \Omega, t \in [t_0, \infty)$ are :

$$\frac{\partial q}{\partial t} + Nq = \frac{\partial q}{\partial t} + \frac{\partial f}{\partial x} + \frac{\partial g}{\partial y} = \frac{\partial}{\partial l} \begin{bmatrix} \rho \\ \rho u \\ \rho v \\ \rho e \end{bmatrix} + \frac{\partial}{\partial x} \begin{bmatrix} \rho u \\ \rho u^2 + p \\ \rho u v \\ \rho u h \end{bmatrix} + \frac{\partial}{\partial y} \begin{bmatrix} \rho u \\ \rho u v \\ \rho v^2 + p \\ \rho v h \end{bmatrix} = 0. \quad (1)$$

An initial flow field $q(x, t_0)$ is given and boundary conditions are defined on $\partial \Omega$.

The Navier-Stokes operator $N^{NS} = N + \epsilon N^d$ consists of the first-order convective Euler part N and a higher-order diffusive part N^d. Since $\lim_{\epsilon \to 0} N^{NS}(\epsilon) = N$ and

---

[1]This author is supported by the Belgian Incentive Program "Information Technology" – Computer Science of the Future, and the Belgian programme on Interuniversitary Poles of Attraction, initiated by the Belgian State – Prime Minister's Service – Science Policy Office.

187

the type of $\mathbf{N}^{NS}$ and $\mathbf{N}$ differ, the Euler equations constitute a singularly perturbed problem. A first impression of the flow field can be obtained by assuming that the flow is convection-dominated. Rather than computing all solutions of the Euler equations, one is interested in the limit of the solutions of a series of Navier-Stokes problems with decreasing viscosity.

2. FINITE VOLUME DISCRETIZATION AND ADAPTIVE REFINEMENT

A first-order spatial discretization of (1) on polygonal meshes has been constructed using the finite volume technique with van Leer flux vector splitting. Adaptive refinement improves the spatial accuracy with a minimal increase of the discrete problem size.

2.1. Spatial discretization

The domain Ω is discretized by a set of polygonal cells, a *grid* $\mathcal{G} = \{\Omega_i\}$. Cell Ω_i is *adjacent* to Ω_j if it has at least one edge in common with Ω_j. $\mathcal{A}(\Omega_j)$ denotes the set of neighbors of cell Ω_j. The area of cell Ω_i is A_i. The interface $\partial\Omega_{ij}$ corresponds to two directed edges $\mathbf{s}_{ij}$ and $\mathbf{s}_{ji}$ with length s_{ij} (see figure 1). Two grids $\mathcal{G}^{(k)}$ and $\mathcal{G}^{(l)}$, $k < l$ are *nested* if $\forall \Omega_j^{(l)} : \exists \Omega_i^{(k)} : \Omega_j^{(l)} \subset \Omega_i^{(k)}$. Cell $\Omega_i^{(k)}$ is the *parent cell* of $\Omega_j^{(l)}$; all cells with this parent form the set $\mathcal{S}(\Omega_i^{(k)})$ of *subcells* of $\Omega_i^{(k)}$.

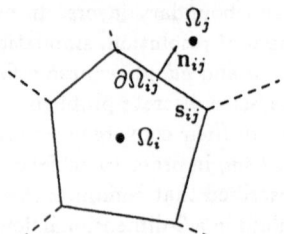

Figure 1: Finite volume discretization with polygonal cells

The first-order operator $\mathbf{N} : E \to \hat{E}$ maps fluid state functions $\mathbf{q}(\mathbf{x})$, $\mathbf{x} \in \Omega$ from the state space E to the residual space $\hat{E}$. The l-th discrete problem is formulated in subspaces $E^{(l)} \subset E$ and $\hat{E}^{(l)} \subset \hat{E}$ by means of projections $\mathbf{R}^{(l)}$ and $\bar{\mathbf{R}}^{(l)}$. First order accuracy is achieved with a piecewise constant (per cell) approximation $\mathbf{q}(\mathbf{x})$:

$$(\mathbf{R}^{(l)}\mathbf{q})_i(t) = \frac{1}{A_i^{(l)}}\int_{\Omega_i^{(l)}}\mathbf{q}(\mathbf{x},t)\,d\Omega, \qquad (\bar{\mathbf{R}}^{(l)}\mathbf{f})_i = \frac{1}{A_i^{(l)}}\int_{\Omega_i^{(l)}}\mathbf{f}(\mathbf{x},t)\,d\Omega.$$

If one defines the flux through an edge by :

$$\mathbf{f}_{ij} = \frac{1}{s_{ij}}\int_{\partial\Omega_{ij}}[\mathbf{f}\ \mathbf{g}]\cdot\mathbf{n}_{ij}\,ds + \mathcal{O}(h) = [\mathbf{f}(\mathbf{q}(\mathbf{x}_{ij}^*))\ \mathbf{g}(\mathbf{q}(\mathbf{x}_{ij}^*))]\cdot\mathbf{n}_{ij} + \mathcal{O}(h).$$

for $\mathbf{x}_{ij}^* \in \partial\Omega_{ij}$ ($\mathbf{n}$ is the unit outward normal), the conservation laws for Ω_i are :

$$A_i\frac{d\mathbf{q}_i}{dt} + \sum_{\Omega_j\in\mathcal{A}(\Omega_i)} s_{ij}\mathbf{f}_{ij} = A_i\mathbf{f}_i, \quad \text{i.e.} \quad (\mathbf{N}^{(l)}\mathbf{q}^{(l)})_i(t) = \frac{1}{A_i^{(l)}}\sum_{\Omega_j^{(l)}\in\mathcal{A}(\Omega_i^{(l)})} s_{ij}\mathbf{f}_{ij}. \tag{2}$$

Constant prolongation is used. The restriction operators are :

$$(\mathbf{R}^{(l \to k)} \mathbf{q}^{(l)})_i = \frac{1}{A_i^{(k)}} \sum_{\Omega_j^{(l)} \in \mathcal{S}(\Omega_i^{(k)})} A_j^{(l)} \mathbf{q}_j^{(l)} \quad \text{and} \quad (\bar{\mathbf{R}}^{(l \to k)} \mathbf{f}^{(l)})_i = \frac{1}{A_i^{(k)}} \sum_{\Omega_j^{(l)} \in \mathcal{S}(\Omega_i^{(k)})} A_j^{(l)} \mathbf{f}_j^{(l)}. \tag{3}$$

The flux through a cell interface $\partial\Omega_{ij}$ is computed in a local coordinate system (x', y') rotated over an angle θ such that $\partial\Omega_{ij}$ is parallel to the y'-axis. Let $\mathbf{M}^{(\theta)}$ denote this rotation in the state space. With *van Leer flux vector splitting* [1] $\mathbf{f}_{VL}(\mathbf{q}_i, \mathbf{q}_j) = \mathbf{f}^+(\mathbf{q}_i) + \mathbf{f}^-(\mathbf{q}_j) = \mathbf{f}(\mathbf{q}(\mathbf{x}_{ij}^*)) + \mathcal{O}(h)$ the flux through $\partial\Omega_{ij}$ is

$$\mathbf{f}_{ij} = \mathbf{M}^{(-\theta_{ij})} \mathbf{f}(\mathbf{M}^{(\theta)} \mathbf{q}(\mathbf{x}_{ij}^*)) = \mathbf{M}^{(-\theta_{ij})} \mathbf{f}_{VL}(\mathbf{M}^{(\theta_{ij})} \mathbf{q}_i, \mathbf{M}^{(\theta_{ij})} \mathbf{q}_j).$$

The consistency condition $\mathbf{f}(\mathbf{q}_i) = \mathbf{f}_{VL}(\mathbf{q}_i, \mathbf{q}_i)$ and the discrete conservation property $\mathbf{f}_{ij} = -\mathbf{f}_{ji}$ are satisfied. The jacobians of the flux $\mathbf{f}_{ij}$ are

$$\mathbf{A}_{ij} = \mathbf{M}^{(-\theta_{ij})} \frac{d\mathbf{f}^+(\mathbf{M}^{(\theta_{ij})} \mathbf{q}_i)}{d\mathbf{q}} \qquad \mathbf{B}_{ij} = \mathbf{M}^{(-\theta_{ij})} \frac{d\mathbf{f}^-(\mathbf{M}^{(\theta_{ij})} \mathbf{q}_j)}{d\mathbf{q}}.$$

Properties $\mathbf{f}_{ij} = -\mathbf{f}_{ji}$ and $\mathbf{A}_{ij} = -\mathbf{B}_{ji}$ allow to compute fluxes and jacobians once per edge.

With each edge along the domain boundary, one associates a *border cell* Ω_B. The boundary conditions impose information about state $\mathbf{q}_B$ and flux $\mathbf{f}^*$ at the border :

$$\mathbf{f}(\mathbf{q}_B, \mathbf{q}_I) = \mathbf{f}^*(\mathbf{q}_B), \qquad \mathbf{b}(\mathbf{q}_B) = 0. \tag{4}$$

2.2. Nested iteration and mesh refinement

A nested sequence of discrete spaces is obtained through mesh refinement. These spaces are characterized by a mesh parameter h_l, such that $h_i > h_j \Leftrightarrow E^{(i)} \subset E^{(j)} \Leftrightarrow \hat{E}^{(i)} \subset \hat{E}^{(j)}$ and $\{E^{(i)}; i \in \mathbb{N}\}$ is dense in E . The approximation found by solving the discrete equations is the exact solution of a Navier-Stokes equation with operator $\mathbf{N} + \epsilon \tilde{\mathbf{N}}^d$. For consistent discretizations the diffusive term $\epsilon \tilde{\mathbf{N}}^d$, also known as the *numerical viscosity*, vanishes as the discretization becomes finer. The singular perturbation problem can therefore be solved arbitrarily accurate by constructing successively finer discretizations. As $\mathbf{N}$ is bounded and stable in a neighborhood of the exact solution, the quality of an approximation is measured by the norm of the residual $\mathbf{r} = \mathbf{f} - \mathbf{Nq}$. The residual norm $\| \cdot \|_{\mathcal{E}} \equiv \|\phi\|_2$ with $\phi = \frac{1}{2\rho}((r_\rho)^2 + (r_{\rho u}/(|u| + c))^2 + (r_{\rho v}/(|v| + c))^2 + (r_{\rho e}/e)^2)^{1/2}$ has been used [10].

A parallel version of the mesh refinement strategy proposed by Van Keirsbilck and Deconinck [11] is used to generate finer discretizations. A refinement criterium must be based on the error made in the operator discretization, i.e. the numerical viscosity. In the exact Euler solution entropy is constant along streamlines in smooth flow regions, and increases over a discontinuity. The streamwise entropy gradient $\mathbf{u} \cdot \nabla s$ therefore yields a robust criterium.

A starting solution $\mathbf{q}^{(l)0}$ for an iterative solver at the next level is obtained from the solution at the previous level. This procedure is known as *nested iteration* [5].

189

3. TIME DISCRETIZATION

Spatial discretization yields a system of ODEs in time. It is assumed that all solution branches $q(x;t), t > t_0$ with an initial value in the neighborhood of $q(x;t_0)$ converge to the same state q^* as $t \to \infty$.

Explicit time-marching schemes are popular in CFD, because of ease of implementation, low memory consumption and limited computational cost per timestep. A well-known first order scheme is Forward Euler (FE) :

$$q^{(l)}(t_{k+1}) = q^{(l)}(t_k) + \Delta t (f^{(l)}(t_k) - N^{(l)} q^{(l)}(t_k)). \tag{5}$$

The main drawback of explicit methods is the existence of a stability condition limiting the timestep. Timesteps are expressed relative to an estimate of the stability limit : the *Courant-Friedrichs-Lewy number* is $CFL = \Delta t / \Delta t^{(CFL)}$ with $\Delta t^{(CFL)} = \min_{\Omega_i} A_i (\sum_{\Omega_j \in A(\Omega_i)} s_{ij} \lambda_{ij})^{-1}$ and $\lambda_{ij} = \max\{0, u_i \cdot n_{ij} + c_i\}$. The stability limit of FE lies at $CFL \approx 0.7$. Multi-stage Runge-Kutta methods have been developed by choosing the coefficients α_j in an n-stage method

$$q^{[0]} = q^{(l)}(t_k); \quad q^{[j]} = q^{[0]} + \alpha_j \Delta t (f^{(l)} - N^{(l)} q^{[j-1]}), j = 1, \ldots, n; \quad q^{(l)}(t_{k+1}) = q^{[n]}$$

so as to improve the smoothing properties, e.g. RK4 ($\alpha_1 = 1/4, \alpha_2 = 1/2, \alpha_3 = 0.55, \alpha_4 = 1; CFL = 1.25$) [7]. Local timestepping variants (LT, as opposed to global timestepping, GT) use a different timestep in each cell. Time-accuracy is lost, but convergence improves significantly.

Unlike explicit methods, implicit techniques are unconditionally stable. Since time-accuracy is of no importance, only first order techniques like Backward Euler are useful :

$$(I^{(l)} + \Delta t \cdot N^{(l)}) q^{(l)}(t_{k+1}) = q^{(l)}(t_k) + \Delta t \cdot f^{(l)}. \tag{6}$$

Linearization around $\tilde{q}^{(l)} = q^{(l)}(t_k)$ gives a block sparse system in the corrections $\Delta q^{(l)}$:

$$L^{(l)}(\tilde{q}^{(l)}; \Delta t) \Delta q^{(l)} = rhs^{(l)}(\tilde{q}^{(l)}), \tag{7}$$

$$L_{ii}^{(l)} = \frac{A_i^{(l)}}{\Delta t} 1 + \sum s_{ij} \tilde{A}_{ij}^{(l)}, \qquad L_{ij}^{(l)} = \sum s_{ij} \tilde{B}_{ij}^{(l)}, \qquad rhs_i^{(l)} = A_i^{(l)} \left(f_i^{(l)} - (N^{(l)} \tilde{q}^{(l)})_i \right).$$

The linearized boundary conditions are :

$$\tilde{A}_{BI}^{(l)} \Delta q_B^{(l)} + \tilde{B}_{BI}^{(l)} \Delta q_I^{(l)} = f^*(\tilde{q}_B^{(l)}) - f(\tilde{q}_B^{(l)}, \tilde{q}_I^{(l)}), \tag{8}$$

$$\tilde{A}_{BI}^{(l)} = M^{(-\theta)} \frac{df^+(M^{(\theta)} \tilde{q}_B^{(l)})}{dq} - \frac{df^*(\tilde{q}_B^{(l)})}{dq}, \qquad \tilde{B}_{BI}^{(l)} = M^{(-\theta)} \frac{df^-(M^{(\theta)} \tilde{q}_I^{(l)})}{dq}.$$

This linearization will be used within a Newton process. Damped Newton iteration is needed when the initial solution is far off the exact solution. A pointwise damping factor of the form $\sigma_i = 2^{-m}, m \in \mathbb{N}$ is chosen, with m as small as possible, such that states with negative ρ or p are avoided [9]. As the iteration proceeds, $\sigma = 1$ always becomes possible and the asymptotically quadratic convergence is preserved.

Iterative solvers for these linear systems require the inversion of 4×4 systems with coefficient matrices $L_{ii}^{(l)}$ which may be ill-conditioned, or $\tilde{A}_{BI}^{(l)}$ which may even be singular. First, Gaussian elimination without pivoting is applied. When small diagonal elements are encountered, elimination with row- and column-pivoting is used. If small pivots are still detected, this implies singularity (occurs only for border cells) : the boundary condition only affects some border state components, while the other ones should remain unchanged. As pivoting slows down Gaussian elimination with a factor 2 to 3, it should be used only when necessary.

4. MULTIGRID SOLVERS

Several multigrid methods have been developed to solve the steady Euler equations [6, 7, 9]. We have implemented a number of them.

4.1. Newton-multigrid

Newton-multigrid methods are derived from the following implicit timestepping procedure : each timestep gives rise to a nonlinear system which is solved by a Newton process. A Newton step consists of a linearization and the solution of the linear system with linear multigrid (LMG).

With a good initial approximation, one timestep with $CFL = \infty$ gives the steady state. The effort expended in each Newton step consists of linearization and a number of LMG cycles. One can speed up the computation by limiting the number n_M of LMG cycles per Newton step or by re-using a linearization in subsequent steps; both yield the same Picard algorithm. One may choose n_M such that the linear problem residual norm is reduced by a prescribed factor [9], or simply put $n_M = 1$ [5]. Convergence no longer is quadratic; it becomes linear. The domain of convergence can be enlarged by using a finite timestep. As $\lim_{\Delta t \to 0} L^{(l)}(q_0^{(l)}; \Delta t) = I^{(l)}$, Δt can be chosen small enough to guarantee diagonal dominance (typically $CFL = 10^5$).

There exist not many parallel linear smoothers for irregular grid discretizations. Colored Gauss-Seidel schemes require graph coloring and have some drawbacks in the context of parallel execution. Therefore collective damped Jacobi relaxation was used.

4.2. Nonlinear multigrid

Nonlinear multigrid methods [5] solve the steady-state equations directly. FAS multigrid is frequently used for solving the Euler equations [6]; it requires a nonlinear smoother.

We have used nonlinear damped collective Jacobi iteration (NLJP) with damping factor $\theta \in (0,1)$, which calls for the solution of a 4×4 nonlinear system in each cell. Linearization around $\tilde{q}_j^{(l)}$ gives a system in $\Delta q_j = q_j^{(l)} - \tilde{q}_j^{(l)}$, which can be solved directly :

$$(\frac{A_j^{(l)}}{\Delta t}1 + \sum_{\Omega_m \in \mathcal{A}(\Omega_j)} s_{jm} A_{jm}(\tilde{q}_j^{(l)}, q_m^{(l)}))\Delta q_j = A_j^{(l)} f_j^{(l)} - \sum_{\Omega_m \in \mathcal{A}(\Omega_j)} s_{jm} f(\tilde{q}_j^{(l)}, q_m^{(l)}).$$

Another type of smoothing iteration is obtained if the full nonlinear system is linearized. One Newton step serves as nonlinear smoother (NLJF); the linear system (7) is solved by linear collective Jacobi relaxation. The Newton process may be damped in order to improve the smoothing properties of NLJF.

4.3. Nonlinear multigrid based on explicit timestepping

There exist several methods that employ a grid hierarchy in order to accelerate explicit time-marching. The method proposed by Jameson has been used for both the regular and the irregular grid case [7, 3]. It consists of the FAS-scheme in which the smoother has been replaced by a multi-stage RK method [8]. Jespersen [6] proved that this method (MGTS) performs a large aggregate timestep with less computation. There is a close relationship between the frozen-τ modified nested iteration technique for continuation in time and a MGTS V-cycle (FE, 0 pre- and 1 post-smoothing step). This explains why the best results are obtained when the same number of timesteps and the same CFL-number are used at each level.

One NLJF and one NLJP step are identical and coincide with one FE timestep if a block-diagonal timestepping matrix is used, with blocks $\Delta t_i = \theta_i (L_{ii}^{(l)})^{-1}$. One may approximate $(L_{ii}^{(l)})^{-1}$ based on physical considerations, as is done in local timestepping (Δt_i is a scalar diagonal matrix) and characteristic preconditioning [12].

5. DISTRIBUTED MEMORY PARALLEL COMPUTATION

For non-adaptive grid hierarchies a fixed data distribution scheme exists that provides both a good calculation load balance and an efficient communication structure on a distributed memory multiprocessor [2, 3]. Adaptive refinement of a well-balanced grid however can result in an unevenly distributed fine grid, requiring load redistribution at run-time [4]. This load balancing problem has been tackled with a two-step approach : first the data set is *partitioned*, subsequently the parts are *mapped* onto the parallel machine architecture.

5.1. Distributed data structures and partitioning

The data parallel programming model was adopted. In this approach the term *phase* is used for each parallel calculation step, consisting of a data exchange between data partitions and the parallel execution of the calculations on them. A phase is characterized by a *process interaction graph*. In the multigrid method there are phases acting on one grid (e.g. smoothing) and phases acting on subsequent grids (e.g. prolongation, restriction). The data dependency relation for smoothing coincides with the $\mathcal{A}$-relation. Smoothing can be carried out in parallel when the grid is partitioned in a set of *parts* $\mathcal{P}^{(l)} = \{p_i^{(l)}\}$. It requires *intra-grid communication* between these parts. The number and volume of these messages should be kept low. A buffer $\mathcal{B}^H$ is associated with each part, storing information concerning part interfaces. Prolongation and restriction act on subsequent grids in the hierarchy. The data dependency relation between cells now coincides with the $\mathcal{S}$-relation. Partitionings $\mathcal{P}^{(l-1)}$ and $\mathcal{P}^{(l)}$ are *nested* if all subcells of $\Omega_k \in \mathcal{G}^{(l-1)}$ belong to the same fine grid part : $\forall \Omega_i, \Omega_j \in \mathcal{S}(\Omega_k) : \Omega_i \in p^{(l)} \Rightarrow \Omega_j \in p^{(l)}$, and if two fine grid cells with parents in different coarse grid parts cannot belong to the same fine grid part : $\forall \Omega_i \in \mathcal{S}(\Omega_m), \Omega_j \in \mathcal{S}(\Omega_n), \Omega_m \in p'^{(l-1)}, \Omega_n \in p''^{(l-1)}, \Omega_i \in p'^{(l)}, \Omega_j \in p''^{(l)} : p'^{(l-1)} \neq p''^{(l-1)} \Rightarrow p'^{(l)} \neq p''^{(l)}$. The $\mathcal{S}$-relation on cells then induces a similar relation on the set of parts. Prolongation and restriction require *inter-grid communication*. A buffer $\mathcal{B}^V$ is associated with a part $p \in \mathcal{S}(p_p)$ if its parent part p_p does not reside in the same processor. This buffer holds a duplicate of the data in p_p. The communication volume is proportional to the number of cells in p_p. Each part has exactly one parent, which limits the number of intergrid messages.

Hierarchical recursive bisection (HRB) is a heuristic that generates nested partitions [4]: the partitioning $\mathcal{P}^{(i)}$ of $\mathcal{G}^{(i)}$ is derived from $\mathcal{P}^{(i-1)}$ by bisecting each part of the latter, and collecting the corresponding subcells. If $\mathcal{G}^{(i)} = \mathcal{G}^{(i-1)} = \ldots = \mathcal{G}^{(0)}$ HRB coincides with recursive bisection for single grid partitioning [13]. Hierarchical inertial recursive bisection (HIRB) implicitly tries to minimize the perimeter of the parts (low intra-grid communication). The number of cells in the partitions depends on where and how much the mesh happened to be refined. A good load balance can be attained when there are more parts than processors. On the other hand, parts should not be too small to avoid a large communication volume or short messages.

5.2. Incremental mapping

A mapping function $m : \mathcal{P} \to \mathcal{Q} : q_k = m(p_i)$ assigning parts to processors is the solution of a minimization problem in which the cost function represents the multigrid cycle time. *Incremental fine grid mapping* assigns the parts of a new fine grid to the processors while taking into account the current assignment of coarse grid parts. Another approach would be to remap parts on the coarser levels too, but this is quite difficult.

Let $\mathcal{P}$ contain the parts of the two finest grids. On this set the relation $\mapsto$ is defined : $\forall p_i, p_j \in \mathcal{P} : p_i \mapsto p_j$ if $p_i \in \mathcal{A}(p_j)$ or $p_j \in \mathcal{A}(p_i)$ or $p_i \in \mathcal{S}(p_j)$ or $p_j \in \mathcal{S}(p_i)$. An $N_{\mathcal{P}} \times N_{\mathcal{P}}$ weight matrix Λ is associated with graph $(\mathcal{P}, \mapsto)$, giving information about the time involved in the corresponding calculation and communication :

$$\Lambda_{ii}^{ts} = a_i t_{calc}, \qquad \forall p_i \in \mathcal{A}(p_j) : \Lambda_{ij}^{ts} = b_{ij} t_{comm}, \qquad \forall p_i \in \mathcal{S}(p_j) : \Lambda_{ij}^{tr} = \Lambda_{ji}^{tr} = c_j t_{comm}.$$

The value of t_{calc} is the time needed to perform one floating point operation; t_{comm} denotes the time for nearest-neighbor communication of one number. a_i is the number of floating point operations required by the calculations for part p_i (proportional to the number of cells in p_i). Intra-grid communication requires exchanging b_{ij} floating point numbers between p_i and p_j (proportional to the number of edges along the interface between both parts). The intergrid transfer message length to or from parent part p_j is c_j (proportional to the number of cells in p_j).

If fine grid smoothing dominates the multigrid cycle, the parallel performance of the smoother is optimized. A suitable cost function is :

$$C_a(m^{(l)}) = \max_{q \in \mathcal{Q}} \{ \sum_{\substack{p_i \in \mathcal{P}^{(l)} \\ m(p_i) = q}} \Lambda_{ii}^{ts} + \max_{\substack{p_j \in \mathcal{A}(p_i) \\ m(p_i) = q}} \Lambda_{ji}^{ts} \Delta_{ji} \}.$$

$\Delta_{ij} = \Delta(m(p_i), m(p_j))$ is the distance between the processors holding p_i and p_j, and accounts for the machine topology.

A cost model for the multigrid cycle can be constructed if one assumes that the work involved in prolongation and restriction (a.o. the residual calculation) is equivalent to ν_0 smoothing steps and if the communication costs for restriction and prolongation are taken to be equal. A model that includes fine grid smoothing and intergrid transfer between the two finest levels is :

$$C_b(m^{(l-1)}, m^{(l)}) = C_a(m^{(l)}) + \theta \cdot \max_{p_i \in \mathcal{S}(p_k)} \Lambda_{ik}^{tr} \Delta_{ik}; \quad \theta = \frac{2\gamma}{\nu_0 + \nu_1 + \nu_2}.$$

Factor θ weighs the relative importance of intergrid and intragrid operations. In cost function

$$C_c(m^{(l-1)}, m^{(l)}) = C_a(m^{(l)}) + \theta \cdot \max_{p_i \in \mathcal{S}(p_k)} \Lambda_{ik}^{tr} \tilde{\Delta}_{ik}; \quad \tilde{\Delta}_{ik} = \Delta_{ik}(1 + (\sum_{p_i \in \mathcal{S}(p_k)} \Lambda_{ik}^{tr} \Delta_{ik} / \Delta_{max} \sum_{p_i \in \mathcal{S}(p_k)} \Lambda_{ik}^{tr})),$$

in which Δ_{max} represents the processor graph diameter, an effective distance $\tilde{\Delta}_{ik}$ is used which depends on the average link occupation during intergrid transfer, thus modelling contention.

Graph $(\mathcal{P}, \mapsto)$ with weight matrix Λ is mapped onto a hypercube architecture by a cost-based heuristic (CRB) [4]. An iterative improvement technique is used to ameliorate the solution. The best results have been obtained with CRB based on cost model C_b, followed by iterative improvement with cost function C_c [4].

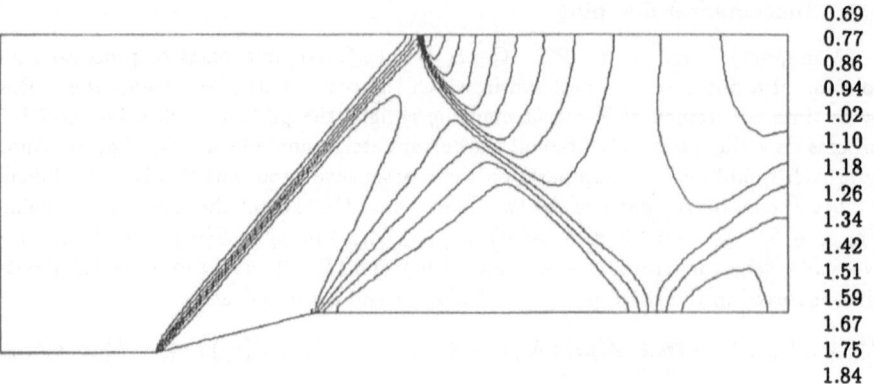

	0.69
	0.77
	0.86
	0.94
	1.02
	1.10
	1.18
	1.26
	1.34
	1.42
	1.51
	1.59
	1.67
	1.75
	1.84

Figure 2: Supersonic flow through a channel : iso-mach lines

6.   EXPERIMENTAL RESULTS

The multigrid algorithms have been implemented using the parallel programming environment LOCO [4] on an Intel iPSC/860 hypercube, a typical distributed memory parallel computer. The flow through a channel with a contraction has been computed; horizontal inflow at Mach 2 of a fluid with density $\rho = 1.271 kg/m^3$ and pressure $p = 101300 Pa$ was specified. Figure 2 shows the iso-mach lines, revealing the presence of a compression shock and its reflections between upper and lower channel walls, and an expension fan.

The mesh hierarchy was built starting from a sufficiently fine coarse mesh, such that the coarse grid solution is not too far off the singular perturbation problem's solution. If not, the adaptation process may be misguided. A mesh ratio $\rho = \#\mathcal{G}^{(l)}/\#\mathcal{G}^{(l-1)} \approx 2$ to 3 was used, which is typical of 2-D irregular mesh hierarchies [3]. In order to evaluate the effectiveness of the mesh adaptation strategy, an initial mesh with 444 cells was refined three times to obtain the solution on a mesh with 4536 cells (figure 3). The partitioning obtained with hierarchical inertial recursive bisection is shown in figure 4. The refinement criterium detects the expansion fan and traces the compression shock and its reflections. The same flow problem was solved with a uniform logically rectangular mesh with approximately the same number of cells. The linear resolution improvement (e.g. thickness of shocks) due to adaptivity was about 2.5.

Figure 5 compares the convergence of single-grid FE (LT, $CFL = 0.7$), a MGTS-FE V-cycle (LT, $CFL = 0.7$, $\nu_1 = \nu_c = 5, \nu_2 = 0$), a MGTS-RK4 V-cycle (LT, $CFL = 1.25$,

Table 1: Relative convergence speed at the fourth multigrid level

Method	seconds/iteration	digits/iteration	digits/second	relative speed
FE	0.189	0.00052	366	1.0
MGTS-FE	1.760	0.036	49	7.5
MGTS-RK4	3.387	0.038	88	4.2
NLMG-NLJF	4.668	0.086	54	6.8
NMG	5.542	0.100	54	6.8

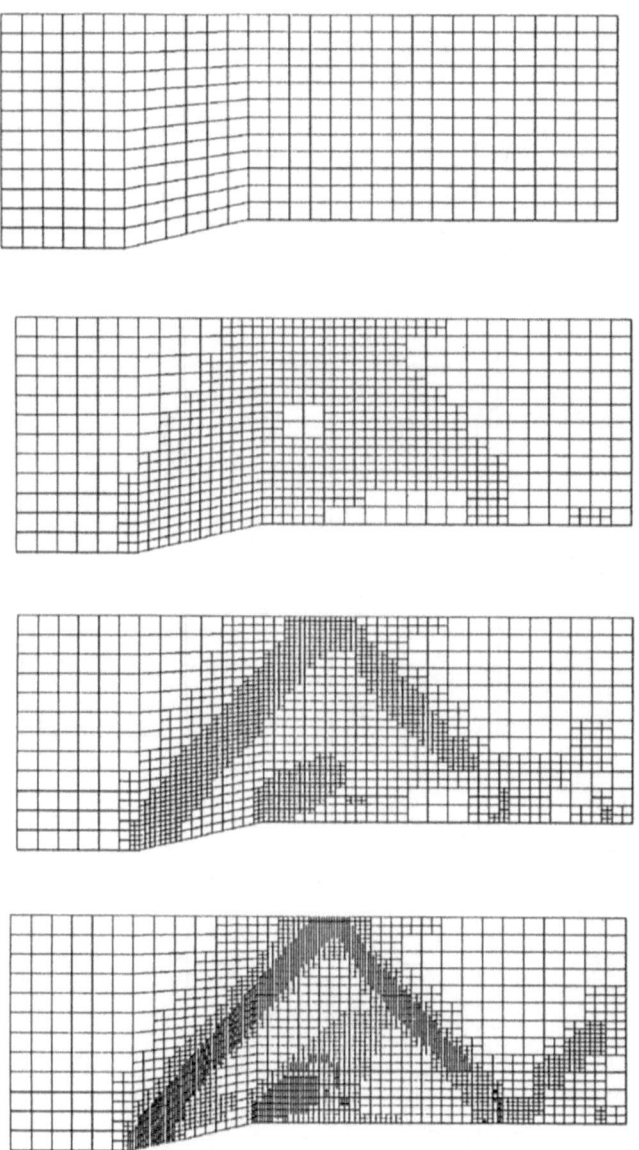

Figure 3: Mesh hierarchy

$\nu_1 = \nu_c = 3$, $\nu_2 = 0$), a NLMG-NLJF V-cycle (LT, $CFL = 10^5$, Newton damping factor $\theta_N = 0.8$, $\nu_1 = 3$, $\nu_2 = 0$, $\nu_c = 15$), and NMG (LT, $CFL = 10^5$, Newton damping factor $\theta_N = 1$) with recomputation of the jacobian every 5 Picard iterations, each of which is solved by 5 LMG V-cycles (Jacobi damping factor $\theta_J = 0.95$, $\nu_1 = 3$, $\nu_2 = 0$, $\nu_c = 5$), i.c. one jacobian update every 25 LMG-cycles. This graph should be interpreted

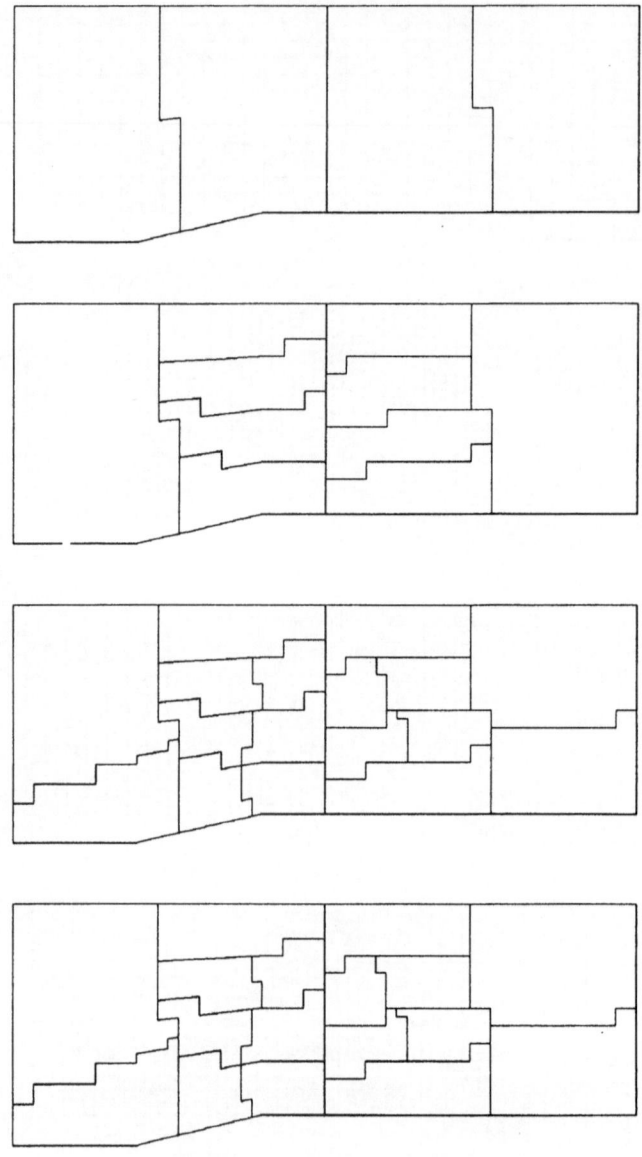

Figure 4: Nested partitioning

with care, as the computational cost for each of these cycles is different. All multigrid methods have a convergence speed which is almost independent of the number of levels (a slight dependence is possible as the mesh ratio is not constant), while single-grid convergence slows down as the discrete problem gets larger. A similar behavior was observed for subsonic flow problems, be it that convergence is a lot slower. Table 1

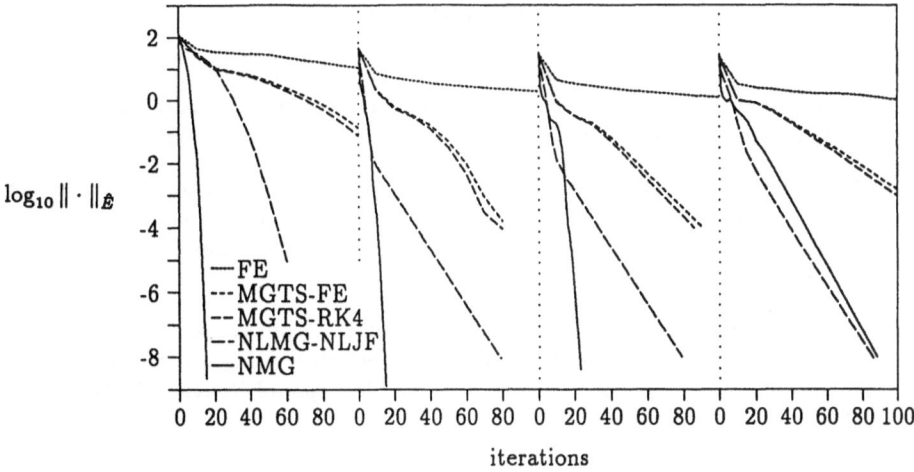

Figure 5: Convergence of single- and multigrid methods at subsequent levels

lists the (sequential) asymptotic execution speed of these methods relative to FE time-stepping. At the fourth grid level the MGTS-FE V-cycle is ≈ 70 times faster than FE in terms of number of iterations, and 7.5 times faster in terms of computation time. The relative multigrid performance improves further as the problem gets larger.

The parallel efficiency of an algorithm is defined as : $\epsilon_p = T(1, S)/P \cdot T(P, S)$, in which $T(P, S)$ denotes the execution time for a problem of size S on a machine with P processors. For the same MGTS-FE V-cycle as before, our current implementation achieved $\epsilon_p \approx 75$ % for $P \leq 16$ with HIRB partitioning and the mapping strategy described earlier. Parallel efficiency losses are due to load imbalance, data exchange communication, and the double calculation of fluxes and jacobians for edges along part interfaces : these are calculated twice, once for either part on both sides of the interface. A load balance between 95-100 % (based on the cost estimates provided by the application) is obtained when 4 to 8 times more parts than processors are available. A difficulty is the inaccuracy of these cost estimates, especially for border cells : the computation time for imposing a particular boundary condition is not always known in advance. The communication and double calculation losses decrease as the problem size per processor is larger, indicating that a sufficiently coarse-grained parallel computer architecture is needed (sufficient amount of memory, communication fast enough) to allow efficient parallel execution.

7. CONCLUSION

Three acceleration techniques have been combined in one steady Euler solver. Incorporating adaptivity and multigrid in a distributed memory parallel computer code calls for specific load balancing techniques. The proposed code proves to be effective on medium-sized parallel computers. For a model problem on a 16 processor machine, a global acceleration factor of the order of $(2.5)^2 \times (7.5) \times (16 \times 0.75) \approx 600$ has been observed with MGTS-FE. For larger problems each of the three factors improves even further.

REFERENCES

[1] W. K. Anderson, J. L. Thomas, and B. van Leer. A comparison of finite volume flux vector splittings for the Euler equations. *AIAA Paper No. 85-0122, presented at the AIAA 23rd Aerospace Sciences Meeting, Reno, Nevada*, 1985.

[2] G. Chesshire and A. Jameson. FLO87 on the iPSC/2 : A parallel multigrid solver for the Euler equations. In J. Gustafson, editor, *Proceedings of the Fourth Conference on Hypercubes, Concurrent Computers and Applications*, pages 957–966. Golden Gate Enterprises, May 1990.

[3] R. Das, D. Mavriplis, J. Saltz, S. Gupta, and R. Ponnusamy. The design and implementation of a parallel unstructured Euler solver using software primitives. In *Proceedings of the 30th Aerospace Sciences Meeting and Exhibit*. AIAA-92-0562, 1992.

[4] J. De Keyser and D. Roose. Incremental mapping for solution-adaptive multigrid hierarchies. In *Proceedings of the Scalable High Performance Computing Conference '92*, pages 401–408. IEEE Computer Society Press, 1992.

[5] W. Hackbush. *Multi-Grid Methods and Applications*. Springer-Verlag, 1985.

[6] P.W. Hemker and B. Koren. Defect correction and nonlinear multigrid for the steady Euler equations. In *Lecture series in computational fluid dynamics*, Rhode-St.-Genese, Belgium, 1988.

[7] A. Jameson. Solution of the Euler equations for two dimensional transonic flow by a multigrid method. *Applied Math. Comp.*, 13:327–355, 1983.

[8] A. Jameson, W. Schmidt, and E. Turkel. Numerical solutions of the Euler equations by finite volume methods using Runge-Kutta time-stepping schemes. *AIAA Paper*, 81-1259-CP:1–14, 1981.

[9] D. C. Jespersen. Design and implementation of a multigrid code for the Euler equations. *Appl. Math. and Computat.*, 13:357–374, 1983.

[10] W. A. Mulder. Multigrid relaxation for the Euler equations. *J. Comp. Phys.*, 60:235–252, 1985.

[11] R. Struijs, P. Van Keirsbilck, and H. Deconinck. An adaptive grid polygonal finite volume method for the compressible flow equations. In *Proceedings of the AIAA 9th CFD Conference*, 1989.

[12] B. van Leer, W.-T. Lee, and P.L. Roe. Characteristic time-stepping or local pre-conditioning of the Euler equations. In *AIAA 10th Computational Fluid Dynamics Conference*, 1991.

[13] R. D. Williams. Performance of dynamic load balancing algorithms for unstructured mesh calculations. *Concurrency: Practice and Experience*, 3:457–481, 1991.

An Object-Oriented Approach for Parallel Self Adaptive Mesh Refinement on Block Structured Grids[1]

Max Lemke[2], Kristian Witsch
Mathematisches Institut, Universität Düsseldorf,
e-mail: {lemke, witsch}@numerik.uni-duesseldorf.de

Daniel Quinlan[2]
Computational Mathematics Group, University of Colorado, Denver,
e-mail: dquinlan@copper.denver.colorado.edu

SUMMARY

Self adaptive mesh refinement dynamically matches the computational demands of a solver for partial differential equations to the activity in the application's domain. In this paper we present two C++ class libraries, P++ and AMR++, which significantly simplify the development of sophisticated adaptive mesh refinement codes on (massively) parallel distributed memory architectures. The development is based on our previous research in this area. The C++ class libraries provide abstractions to separate the issues of developing parallel adaptive mesh refinement applications into those of parallelism, abstracted by P++, and adaptive mesh refinement, abstracted by AMR++. P++ is a parallel array class library to permit efficient development of architecture independent codes for structured grid applications, and AMR++ provides support for self adaptive mesh refinement on block-structured grids of rectangular non overlapping blocks. Using these libraries the application programmers' work is greatly simplified to primarily specifying the serial single grid application, and obtaining the parallel and self adaptive mesh refinement code with minimal effort.

First results for simple singular perturbation problems solved by self adaptive multilevel techniques (FAC, AFAC), being implemented on the basis of prototypes of the P++/AMR++ environment, are presented. Singular perturbation problems frequently arise in large applications, e.g. in the area of computational fluid dynamics. They usually have solutions with layers which require adaptive mesh refinement and fast basic solvers in order to be resolved efficiently.

---

[1]This research has been supported by the National Aeronautics and Space Administration under grant number NASI-18606 and the German Federal Ministry of Research and Technology (BMFT) under PARANUSS, grant number ITR 900689.
[2]Part of this work belongs to the author's dissertation.

INTRODUCTION

The purpose of local mesh refinement during the solution of partial differential equations (PDEs) is to match the computational demands to an application's activity: In a fluid flow problem this means that only regions of high local activity (shocks, boundary layers, etc.) can demand increased computational effort; regions of little flow activity (or interest) are more easily solved using only relatively little computational effort. In addition, the ability to adaptively tailor the computational mesh to the changing requirements of the application problem at runtime (e.g. moving fronts in time dependent problems) provides for much faster solution methods than static refinement or even uniform grid methods. Combined with increasingly powerful parallel computers that are becoming available, such methods allow for much larger and more comprehensive applications to be run. With local refinement methods, the greater disparity of scale introduced in larger applications can be addressed locally. Without local refinement, the resolution of smaller features in the applications domain can impose global limits either on the mesh size or the time step. The increased computational work associated with processing the global mesh cannot be readily offset even by the increased computational power of advanced parallel computers. Thus, local refinement is a natural part of the use of advanced massively parallel computers to process larger and more comprehensive applications.

Our experiments with different local refinement algorithms for the solution of the simple potential flow equation on parallel distributed memory architectures (e.g. [2]) demonstrates that, with the correct choice of solvers, performance of local refinement codes shows no significant sign of degradation as more processors are used. In contrast to conventional wisdom, the fundamental techniques used in our adaptive mesh refinement methods do not oppose the requirements for efficient vectorization and parallelization. However, the best choice of the numerical algorithm is highly dependent on its parallelization capabilities, the specific application problem and its adaptive grid structure, and, last not least, the target architectures' performance parameters. It occurs that algorithms, which are expensive on serial and vector architectures, but are highly parallelizable, can be superior on one or several classes of parallel architectures.

Our previous work with parallel local refinement, which was done in the C language to better allow access to dynamic memory management, has permitted only simplified application problems on non block structured composite grids of rectangular patches. The work was complicated by the numerical properties of local refinement including self adaptivity and their parallelization capabilities like, for example, static and dynamic load balancing. In particular the explicit introduction of parallelism in the application code is very cumbersome. Software tools for simplifying this are not available, e.g., existing grid oriented communication libraries (as e.g. used in [2]) are far too restrictive to be efficiently applied to this kind of dynamic problem. Thus, extending this code for the solution of more general complex fluid flow problems on complicated block structured grids is limited by the software engineering problem of managing the large complexities of the application problem, the numerical treatment of self adaptive mesh refinement, complicated grid structures and explicit parallelization. The development of codes that are portable across different target architectures and that are applicable to not just one problem and algorithm but to a larger class is impossible under these conditions.

These experiences with parallel adaptive mesh refinement have motivated our work to simplify the development of such codes for parallel distributed memory computers:

Our solution to this software difficulty presents abstractions as a means of handling the combined complexities of adaptivity, mesh refinement, the application specific algorithm, and parallelism. These abstractions greatly simplify the development of algorithms and codes for complex applications. As an example, the abstraction of parallelism permits the development of application codes (necessarily based on parallel algorithms as opposed to serial algorithms, whose data and computation structures do not allow parallelization) in the simplified serial environment and the same code to be executed in a massively parallel distributed memory environment.

This paper introduces an innovative set of software tools to simplify the development of parallel adaptive mesh refinement code for difficult algorithms. The tools are present in two parts, which form C++ class libraries and allow for the management of the great complexities described above. The first class library, P++ (Section 2), forms a data parallel superset of the C++ language with the commercial C++ array class library M++ (Dyad Software Corporation). A standard C++ compiler is used with no modifications of the compiler required. The second set of class libraries, AMR++ (Section 3), forms a superset of the C++/M++, or P++, environment and further specifies the combined environment for local refinement (or parallel local refinement). In Section 4 we introduce multilevel algorithms, that allow for the introduction of self adaptive mesh refinement (Asynchronous) Fast Adaptive Composite Methods (FAC and AFAC)). In Section 5, we present first results for a simple singular perturbation problem that has been solved using FAC and AFAC algorithms being implemented on the bases of AMR++ and P++ prototypes. This problem serves as a good model problem for complex fluid flow applications, because several of the properties that are related to self adaptive mesh refinement are already present in this simple problem. We would like to thank everybody who discussed the P++ or AMR++ development with us or in any other way supported our work. We are particularly grateful to Steve McCormick, without whose support this joint work would not have been possible.

P++ — A PARALLEL ARRAY CLASS LIBRARY FOR STRUCTURED GRIDS

P++ is an innovative, robust and architecture-independent array class library that simplifies the development of efficient, parallel programs for large scale scientific applications by abstracting parallelism ([2]). The target machines are current and evolving massively parallel distributed memory multiprocessor systems (e.g. Intel iPSC/860 and PARAGON, Connection Machine 5, Cray MPP, IBM RS 6000 networks) with different types of node architectures (scalar, vector, or superscalar). Through the use of portable communication and tools libraries (e.g. EXPRESS, ParaSoft Corporation), the requirements of shared memory computers are also addressed. The P++ parallel array class library is implemented in standard C++ using the serial M++ array class library, with absolutely no modification of the compiler. P++ allows for software development in the preferred serial environment, and such software to be efficiently run, unchanged, in all target environments. The runtime support for parallelism is both completely hidden and dynamic so that array partitions need not be fixed during execution. The added degree of freedom presented by parallel processing is exploited by use of an optimization module within the array class interface.

Application class: The P++ application class is currently restricted to structured grid-oriented problems, which form a primary problem class currently represented in scientific supercomputing. This class is represented by dimensionally independent block structured grids (1D - 4D) with rectangular or logically rectangular grid blocks. The M++ array interface, which is also used as the P++ interface and whose functionality is similar to the array features of Fortran 90, is particularly well suited to express operations on grid blocks to the compiler and to the P++ environment at runtime. Grid blocks and indices to access groups of their elements (*double(_VSG_)Array, (VSG_)Index*) form the major object classes of M++ and P++.

```
#include "header.h"
#ifdef PPP
#define doubleArray double_VSG_Array
#define Index       VSG_Index
#endif
//-------------------------------------------------------------------
void MacCormack (Index I, double Time_Step, doubleArray & F,
                doubleArray & U_new, doubleArray &U_old)
   { F          = (U_old * U_old) / 2;
// array expression
     U_new (0) = U_old (0) - Time_Step * (F ( 1 ) - F (0));
// scalar expression
     U_new (I) = U_old (I) - Time_Step * (F (I+1) - F (I));
// indexed array expression
     F          = (U_new * U_new) / 2;
// array expression
     U_new (I) =   0.5 * (U_old (I) + U_new (I))
                 - 0.5 * Time_Step * (F (I) - F (I-1));        }
// indexed array expression
//-------------------------------------------------------------------
void main()
   { ...
     int Start_Position = 1;
     int Count        = N-2;
     int Stride       = 1;
     Index Interior (Start_Position, Count, Stride);
     doubleArray U_old (N,N) ,U_new (N,N), F(N,N);
     ...
     MacCormack (Interior, Time_Step, F, U_new, U_old);
     ...                                                       }
```

Figure 1: Sample of M++/P++ code.

Object-Oriented Design and User-Interface: In C++, an object is a user defined type that contains both data and member functions for manipulation of that data. The basic structure of P++ consists of multiple interconnected classes, namely the P++ kernel class double_VSG_Array as the grid block object, the VSG_Index class, and Communication-, I/O-, Parallel-Data-, Diagnostics- and Optimization-Manager classes.

The P++ user interface consists of a combination of the M++ array language interface and a set of functions for parallel optimization, the Optimization Manager. Standard C++ I/O functions are overloaded by P++ parallel I/O-Manager functions. The P++ user interface provides for switching between M++ in the serial environment to P++ in the serial or parallel environment. The Optimization Manager allows for overriding defaults for user control of partitioning, communication, array to processor mappings, communication models of VSG, parallel I/O system, scheduling, etc. Optimizations of this kind only have significance in a parallel environment. The Optimization Manager is the only means by which the user can affect the parallel behavior of the code. It provides a consistent means of tailoring the parallel execution and performance.

The code segment in Figure 1 shows the combined use of array and index objects in the user interface for a MacCormack scheme to solve Burger's equation (here F, U, and U_old are arrays; I is an Index object containing the position, count, and stride of the interior of the problem domain; and Time_Step is a scalar). This serial code executes on both serial (using M++ or P++) and parallel machines (using P++), with no changes required.

Programming Model and Parallelism: P++ is based on a Single Program Multiple Data Stream (SPMD) programming model, which consists of executing one single program source on all nodes of the parallel system. Its combination with the Virtual Shared Grids (VSG) model of data parallelism (a restriction of virtual shared memory to structured grids, whose communication is controlled at runtime) is essential for the simplified representation of the parallel program using the serial program and hiding communication within the grid block classes. Besides data parallelism, control flow oriented functional parallelism is essential for obtaining high performance. To date only natural functional parallelism, which results from the execution of operations and expressions with different distributed grid variables on different processors, is available within P++. Two different principles for providing very efficient additional functional parallelism through asynchronous scheduling are currently under investigation: Parallel blocks or a *for all*-like functionality at compile time (e.g., CC++ ([2]); and deferred expression evaluation (across multiple expressions) with runtime dependency analysis.

Data Partitioning: Currently default partitioning of the grid data within P++ consists of even partitioning of each grid variable across all nodes, based on 1D partitioning of the last array dimension (grid block 1 in Figure 3 (b), [2]). In addition, we provide associated partitioning for consistently aligned partitionings of grid variables (same size, coarser, or finer) with others (e.g. for multigrid algorithms); user defined partitioning, which is based on a user filled mapping structure; and application based partitioning, which allows for the introduction of user specified load balancing algorithms to handle the partitioning of one or more specified grid variables (Figure 3, [2]), are available as user supported choices that can override the default. In combination with VSG, this allows partitioning to be a parameter of optimization. The restriction to 1D partitionings has been chosen for ease of implementations. It is straight forward to extend the code to multidimensional partitionings as they are advantageous on massively parallel architectures, e.g. in multigrid context.

Data Parallelism Based on Virtual Shared Grids (VSG): The VSG concept gives the appearance of virtual shared memory restricted to array/grid variables. The M++ array syntax provides the means for the user to express the algorithm's data and operation structure to the compiler and runtime system. Computations are based on global indexing. Communication patterns are derived at runtime, and the appropriate send and receive messages of grid portions are automatically generated by P++ selecting the most efficient communication models for each operation: overlap update for computations on closely aligned grid portions or VSG update for general operations between non aligned grids. The ability of P++ to optimally schedule communication, for each array expression, guarantees that the same minimum amount of data is communicated in the same minimum number of messages as in codes based on the traditional explicit message passing programming model. This is a great advantage over the more general virtual shared memory model, where the operating system triggers communication without having information about the algorithm's use of data. In this case, either full memory pages are communicated for each data item of basic type (float, int, ...), in the hope that the succeeding non local data item is located on the same page as the preceding item, or expensive element by element communication is done.

- *VSG Update:* In the implementation of the general VSG concept, within the VSG update communication model, we restrict the classical Owner Computes rule to binary subexpressions and define the owner arbitrarily to be the left operand. This simple rule handles the communication required in the parallel environment; specifically, the designated owner of the left operand receives all parts of the distributed array necessary to perform the given binary operation. Thus the temporary result and the left operand are partitioned similarly.

- *Overlap Update:* As an optimization for a broad range of iterative methods we have implemented nearest neighbor access to array elements through the widely used technique of grid partitioning with overlap (currently restricted to width one, see [2]). In this way, the most common type of nearest neighbor grid accesses can be handled by complicated expressions, and the communication can be limited to one overlap update in the "=" operator (overloaded) for grids. To minimize communication, the necessity of updating the overlapping boundaries is detected at runtime, based on how the overlap has been modified.

Grids are constructed in a distributed fashion across the processors of the parallel system. All information required for evaluating expressions is available from the array syntax and a partitioning table, which stores partitioning information and processor mappings. The number of entries in the table is greatly reduced by grouping associated grids and providing for storage of only the required data on a processor by processor basis. This is necessary due to the large sizes that these tables can be on massively parallel systems. All necessary global information is locally available on each processor with no communication required, again a significant advantage of P++ compared to other developments.

In addition to the above two basic communication models, whose combination allows for the optimization of the communication structure of application codes, the principle of deferred expression evaluation (lazy evaluation), which has not been implemented in the current prototype, can contribute to a greatly enhanced single node performance: This evaluation rule allows for significant expression optimization w.r.t. overlapping functional hardware units by aggregate operators and, most important, intermediate array storage minimization (e.g. through chaining and optimal vector register use) or cache utilization.

Summary of P++ Features: The major features of the P++ array class library as a programming environment for structured grid applications are:

- Object oriented indexing of the array objects simplifies development of serial codes by removing error prone explicit indexing common to *for* or *do* loops;

- Algorithm and code development takes place in a serial environment;

- Serial codes are re-compilable to run in parallel without modification;

- P++ codes are portable between different architectures;

- Vectorization, parallelization and data partitioning are hidden from the user, except for optimization switches;

- P++ application codes exhibit communication as efficiently as codes with explicit message passing;

- With improved C++ compilers and an optimized implementation of M++, single node performance of C++ with array classes has the potential to approximate that of Fortran.

Current State, Performance Issues and Related Work: The P++ prototype is currently implemented on the bases of the AT&T C++ C-Front precompiler using the Intel NX-2 communication library (or, on an experimental basis, an EXPRESS-like portable communication library from Caltech). Current versions are running on the Intel iPSC/860[3] Hypercube, the Intel Simulator, SUN workstations, the Cray 2 and IBM PCs. The prototype contains all major concepts described above. At several points, without loss of generality, its functionality is restricted to the needs within our own set of test problems (3D multigrid codes, FAC and AFAC codes implemented on the basis of AMR++).

The feasibility of the approach has been proven by the successful implementation and test of our set of test problems on the basis of P++, in particular the very complex AMR++ class library. The results that have been obtained with respect to parallel efficiency, which to optimize was one of the major goals of the P++ development, are also very satisfying: Comparisons for P++ and Fortran with message passing based test codes, respectively, have shown that the number of messages and the amount of communicated data is roughly the same. Thus, besides a negligible overhead, similar parallel efficiency can be achieved. With respect to single node performance, only little optimization has been done. The major reason is that the used system software components (AT&T C++ C-Front precompiler 2.1, M++) are not very well optimized for the target machines. However, our experiences with C++ array language class libraries on workstations and on the Cray Y-MP (in collaboration with Sandia National Laboratories: about 90% of the Fortran vector performance is achieved) are very promising: With new optimized system software versions, Fortran performance can be approximated. Therefore, altogether, we expect the parallel performance for P++ based codes to be similar to that obtained for optimized Fortran codes with explicit message passing.

To our knowledge the P++ approach is unique. There are several other developments that contain distinct components of the P++ concepts (e.g. High Performance Fortran (HPF), the CC++ declarative concurrent object oriented programming notation developed at California Institute of Technology, the PARAGON programming environment

---

[3]Thanks to the people at the Federal German Research Center Jülich (KFA) for their generous support in letting us use their iPSC/860 environment.

developed at Cornell University, the Parallel Runtime Tools developed at NASA/ICASE or several virtual shared memory implementations). However, their major target application classes or their primary goals are different (for more details see [2]).

P++ is used within the AMR++ class libraries (detailed below) to abstract the parallel issues from the local refinement issues that are abstracted into AMR++. Thus the AMR++ class libraries represent a P++ application that has been fully developed in a serial environment, but which also runs in a parallel environment. The nested set of abstractions provided by AMR++ uses P++ at its lowest level to provide architecture independent support.

AMR++ — AN ADAPTIVE MESH REFINEMENT CLASS LIBRARY

AMR++ is a C++ class library which simplifies the details of building self adaptive mesh refinement applications. The use of this class library significantly simplifies the construction of local refinement codes for both serial and parallel architectures. AMR++ has been developed in a serial environment using C++ and the M++ array class interface. It runs in a parallel environment, because M++ and P++ share the same array interface. Therefore AMR++ inherits the machine targets of P++, and, thus, has a broad base of machines on which to run. The efficiency and performance of AMR++ is mostly dependent on the efficiency of M++ and P++, in the serial and parallel environments respectively. In this way, the P++ and AMR++ class libraries separate the abstractions of local refinement and parallelism to significantly ease the development of parallel adaptive mesh refinement applications in an architecture independent manner.

The AMR++ class library represents work which combines complex numerical, computer science and engineering application requirements. Therefore, the work naturally involves compromises in its initial development. In the following sections the features and current restrictions of the AMR++ class library are summarized.

Block Structured Grids — Features and Restrictions: The target grid types of AMR++ are 2D and 3D block structured grids with rectangular or logically rectangular grid blocks. On the one hand, they allow for a very good representation of complex internal geometries which are introduced through local refinement in regions with increased local activity. This flexibility of local refinement block structured grids equally applies to global block structured grids that allow to match complex external geometries. On the other hand, the restriction to structures of rectangular blocks, as opposed to fully unstructured grids, allows for the application of the VSG programming model of P++ and therefore is the foundation for good efficiency and performance in distributed environments, which is one of the major goals of the P++/AMR++ development. Thus, we believe that block structured grids are the best compromise between full generality of the grid structure and efficiency in a distributed parallel environment. The application class forms a broad cross section of important scientific applications.

In the following, the global grid is the finest uniformly discretized grid, that covers the whole physical domain. Local refinement grids (level $i+1$) are formed from the global grid (level $i = 0$) or recursively from refinement grids (discretization level i) by standard refinement with $h_{i+1} = \frac{1}{2}h_i$ in each coordinate direction. Thus, boundary lines of block structured refinement grids always match grid lines on the underlying discretization level.

The construction of block structured grids in AMR++ has some practical limitations that simplify the design and use of the class libraries. Specifically, grid blocks at the same level of discretization cannot overlap. Block structures are formed by distinct or connected rectangular blocks that share their boundary points (block interfaces) at those places where they adjoin each other. Thus, a connected region of blocks forms a block structured refinement grid. It is possible that one refinement level consists of more than one disjunct block structured refinement grid. In the dynamic adaptive refinement procedure, refinement grids can be automatically merged, if they adjoin each other.

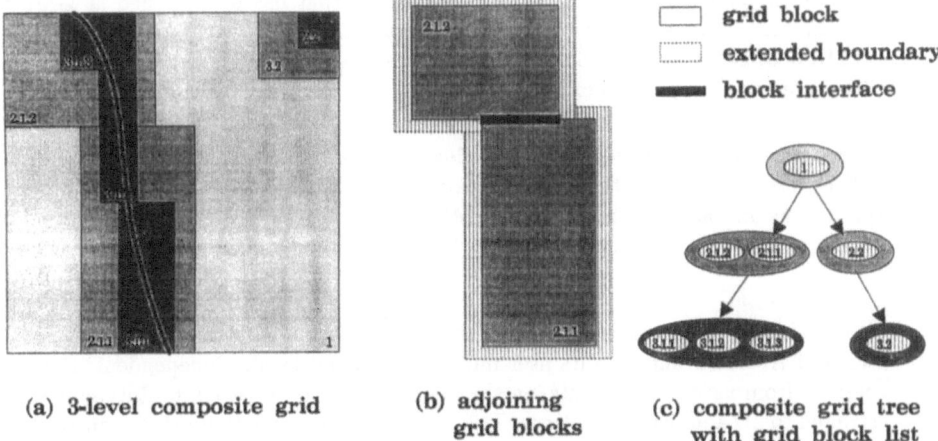

(a) 3-level composite grid

(b) adjoining grid blocks

(c) composite grid tree with grid block list

grid block
extended boundary
block interface

Figure 2: Example of a composite grid, its composite grid tree and a cut out of 2 blocks with their extended boundaries and their interface.

In Figure 2 (a) an example for a composite grid is illustrated: The composite grid shows a rectangular domain within which we center a curved front and a corner singularity. The gird blocks are ordered lexicographically; the first digit represents the level, the second digit the connected block structured refinement grid, and the third digit the grid block. Such problems could represent the structure of shock fronts or multi fluid interfaces in fluid flow applications: In oil reservoir simulations, for example, the front could be an oil water front moving with time and the corner singularity could be a production well. In this specific example the front is refined with two block structured refinement grids; the first grid on refinement level 2 is represented by grid blocks 2.1.1 and 2.1.2, the second grid on level 2 by grid blocks 3.1.1, 3.1.2 and 3.1.3. In the corner on each of the levels a single refinement block is introduced.

For ease of implementation, in the AMR++ prototype the global grid has to be uniform. This simplification of the global geometry was necessary in order to be able to concentrate on the major issues of this work, namely to implement local refinement and self adaptivity in an object-oriented environment. The restriction is no general constraint and can be more or less easily raised in a future version of the prototype. Aside from implementation issues, some additional functionality has to be made available:

- For implicit solvers the resulting domain decomposition of the global grid may require special capabilities within the single grid solvers (e.g. multigrid solvers for block structured grids with adequate smoothers, such as inter block line or plane relaxation methods).

- The block structures in the current AMR++ prototype are defined only by the needs of local refinement of a uniform global grid. This restriction allows them to be cartesian. More complicated structures as they result from difficult non cartesian external geometries (e.g. holes or spliss points, see [2]) currently are not taken into consideration. An extension of AMR++ however, is principally possible. The wide experience for general 2D block structured grids, that has been gained at GMD [2], can form a basis for these extensions. Whereas our work is comparably simple in 2D, because no explicit communication is required, extending the GMD work to 3D problems is very complex.

Some Implementation Issues: In the following some implementation issues are detailed. They also demonstrate the complexity of a proper and efficient treatment of block structured grids and adaptive refinement. AMR++ takes care of all these issues, which would have to be handled explicitly, if everything had to be handled at the application level.

- *Dimensional independence and multi indexing:* The implementation of most features of AMR++ and also its user interface is dimensionally independent. Being derived from user requirements, on the lowest level, the AMR++ prototype is restricted to 2D and 3D applications. This, however, is a restriction, which can easily be removed.

  One important means by which dimensional independence is reached, are multi-dimensional indices (multi indices), which contain one index for each coordinate direction. On top of these multi indices there are index variants defined for each type of sub block (interior, interior and boundary, boundary only, ...), which contain multiple multi indices. For example, for addressing the boundary of a 3D block (non convex), one multi index is needed for each of the six planes. In order to avoid special treatment of physical boundaries, all index variants are defined twice, including and excluding the physical boundary, respectively. All index variants, several of them also including extended boundaries (see below), are precomputed at the time when a grid block is allocated. In the AMR++ user interface and in the top level classes, only index variants or indicators are used and therefore allow a dimensionally independent formulation, except for the very low level implementation.

- *Implementation of block structured grids:* The AMR++ grid block objects consist of the interior, the boundary and an extended boundary of a grid block and, in addition, links that are formed between adjacent pairs of grid block objects. The links contain P++ array objects that do not consist of actual data but serve as views (subarrays) of the overlapping parts of the extended boundary between adjacent grid block objects. The actual boundaries that are shared between different blocks (block interfaces) are very complex structures, that are represented properly in the grid block objects. For example in 3D, interfaces between blocks are 2D planes. Those between plane-interfaces are 1D-line interfaces, and, one step further, those between line-interfaces are points (zero-dimensional).

  In Figure 2 (b), grid blocks 2.1.1 and 2.1.2 of the composite grid in Figure 2 (a) are depicted including their block interface and their extended boundary. The regular lines denote the outermost line of grid points of each block. Thus, with an extended boundary of two, there is one line of points between the block boundary line and the dashed line for the extended boundary. In its extended boundary, each grid block has views of the values of the original grid points of its adjoining neighboring block. This way it is possible to evaluate stencils on the interface and, with an extended boundary width of two, to also define a coarse level of the block structured refinement grid in multigrid sense.

- *Data structures and iterators:* In AMR++, the composite grid is stored as a tree of all refinement grids, with the global grid being the root. Block structured grids are stored as lists of blocks (for ease of implementation; collections of blocks would be sufficient in most cases). In Figure 2 (c), the composite grid tree for the example composite grid in Figure 2 (a) is illustrated.

  The user interface for doing operations on these data structures are so-called iterators. For example, for an operation on the composite grid (e.g. zeroing each level or interpolating a grid function to a finer level) an iterator is called, that traverses the tree in the correct order (preorder, postorder, no order). This iterator takes the function to be executed and two indicators that specify the physical boundary treatment and the type of sub grid to be treated as an argument. The iteration starts at the root and recursively traverses the tree. For doing an operation (e.g. Jacobi relaxation) on a block structured grid, iterators are available, that process the list of blocks and all block interface lists. They take similar arguments as the composite grid tree iterators.

Object-Oriented Design and User Interface: The AMR++ class libraries are customizable by using the object oriented features of C++. For example, in order to obtain efficiency in the parallel environment, it may be necessary to introduce alternate iterators that traverse the composite grid tree or the blocks of a refinement region in a special order. This is implemented by alternate usage of different base classes in the serial and parallel environment. The same is true for alternate composite grid cycling strategies as for example needed in AFAC on the contrary to FAC algorithms (Section 4).

Application specific parts of AMR++, such as the single grid solvers or criteria for adaptivity, which have to be supplied by the user, are also simply specified through substitution of alternate base classes: A preexisting application (e.g. problem setup and uniform grid solver) uses AMR++ to extend its functionality and to build an adaptive mesh refinement application. Thus, the user supplies a solver class and some additional required functionality (refinement criteria, ...) and uses the functionality of the highest level AMR++ ((Self_)Adaptive_)Composite_Grid class to formulate his special algorithm or to use one of the supplied PDE solvers. In the current prototype of AMR++, FAC and AFAC based solvers (Section 4) are supplied. If the single grid application is written using P++, then the resulting adaptive mesh refinement application is architecture independent, and so can be run efficiently in a parallel environment.

The design of AMR++ is object-oriented and the implementation of our prototype extensively uses features like encapsulation and inheritance: The abstraction of self adaptive local refinement, which involves the handling of many issues including memory management, interface for application specific control, dynamic adaptivity and efficiency, is reached through grouping these different functionalities in several interconnected classes.

For example, memory management is greatly simplified by the object oriented organization of the AMR++ library: Issues such as lifetime of variables are handled automatically by the scoping rules for C++, thus memory management is automatic and predictable.

As the AMR++ interface is object oriented, the control over construction of the composite grid is intuitive and natural: The creation of composite grid objects is similar to the declaration of floating point or integer variables in procedural languages like Fortran and C. The user basically formulates his solver by allocating one of the predefined composite grid solver objects or by formulating his own solver on the basis of the composite grid objects and the associate iterators and by additionally supplying the single grid solver class.

Although not a part of the current implementation of AMR++, C++ introduces a template mechanism in the latest standardization of the language, which only now starts to be part of commercial products. The general purpose of this template language feature is to permit class libraries to use user specified base types. For AMR++ for example, the template feature could be used to allow the specification of the base solver and adaptive criteria for the parallel adaptive local refinement implementation. In this way, the construction of adaptive local refinement code from the single grid application on the basis of the AMR++ class library can become even simpler and cleaner.

In this paper the object-oriented design of interconnected classes is not further detailed. The reader is referred to [2] and [2].

Static and Dynamic Adaptivity, Grid Generation: In the current AMR++ prototype, static adaptivity is fully implemented. The user can specify his composite grid either interactively or by some input file: For each grid block AMR++ needs its global coordinates and the parent grid block. Block structured local refinement regions are formed automatically by investigating neighboring relationships. In addition, the functionalities for adding and deleting grid blocks under user control are available within the Adaptive_Composite_Grid object of AMR++.

Recently, dynamic adaptivity has been a subject of intensive research. First results are very promising and some basic functionality has been included in the AMR++ prototype: Given a global grid, a flagging criteria function and some stopping criteria, the Self_Adaptive_Composite_Grid object contains the functionality for iteratively solving on the actual composite grid and generating a new discretization level on top of the respective finest level. Building a new composite grid level works as follows:

1. The flagging criteria delivers an unstructured collection of flagged points in each grid block.

2. For representing grid block boundaries, all neighboring points of flagged points are also flagged.

3. The new set of grid blocks to contribute to the refinement level (gridding) is built by applying a smart recursive bisection algorithm similar to the one developed in [2]: If building a rectangle around all flagged points of the given grid block is too inefficient, it is bisected in the longer coordinate direction and new enclosing rectangles are computed. The efficiency of the respective fraction is measured by the ratio of flagged points to all points of the new grid block. In the subsequent tests 75% is used. This procedure is repeated recursively if any of the new rectangles is also inefficient. Having the goal of building the rectangles as large as possible within the given efficiency constraint, the choice of the bisection point (splitting in

halves is too inefficient because it results in very many small rectangles) is done by a combination of signatures and edge detection. As a detailed description of this method reaches beyond the scope of this paper, the reader is referred to [2] or [2].

4. Finally, the new grid blocks are added to the composite grid to form the new refinement level. Grouping these blocks into connected block structured grids is done the same way as it is done in the static case.

This flagging and gridding algorithm has the potential for further optimization: The bisection method can be further improved and also a clustering and merging algorithm could be applied. This is especially true for refinement blocks of different parent blocks that could form one single block with more than one parent. Internal to AMR++, this kind of parent / child relationship is supported. The results in Section 5 however show, that the gridding already is quite good. The number of blocks that are constructed automatically is only slightly larger ($< 10\%$) than a manual construction would deliver.

A next step in self adaptive refinement would be to support time dependent problems whose composite grid structure changes dynamically with time (e.g. moving fronts). In this case, in addition to adding and deleting blocks, e.g. enlarging and diminishing blocks has to be supported. Though some basic functionality and the implementation of the general concept is already available, this problem has not yet been further tackled.

Current State and Related Work: The AMR++ prototype is implemented using M++ and the AT&T Standard components class library to provide standardized classes (e.g. linked list classes). Through the shared interface of M++ and P++, AMR++ inherits all target architectures of P++. AMR++ has been successfully tested on SUN workstations and on the Intel iPSC/860.

Taking into account the large functionality of AMR++, there are still several insufficiencies and restrictions and also a large potential for optimization in the current prototype (as already pointed out in the preceding description). Until now, AMR++ has been successfully used as a research tool for the algorithms and model problems described in the next two sections. However, AMR++ provides the functionality to implement much more complicated application problems.

Concerning parallelization, running AMR++ under P++ on the Intel iPSC/860 has proven its full functionality. Intensive optimization however, has only been done within P++. AMR++ itself offers a large potential for optimization. For example, efficiently implementing self adaptivity including load (re)balancing in a parallel environment requires further research. In addition the iterators that are currently available in AMR++, though working in a parallel environment, are best suited for serial environments. Special parallel iterators that for example support functional parallelism on the internal AMR++ level would have to be provided.

To the authors' knowledge, the AMR++ approach is unique. There are several other developments in this area (e.g. [2]), but they either address a more restricted class of problems or they are still restricted to serial environments.

MULTILEVEL ALGORITHMS WITH ADAPTIVE MESH REFINEMENT

The fast adaptive composite grid method (FAC, [2], [2]), which was originally developed from and is very similar to the Multi-Level Adaptive Technique (MLAT, [2]), is an algorithm that uses uniform grids, both global and local, to solve partial differential equations. This method is known to be highly efficient on scalar or single processor vector computers, due to its effective use of uniform grids and multiple levels of resolution of the solution. On distributed memory multiprocessors, methods like MLAT or FAC benefit from their tendency to create multiple isolated refinement regions, which may be effectively treated in parallel. However, for several problem classes they suffer from the way in which the levels of refinement are treated sequentially in each region. Specifically, the finer levels must wait to be processed until the coarse-level approximations have been computed and passed to them; conversely, the coarser levels must wait until the finer level approximations have been computed and used to correct their equations. Thus, the parallelization potential of these "hierarchical" methods is restricted to intra-level parallelization.

The asynchronous fast adaptive composite method (AFAC) eliminates this bottleneck of parallelism. Through a simple mechanism used to reduce inter-level dependencies, individual refinement levels can be processed by AFAC in parallel. The result is that the convergence rate for AFAC is the square root of the one for FAC. Therefore, since both AFAC and FAC have roughly the same number of floating point operations, AFAC requires twice the serial computational time as FAC, but AFAC allows for the introduction of inter-level parallelization.

As opposed to the original development of FAC and AFAC, in this paper modified algorithms, which are also known as FACx and AFACx are discussed and used. They differ in the treatment of the refinement levels. Whereas in FAC and AFAC, a rather accurate solution is computed (e.g. one MG V-cycle), in FACx only a couple of relaxations are applied (smoothing). In AFACx, a two grid procedure (of FMG-type) on the refinement level and its standard coarsening with several relaxations on each of these levels is used. Experiments and some theoretical observations show, that all of the results that have been obtained for FAC and AFAC also hold for FACx and AFACx (see [2] and [2]). In the following, FAC and AFAC always denote the modified versions (FACx and AFACx).

Numerical algorithms: Both FAC (MLAT) and AFAC consist of two basic steps which are described loosely as follows:

1. *Smoothing phase:* Given the solution approximation and composite grid residuals on each level, use relaxation or some restricted multigrid procedure to compute a correction local to that level (a better approximation is required on the global grid, the finest uniform discretization level).

2. *Level transition phase:* Combine the local corrections with the global solution approximation, compute the global composite grid residual, and transfer the local components of the approximation and residual to each level.

The difference between MLAT and FAC on the one hand and AFAC on the other hand is in the order in which the levels are processed and in the details of how they are combined:

- *MLAT and FAC:* FAC and MLAT can roughly be viewed as standard multigrid methods with mesh refinement and a special treatment of the interfaces between the refinement levels and the underlying coarse level. In FAC and MLAT the treatment of the refinement levels is hierarchical. Theory on FAC is based on its interpretation as a multiplicative Schwarz Alternating Method or as a block relaxation method of Gauss-Seidel type.

FAC and MLAT mainly differ by their motivation. Whereas it is the goal of FAC to compute a solution for the composite grid (grid points of the composite grid are all the interior points of the respective finest discretization level), the major goal of MLAT is to get the best possible solution on a given uniform grid (with using local refinement). Thus, in FAC coarse levels of the composite grid serve for the computation of corrections. Therefore FAC was originally formulated as a correction scheme (CS). The MLAT formulation requires a full approximation scheme (FAS), because coarse levels serve as correction levels in those points, that are covered by a finer level and as the approximation grid for the solution in those points that have discretized finer. MLAT was first developed using finite difference discretization, whereas for FAC finite volume discretizations were used. However, they are closely related and in many problems lead to the same stencil representation. This is true except for the interface points, where finite volume discretizations generally lead to conservative discretizations, whereas finite difference discretizations do not. Because of the modification of the original FAC algorithm as discussed above, there is no difference in the treatment of the refinement levels between the original MLAT algorithm and the modified FAC algorithm, that are discussed in this paper. It can be shown ([2]) that a FAS version of FAC with a special choice of the operators on the interface is equivalent to the originally developed Multilevel Adaptive Technique (MLAT).

- *AFAC:* AFAC on the other hand consists of the same discretization and operators as FAC, but a decoupled and asynchronous treatment of the refinement levels in the solution phase, which dominates the arithmetic work in the algorithm. Theory on AFAC can be based on its interpretation as an additive Schwarz Alternating Method or as a block relaxation method of Jacobi type.

Convergence theory in [2] shows that, under appropriate conditions, the convergence rates of FAC and AFAC have the relation $\rho_{AFAC} = \sqrt{\rho_{FAC}}$. This implies that two cycles of AFAC are roughly equivalent to one cycle of FAC. If the algorithmic components are chosen slightly different than for the convergence analysis or if applied to singular perturbation problems as discussed in the next section, experiences show that AFAC is usually better as it is represented in the above formula: In several cases, the convergence rate of AFAC shows only a slight degradation of the FAC rate (Section 5).

Parallelization — an Example for the Use of P++/AMR++: By example we demonstrate some of the features of AMR++ and examples for the support of P++ for the design of parallel block structured local refinement applications on the basis of FAC and AFAC algorithms.

In a parallel environment, partitioning the composite grid levels becomes a central issue in the performance of composite grid solvers. In Figure 3, two different partitioning strategies that are supported within P++/AMR++ are illustrated for the composite grid in Figure 2. For ease of illustration, grid blocks 2.2 and 2.3 are not included. The so-called FAC partitioning in Figure 3 (b) is typical for implicit and explicit algorithms, where the local refinement levels have to be treated in a hierarchical manner (FAC, MLAT,...). The so-called AFAC partitioning in Figure 3 (c) can be optimal for implicit

213

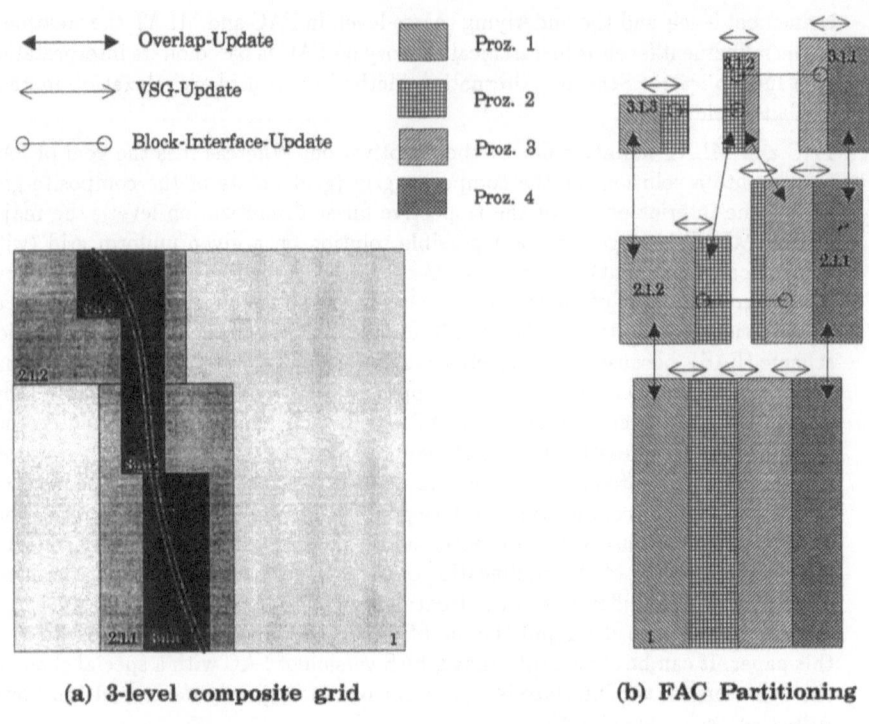

◄───────► Overlap-Update	▨ Proz. 1
◄───────► VSG-Update	▨ Proz. 2
⊙───────⊙ Block-Interface-Update	▨ Proz. 3
	▨ Proz. 4

(a) 3-level composite grid   (b) FAC Partitioning

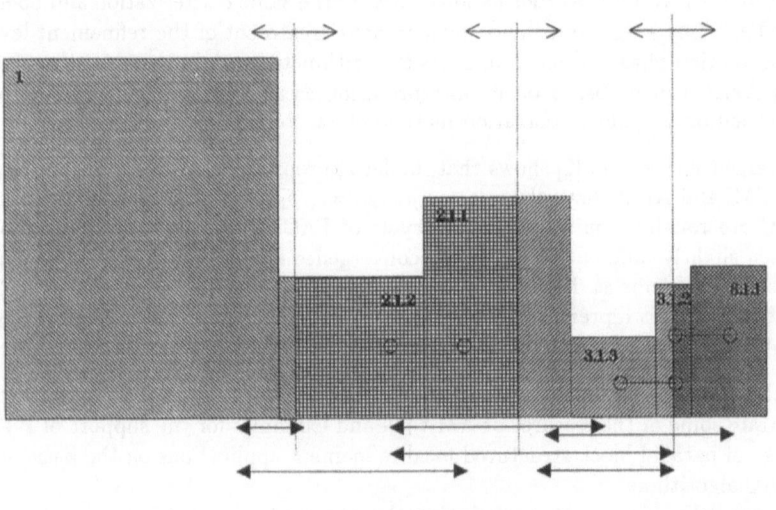

(c) AFAC Partitioning

Figure 3: Parallel multilevel local refinement algorithms on block structured grids — an example for AMR++ use and hidden interaction of the P++ communication models.

algorithms, that allow an independent and asynchronous treatment of the refinement levels. In case of AFAC however, it has to be taken into consideration that this partitioning is only optimal for the solution phase, which dominates the arithmetic work of the algorithm. The efficiency of the level transition phase, which is based on the same hierarchical structure as FAC and which can eventually dominate the aggregate communication work of the algorithm, highly depends on the architecture and the application (communication / computation ratio, single node (vector) performance, message latency, transfer rate, congestion, ...). For determining whether AFAC is better than FAC in a parallel environment, the aggregate efficiency and performance of both phases and the relation of the convergence rates have to be properly evaluated. For more details see [2] and [2]. Both types of partitioning are supported in the P++/AMR++ environment.

Solvers used on the individually partitioned composite grid levels make use of overlap updates within P++ array expressions which automatically provide communication, as needed. The inter grid transfers between local refinement levels, typically located on different processors, rely on VSG updates. The VSG updates are also provided automatically by the P++ environment. Thus the underlying support of parallelism is isolated in P++ through either overlap update or VSG update or a combination of both, and the details of parallelism are isolated away from the AMR++ application. The block structured interface update is handled in AMR++. However communication is hidden in P++ (mostly VSG update).

RESULTS FOR SINGULAR PERTURBATION PROBLEMS

The use of the tools described above is now demonstrated with first examples. The adaptivity provided by AMR++ is necessary in case of large gradients or singularities in the solution of the PDE. They may be due to rapid changes in the right hand side or the coefficients of the PDE, corners in the domain, or singular perturbations. Here, the first and the last case will is examined on the basis of model problems.

Singularly perturbed PDEs represent the modelling of physical processes with relatively small diffusion (viscosity) and dominating convection. They may occur as a single equation or within systems of complex equations, e.g. as the momentum equations within the Navier-Stokes or, in addition, as supplementary transport equations in the Boussinesq system of equations. Here we merely treat a single equation. However, we only use methods which generalize directly to more complex situations. Therefore we do not rely on the direct solution methods provided by downstream or ILU relaxations for simple problems with pure upstream discretization. The latter are no direct solution methods for systems of equations, cf. [2]. Further on, these types of flow direction dependent relaxations are not efficiently parallelizable in the case of only few relaxations as usually used in multilevel methods. This in particular holds on on massively parallel systems.

Model Problem and Solvers: The numerical results have been obtained for the model problem

$$-\varepsilon\Delta u + au_x + bu_y = f \qquad \text{on } \Omega = (0,1)^2$$

with Dirichlet boundary conditions on $\partial\Omega$ and $\varepsilon = 0.00001$. This problem serves as a good model problem for complex fluid flow applications, because several of the properties that are related to self adaptive mesh refinement are already present in this simple problem.

The equation is discretized using isotropic artificial viscosity (diffusion):

$$L_h := -\varepsilon_h \Delta_h + a D_{2h,x} u + b D_{2h,y} u \text{ with } \Delta_h = D_{h,x}^2 + D_{h,y}^2$$

$$\varepsilon_h := \max\{\varepsilon, \beta h \max\{|a|, |b|\}/2\}$$

For comparison, some results with a nine point stencil discretization corresponding to bilinear finite elements for the Laplace operator have also been obtained.

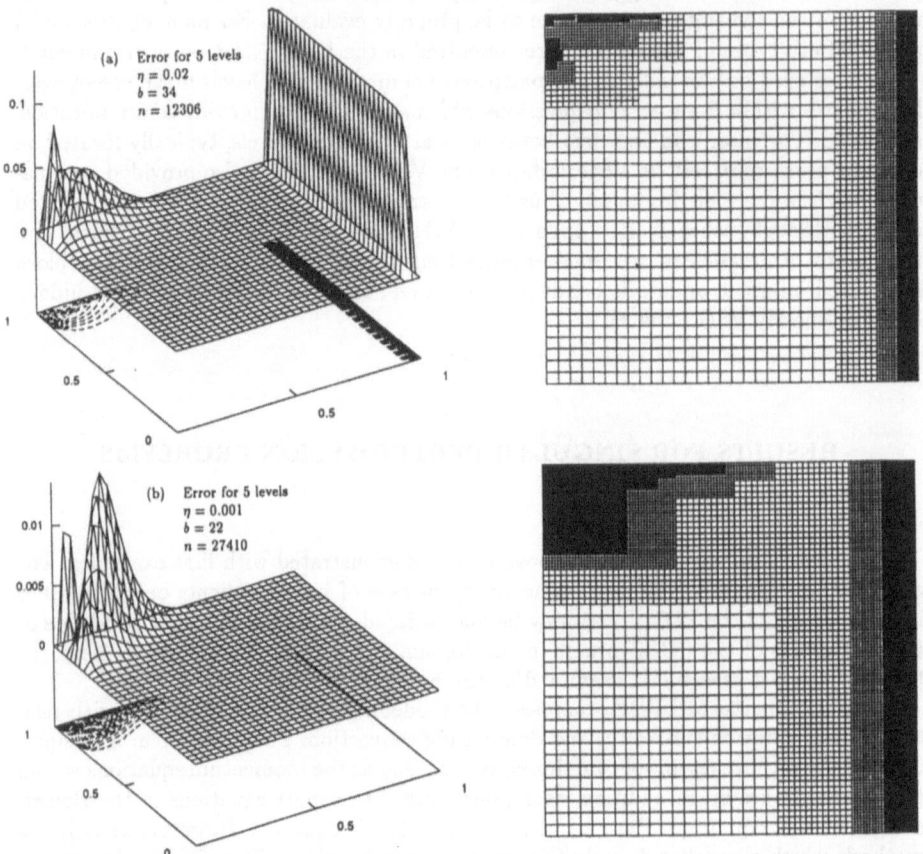

Figure 4: Results for a singular perturbation problem: Plots of the error and the composite grid with two different choices of the accuracy η in the self adaptive refinement process.

The discrete system is solved by multilevel methods: MG on the finest global grid and FAC or AFAC on composite grids with refinement. For the multigrid method it is known that, with artificial viscosity, the two-grid convergence rate (spectral radius of the corresponding iteration matrix) is bounded below by 0.5 (for $h \to 0$). Therefore, multilevel convergence rates tend to 1.0 with an increasing number of levels (see e.g. examples in [2]). In [2] a multigrid variant which shows surprisingly good convergence behavior has been developed: MG convergence rates stay far below 0.5 (with three relaxations on each level). Here, essentially this method is used, which is described as follows:

216

- Discretization with isotropic artificial viscosity using $\beta = 3$ on the finest grid m and $\beta_{l-1} = 1/2\,(\beta_l + 1/\beta_l)$ for coarser grids $l = m - 1, m - 2, \ldots$,

- Standard MG components (odd even relaxation, full weighting and bilinear interpolation).

Anisotropic artificial viscosity may also be used, but generally requires (parallel) zebra line relaxation, which has not yet been fully implemented.

For FAC and AFAC the above MG method with V(2,1) cycling is used as a global grid solver. On the refinement levels three relaxations are performed.

Convergence Results: In Table 1 several convergence rates for FAC, AFAC and, for comparison, for MG are shown. The finest grids have mesh sizes of $h = 1/64$ or $h = 1/512$, respectively. For FAC and AFAC the global grid has the mesh size $h = 1/32$, the (predetermined) fine block always covers 1/4 of the parent coarse block along the boundary layer. The following conclusions can be drawn:

Table 1: Convergence rates for a singular perturbation problem and, for comparison, for Poisson's equation.

Convergence Rate ρ	h	5-point stencil		9-point stencil	
		1/64	1/512	1/64	1/512
Poisson	**MG-V**	0.08	0.08	0.02	0.03
$a = b = 0$	**FAC**	0.33	0.33	0.10	0.10
$\beta_0 = 1, \varepsilon = 1$	**AFAC**	0.41	0.41	0.31	0.32
SPP	**MG-V**	0.14	0.30	0.19	0.48
$a = b = 1$	**FAC**	0.65	0.66	0.60	0.70
$\beta_0 = 3, \varepsilon = 0.00001$	**AFAC**	0.67	0.67	0.60	0.75
SPP	**MG-V**	0.81	1.0	0.20	0.52
$a = b = 1$	**FAC**	0.65	0.66	0.53	0.53
$\beta_0 = 1.1, \varepsilon = 0.00001$	**AFAC**	0.65	0.70	0.55	0.70

- In the case of FAC and AFAC, both for the 5 point and the 9 point Laplacian, the choice of β has to be further investigated.

- For MG and the 5 point stencil the results are as expected, the 9 point stencil gives better but also deteriorating results.

- V cycles are used; W or F cycles would yield better convergence rates but worse parallel efficiency.

- The 9 point stencil for the Laplacian fulfills the Galerkin condition with respect to the level transfers used and shows better convergence rates than the 5 point stencil.

- If $\rho(FAC)$ is small, the expected result $\rho(AFAC) \approx \sqrt{\rho(FAC)}$ can be observed, otherwise $\rho(FAC) \approx \rho(AFAC) \ll \sqrt{\rho(FAC)}$.

Self Adaptive Mesh Refinement Results: More interesting for the goal of this paper are applications of the self adaptive process. As opposed to the convergence rates, they do not only depend on the PDE but also on the particular solution. The results in this paper have been obtained for the the exact solution

$$u(x) = \frac{e^{(x-1)/\varepsilon} - e^{-1/\varepsilon}}{1 - e^{-1/\varepsilon}} + \frac{1}{2} e^{-100(x^2+(y-1)^2)} \, .$$

This solution has a boundary layer for $x = 1, 0 \leq y \leq 1$ and a steep hill around $x = 0, y = 1$ (see Figure 4 (c)). In order to measure the error of the approximate solution, a discrete approximation to the L_1 error norm is used. This is appropriate for this kind of problem: For solutions with discontinuities of the above type one can observe 1st order convergence only with respect to this norm (no convergence in the L_∞ norm, order 0.5 in the L_2 norm).

The result have been obtained with the flagging criteria

$$h^f \left[\beta h \max\{|a|, |b|\} \left(|D_{h,x}^2 u| + |D_{h,y}^2 u| \right) \right] \geq \eta$$

with a given value of η. For $\varepsilon < \varepsilon_h$, the second factor is an approximation to the lowest order error term of the discretization. Based on experiments, $f = 1$ is a good choice. Starting with the global grid, the composite grid is self adaptively built on the basis of the flagging and gridding algorithm described in Section 3.

Table 2: Accuracy (L1-norm e) vs. the number of grid points (n) and the number of blocks (b) for MG-V on a uniform grid and FAC on self adaptively refined composite grids.

	MG-V uniform		FAC $\eta = 0.02$			$\eta = 0.01$			$\eta = 0.001$		
h	e	n	e	n	b	e	n	b	e	n	b
1/32	0.0293	961	0.0293	961	1	0.0293	961	1	0.0293	961	1
1/64	0.0159	3969	0.0160	1806	4	0.0160	1967	4	0.0159	2757	3
1/128	0.0083	16129	0.0089	3430	10	0.0087	3971	10	0.0083	6212	7
1/256	0.0043	65025	0.0056	6378	19	0.0051	7943	16	0.0043	13473	12
1/512	0.0023	261121	0.0073	12306	34	0.0044	15909	30	0.0023	27410	22

In Table 2, the results for MG and FAC are presented for three values of η. In Figure 4, two of the corresponding block structured grids are displayed. The corresponding error plots give an impression of the error distribution restricted from the composite grid to the global uniform grid. Thus, larger errors near the boundary layer are not visible. These results allow the following conclusions:

- In spite of the well known difficulties in error control of convection dominated problems, the grids that are constructed self adaptively are reasonably well suited to the numerical problem.

- As long as the accuracy of the finest level is not reached, the error norm is approximatively proportional to η. As usual in error control by residuals, with the norm of the inverse operator being unknown, the constant factor is not known.

- If the refinement grid does not properly match the local activity, convergence rates significantly degrade and the error norm may even increase.

- Additional tests have shown that, if the boundary layer is fully resolved with an increased number of refinement levels, the discretization order, as expected, changes from one to two.

- The gridding algorithm is able to treat very complicated refinement structures efficiently: The number of blocks that are created is nearly minimal (compared to hand coding).

- Though this example needs relatively large refinement regions, the overall gain by using adaptive grids is more than 3.5 (taking into account the different number of points and the different convergence rates). For pure boundary layer problems, factors larger than 10 have been observed.

- These results have been obtained in a serial environment. AMR++ however, has been successfully tested in parallel. For performance and efficiency considerations see Sections 2 and 3.

REFERENCES

[1]   **Balsara, D.; Lemke, M.; Quinlan, D.:** *AMR++, a parallel adaptive mesh refinement object class library for fluid flow problems*; Symposium on Adaptive, Multilevel and Hierarchical Strategies, ASME Winter Annual Meeting, Anaheim, CA, Nov. 8 - 13, 1992; accepted for publication in the Proceedings.

[2]   **Bell, J; Berger, M.; Saltzman, J., Welcome, M.:** *Three dimensional adaptive mesh refinement for hyperbolic conservation laws*; Internal Report, Los Alamos National Laboratory.

[3]   **Brandt, A.:** *Multi-level adaptive solutions to boundary value problems*; Math. Comp., 31, 1977, pp. 333-390.

[4]   **Chandy, K.M.; Kesselman, C.:** *CC++: A Declarative Concurrent Object Oriented Programming Notation*; California Institut of Technology, Report, Pasadena, 1992.

[5]   **Dörfer, J.:** *Mehrgitterverfahren bei singulären Störungen*; Dissertation, Heinrich-Heine Universität Düsseldorf, 1990.

[6]   **Dowell B.; Govett M.; McCormick, S.; Quinlan, D.:** *Parallel Multilevel Adaptive Methods*; Proceedings of the 11th International Conference on Computational Fluid Dynamics, Williamsburg, Virginia, 1988.

[7]   **Frohn-Schauf:** *Flux-Splitting-Methoden und Mehrgitterverfahren für hyperbolische Systeme mit Beispielen aus der Strömungsmechanik*; Dissertation, Heinrich-Heine Universität Düsseldorf, 1992.

[8] **Hart, L.; McCormick, S.F.**: *Asynchronous multilevel adaptive methods for solving partial differential equations on multiprocessors: Basic ideas*; Parallel Computing 12, 1989, pp 131-144.

[9] **Hempel, R.; Lemke, M.**: *Parallel black box multigrid*; Proceedings of the Fourth Copper Mountain Conference on Multigrid Methods, 1989, SIAM, Philadelphia.

[10] **Lemke, M.**: *Multilevel Verfahren mit selbst-adaptiven Gitterverfeinerungen für Parallelrechner mit verteiltem Speicher;* Dissertation, Heinrich-Heine-Universität Düsseldorf, to appear in 1993.

[11] **Lemke, M.; Quinlan, D.**: *Fast adaptive composite grid methods on distributed parallel architectures*; in Communications in Applied Numerical Methods, Vol. 8, No. 9, Wiley, 1992.

[12] **Lemke, M.; Quinlan, D.**: *P++, a C++ Virtual Shared Grids Based Programming Environment for Architecture-Independent Development of Structured Grid Applications*; CONPAR/VAPP V, September 1992, Lyon, France; Lecture Notes in Computer Science, No. 634, Springer Verlag, September 1992.

[13] **Lonsdale, G; Schüller, A.**: *Multigrid efficiency for complex flow simulations on distributed memory machines*; Parallel Computing 19, 1993, pp23 - 32.

[14] **Lemke, M.; Schüller, A.; Solchenbach, K.; Trottenberg, U.**: *Parallel processing on distributed memory multiprocessors*; Proceedings, GI-20. Annual meeting 1990, Informatik Fachberichte Nr. 257, Springer, 1990.

[15] **McCormick, S.**: *Multilevel Adaptive Methods for Partial Differential Equations*; Society for Industrial and Applied Mathematics, Frontiers in Applied Mathematics, Vol. 6, Philadelphia, 1989.

[16] **McCormick, S.; Thomas, J.**: *The fast adaptive composite grid method for elliptic boundary value problems*; Math. Comp. 46, (1986), pp. 439-456.

[17] **McCormick, S., Quinlan, D.**: *Asynchronous multilevel adaptive methods for solving partial differential equations on multiprocessors: Performance results*; Parallel Computing 12, 1989, pp 145-156.

[18] **McCormick, S., Quinlan, D.**: *Dynamic Grid Refinement for Partial Differential Equations on Parallel Computers*; Proceedings of the Seventh International Conference on Finite Element Methods in Flow Problems, 1989.

[19] **McCormick, S.; Quinlan, D.**: *Idealized analysis of asynchronous multilevel methods*; Symposium on Adaptive, Multilevel and Hierarchical Strategies, ASME Winter Annual Meeting, Anaheim, CA, Nov. 8 - 13, 1992; accepted for publication in the Proceedings.

[20] **Peery, J.; Budge, K.; Robinson, A.; Whitney, D.**: *Using C++ as a scientific programming language*; Report, Sandia National Laboratories, Albuquerque,NM, 1991

[21] **Quinlan, D.**: Dissertation, University of Colorado, Denver, to appear in 1993.

[22] **Ritzdorf, H.**: *Lokal verfeinerte Mehrgitter-Methoden für Gebiete mit einspringende Ecken*; Diplomarbeit, Institut für Angewandte Mathematik der Universität Bonn, 1984.

A posteriori error estimates for the cell-vertex finite volume method

J.A. Mackenzie[1], E. Süli[2] and G. Warnecke[3]

[1]Department of Mathematics, University of Strathclyde, Glasgow, UK
[2]Computing Laboratory, University of Oxford, Oxford, UK
[3]Department of Mathematics, University of Stuttgart, Stuttgart, Germany

Summary

In this paper we develop and analyse a posteriori error estimates for the cell-vertex finite volume method for steady two-dimensional linear advection problems. The error estimates are based on appropriate norms of the finite element residual corresponding to the cell-vertex solution, which has been obtained from a Petrov-Galerkin formulation of the method. Local estimates are obtained by decomposing the error into its locally generated part and a propagated part. Numerical experiments are given which demonstrate the sharpness of the error estimates for model smooth and non-smooth problems.

1 Introduction

One of the major challenges today in the field of computational fluid dynamics is the efficient numerical solution of partial differential equations describing complex multidimensional fluid problems. Such problems invariably have solutions which have localised structures such as shocks and boundary layers. For the accurate and efficient resolution of these local features one needs to use a suitable locally refined computational grid. Such grids are usually termed adaptive grids.

Fundamental to the success of adaptive algorithms is the availability of sharp a posteriori error estimates. The use of such an estimate ensures that the adaptive algorithm is reliable, in the sense that a norm of error is guaranteed to be below a given tolerance when the error estimate is below the tolerance. The use of sharp error estimates ensures that the adaptive algorithm is efficient in that it does not produce a grid which is overly refined.

As a first step towards developing successful adaptive algorithms for the compressible Euler and Navier–Stokes equations we develop here a posteriori error estimates for the cell-vertex finite volume approximation of steady two-dimensional linear advection problems. Due to their inherent conservation properties, finite volume methods are presently the most popular discretisation schemes for hyperbolic conservation laws. Finite volume methods on quadrilateral grids are often interpretated as finite difference methods. However, the cell-vertex finite volume method presented here, has a natural interpretation as a non-conforming Petrov–Galerkin finite element method [3] which allows us to apply error estimation techniques similar to those successfully used with finite element approximations of elliptic problems. In particular we make use of the finite element residual which is easily computable from the approximate solution.

Obtaining localisation results for hyperbolic problems can be difficult and we attempt to solve this problem by splitting the global error into an error which is generated locally in a cell and an error which is simply transmitted through the cell. Error estimates are then derived for both errors.

The rest of this paper is structured as follows: in Section 2 we introduce some notation which will be used throughout the rest of the paper. In Section 3 we state the two-dimensional linear advection model problem and in Section 4 we present its cell-vertex finite volume approximation. The a posteriori error estimates are given in Section 5 and numerical experiments investigating the theory are presented in Section 6. Finally, we make some conclusions in Section 7.

2 Notation

Let Ω denote an open, bounded subset of $\mathbb{R}^2$ with a Lipschitz-continuous boundary Γ. Let $\mathcal{E}(\Omega)$ denote the space of real valued, infinitely differentiable functions on Ω for which derivatives of all orders have continuous extensions on $\overline{\Omega}$ and let $\mathcal{D}(\Omega)$ denote the subspace of $\mathcal{E}(\Omega)$ of functions which have a compact support in Ω.

For each integer $m \geq 0$ and $1 \leq p \leq \infty$, we define the Sobolev space, $W^{m,p}(\Omega)$ as the completion of $\mathcal{E}(\Omega)$ in the norm given by

$$\|u\|_{m,p,\Omega} = \begin{cases} \left\{\sum_{k \leq m} |u|^p_{k,p,\Omega}\right\}^{1/p}, & p \in [1, \infty), \\ \max_{k \leq m} |u|_{k,p,\Omega}, & p = \infty, \end{cases}$$

where $|\cdot|_{k,p,\Omega}$ is the Sobolev semi-norm given by

$$|u|_{k,p,\Omega} = \begin{cases} \left(\sum_{|\alpha|=k} \int_\Omega |D^\alpha u(\mathbf{x})|^p d\mathbf{x}\right)^{1/p}, & p \in [1, \infty), \\ \max_{|\alpha|=k}(\text{ess. sup}_{\mathbf{x} \in \Omega} |D^\alpha u(\mathbf{x})|), & p = \infty \; ; \end{cases}$$

here α is a multi-index and $D^\alpha = \partial_1^{\alpha_1} \partial_2^{\alpha_2}$ denotes a generalised derivative.

The completion of $\mathcal{D}(\Omega)$ in the norm on $W^{m,p}(\Omega)$ is denoted by $W_0^{m,p}(\Omega)$. When $p = 2$, $W^{m,2}(\Omega)$ and $W_0^{m,2}(\Omega)$ will be denoted by $H^m(\Omega)$ and $H_0^m(\Omega)$ respectively. We denote the dual space $(H_0^m(\Omega))' = H^{-m}(\Omega)$. Finally, on the boundary $\partial\Omega$ we define

$$H^{1/2}(\partial\Omega) = \{u \in L_2(\partial\Omega) : \exists v \in H^1(\Omega) : \text{tr}(v) = u, \; \mathbf{x} \in \partial\Omega\},$$

where $\text{tr}(\cdot)$ is the trace operator.

3 The model problem

We begin by introducing the necessary notation. Let Ω denote the unit square $(0, 1) \times (0, 1)$ in $\mathbb{R}^2$. For $\mathbf{a} = (a_1, a_2)$, a two-component real vector-function with continuously differentiable and positive entries defined on $\overline{\Omega}$, we define the following subsets of $\partial\Omega$:

$$\partial_-\Omega = \{\mathbf{x} \in \partial\Omega : \mathbf{a}(\mathbf{x}).\mathbf{n}(\mathbf{x}) < 0\},$$

$$\partial_+\Omega = \{\mathbf{x} \in \partial\Omega : \mathbf{a}(\mathbf{x}).\mathbf{n}(\mathbf{x}) \geq 0\},$$

where $\mathbf{n}(\mathbf{x})$ denotes the unit outward normal to $\partial\Omega$ at $\mathbf{x} \in \Omega$. Suppose that $b \in C(\overline{\Omega})$, $f \in L_2(\Omega)$ and $g \in H^{1/2}(\partial\Omega)$.

We consider the boundary value problem:

$$\nabla.(\mathbf{a}u) + bu = f, \quad \mathbf{x} \in \Omega, \tag{1}$$
$$u = g \quad \mathbf{x} \in \partial_-\Omega. \tag{2}$$

To transform this into its weak formulation, we define the bilinear form $B : H^1(\Omega) \times L_2(\Omega) \to \mathbb{R}$ associated with the hyperbolic operator appearing in (1) by

$$B(u, p) = (\nabla.(\mathbf{a}u) + bu, p),$$

where $(\cdot, \cdot)$ denotes the L_2 inner product. The boundary value problem (1) and (2) can then be reformulated as follows: find $u \in H^1_-(\Omega)$ satisfying

$$B(u, p) = (f, p) \quad \forall p \in L_2(\Omega), \tag{3}$$

where $H^1_-(\Omega)$ consists of all functions in $H^1(\Omega)$ whose trace on the inflow boundary $\partial_-\Omega$ is equal to g.

4 The cell vertex method

The finite volume discretisation of the boundary value problem (1) and (2) is based on the weak formulation (3). First we define $\mathcal{F} = \{T^h\}$, $h > 0$, a family of partitions of Ω, as follows. For a pair of non-negative integers, $M = M(h)$ and $N = N(h)$, we consider the tensor product nonuniform grid

$$\mathcal{G}^h = \{\mathbf{x}_{ij} = (x^1_i, x^2_j) \in \overline{\Omega} \; : \; 0 = x^1_0 < x^1_1 < \cdots < x^1_M = 1,$$
$$0 = x^2_0 < x^2_1 < \cdots < x^2_N = 1\},$$

and define the 'finite volume'

$$K_{ij} = (x^1_{i-1}, x^1_i) \times (x^2_{j-1}, x^2_j), \quad i = 1, \ldots, M, \; j = 1, \ldots, N.$$

The discretisation of (3) is performed on the partition $T^h = \{K_{ij}\}$. With $h_{K_{ij}}$ denoting the diameter of K_{ij}, we shall assume that $h = \max\{h_{K_{ij}} : K_{ij} \in T^h\}$ approximates zero.

In order to introduce the relevant finite element spaces, we define the reference square $\hat{K} = (0, 1)^2$, and denote by $F_{K_{ij}}$ the bilinear function which maps $\hat{K}$ onto K_{ij}. Let $Q_1(\hat{K})$ be the set of bilinear functions on $\hat{K}$ and $Q_0(\hat{K})$ the set of constant functions on $\hat{K}$; we define

$$\mathcal{U}^h = \{v \in H^1(\Omega) : \; v = \hat{v} \circ F^{-1}_{K_{ij}}; \; \hat{v} \in Q_1(\hat{K}), \; K_{ij} \in T^h\}$$

and

$$\mathcal{M}^h = \{p \in L_2(\Omega) : \; p = \hat{p} \circ F^{-1}_{K_{ij}}; \; \hat{p} \in Q_0(\hat{K}), \; K_{ij} \in T^h\},$$

as well as $\mathcal{U}^h_- = \mathcal{U}^h \cap H^1_-(\Omega)$. Let $I^h : (H^1_-(\Omega) \cap C(\overline{\Omega}))^2 \to (\mathcal{U}^h_-)^2$ be the interpolation projector onto $(\mathcal{U}^h_-)^2$. The discrete analogue of the bilinear form B is defined by

$$B^h(v, p) = (\nabla.I^h(\mathbf{a}v) + I^h(bv), p) \quad \forall v \in \mathcal{U}^h_-, \; \forall p \in \mathcal{M}^h.$$

The finite volume approximation of (3) is defined as follows: find $u^h \in \mathcal{U}^h_-$ satisfying

$$B^h(u^h, p) = (f, p) \quad \forall p \in \mathcal{M}^h. \tag{4}$$

5 A posteriori error estimates

A key feature of the a posteriori error estimates which we are about to develop is the role played by the finite element residual which is defined by

$$r = \nabla.(au^h) + bu^h - f.$$

The proofs of all the following theorems will be reported elsewhere. We start with the following result.

Theorem 5.1 *For each cell $K_{ij} \in T^h$ there exists a positive constant $c_1(K_{ij})$ and a positive function $m(\mathbf{x})$ such that for the error $e = u - u^h$*

$$\|me\|_{L_2(K_{ij})}^2 + c_1 \|me\|_{L_2(\partial_+ K_{ij})}^2 \le c_1 \|me\|_{L_2(\partial_- K_{ij})}^2 + c_1^2 \|mr\|_{L_2(K_{ij})}^2. \tag{5}$$

Remark 1. An example of a function satisfying the stability requirements of the theorem is $m(\mathbf{x}) = \exp(-\alpha.\mathbf{x})$ where we choose α such that

$$\alpha.\mathbf{a} + (b + \frac{1}{2}\nabla.\mathbf{a}) \ge \frac{1}{c_1}. \tag{6}$$

Note that m can be defined in an element-by-element fashion.

Remark 2. To determine how sharp the error estimate is, we have to consider the a priori error estimates which are available for the method. In Süli [5] and Morton and Süli [3], it is shown that an appropriate mesh-dependent l_2 norm of the error converges as $O(h^2)$ when the solution is sufficiently smooth, on arbitrarily stretched tensor product non-uniform grids. However, at present, there are no convergence results available in the L_2-norm of the error although numerical evidence in Section 6, suggests that $\|e\|_{L_2(\Omega)}$ converges as $O(h^2)$ for a sufficiently smooth solution. Since $r = \nabla.(ae) + be$, we would expect $\|r\|_{L_2(\Omega)}$ to converge at a rate no better than $O(h)$, since this would the optimal rate of convergence for the streamwise gradient error, obtained for bilinear functions. Therefore it appears that this estimate is non-optimal and that a multiplicative factor of h is needed in front of the L_2 norm of the residual. A similar conclusion is reached by Sonar [4] in his numerical investigations of error indicators for the compressible Euler equations. Interestingly, Eriksson and Johnson [1] have a theoretical justification for the presence of a factor of h for their streamline diffusion method which is, however, only $O(h^{3/2})$ convergent for smooth solutions.

Remark 3. The application of this theorem requires the knowledge of the error at the inflow boundary of a cell in order to bound the error in the cell and at its outflow boundary. Since the entries of the velocity field, $\mathbf{a}$, are of constant sign, this can be achieved by sweeping through the mesh from left to right and from bottom to top. However, this process could be expected to badly overestimate the error. What is needed is a localisation result which bounds the error by a locally computable quantity.

5.1 Error localisation

Consider the exact solutions $\tilde{u}$ and $\tilde{u}^h$ to the local boundary value problems

$$L\tilde{u} := \nabla \cdot (a\tilde{u}) + b\tilde{u} = f \quad \text{on } K, \qquad \tilde{u}|_{\partial_- K} = u|_{\partial_- K}, \tag{7}$$

$$L\tilde{u}^h = f \quad \text{on } K, \qquad \tilde{u}^h|_{\partial_- K} = u^h|_{\partial_- K}. \tag{8}$$

Due to the unique solvability of the boundary value problem we have $\tilde{u} = u|_K$. For $\tilde{u}^h$ we have the following interpretation. Suppose that we have determined the numerical solution in the region between $\partial_-\Omega$ and ∂_-K. Then the numerical data $u^h|_{\partial_-K}$ are distorted initial data due to the numerical error upwind of the cell K. In the second problem (8) above we determine the exact solution to these distorted data. Then the quantity $e_c := \tilde{u}^h - u^h$ is the error produced by the scheme on this cell. We call it the *cell error*. The complementary quantity $e_{tr} := u - \tilde{u}^h$ reflects the component of the error which is created upwind of the cell which is just advected through the cell. We call this error the *transmitted error*. Therefore we have a decomposition on each cell of the error

$$e|_K = (u - u^h)|_K = e_c + e_{tr}.$$

which satisfy the equations

$$\begin{aligned} Le_c &= -r & \mathbf{x} \in K \\ e_c &= 0 & \mathbf{x} \in \partial_-K \end{aligned}$$

and

$$\begin{aligned} Le_{tr} &= 0 & \mathbf{x} \in K \\ e_{tr} &= e & \mathbf{x} \in \partial_-K. \end{aligned}$$

We have the following bounds for the cell error.

Theorem 5.2 *For each cell K_{ij}, there exists two positive constants $c_0(K_{ij})$ and $c_1(K_{ij})$ such that with $m = e^{-2\alpha.\mathbf{x}}$,*

$$c_0 \|mr\|_{H^{-1}(K_{ij})} \le \|me_c\|_{L_2(K_{ij})} \le c_1 \|mr\|_{L_2(K_{ij})}. \tag{9}$$

Remark 4. Using this theorem we define an upper cell error *indicator* $\bar{\epsilon}_{ij}$ and a global upper error *estimate* $\bar{\epsilon}$ by noting that

$$\begin{aligned} \|me_c\|^2_{L_2(\Omega)} &= \sum_{i=1}^{M}\sum_{j=1}^{N} \|me_c\|^2_{L_2(K_{ij})} \\ &\le \sum_{i=1}^{M}\sum_{j=1}^{N} c_1^2(k_{ij}) \|mr\|^2_{L_2(K_{ij})} \\ &\equiv \sum_{i=1}^{M}\sum_{j=1}^{N} \bar{\epsilon}_{ij}^2 = \bar{\epsilon}^2. \end{aligned}$$

Similarly we define a lower cell error indicator $\underline{\epsilon}_{ij}$ and a lower cell error estimate $\underline{\epsilon}$ by noting that

$$\begin{aligned} \|e_c\|^2_{L_2(\Omega)} &= \sum_{i=1}^{M}\sum_{j=1}^{N} \|e_c\|^2_{L_2(K_{ij})} \\ &\ge \sum_{i=1}^{M}\sum_{j=1}^{N} c_0^2(K_{ij}) \|r\|^2_{H^{-1}(K_{ij})} \\ &\equiv \sum_{i=1}^{M}\sum_{j=1}^{N} \underline{\epsilon}_{ij}^2 = \underline{\epsilon}^2. \end{aligned}$$

Remark 5. Calculation of $\bar{\varepsilon}$ and $\underline{\varepsilon}$ requires the numerical evaluation of the L_2 and H^{-1} norms of the residual. The L_2-norm of the residual will be approximated by numerical quadrature the effect of which will be discussed in the next section. The difficulty in evaluating the H^{-1} norm lies in its definition

$$\|mr\|_{H^{-1}(K_{ij})} \equiv \sup_{\phi \in H_0^1(K_{ij})} \frac{|\int_{K_{ij}} mr\phi \mathrm{dx}|}{\|\phi\|_{H^1(K_{ij})}}.$$

For practical purposes we cannot test the residual against all functions in $H_0^1(K_{ij})$ and instead we use the approximation

$$\|mr\|_{H^{-1}(K_{ij})} \approx \max_{l,m=1,2} \frac{|\int_{K_{ij}} mr\psi_{lm} \mathrm{dx}|}{\|\psi_{lm}\|_{H^1(K_{ij})}}, \tag{10}$$

where $\{\psi_{lm}\}_{l,m=1}^2$ is a set of four bilinear functions lying in $H_0^1(K_{ij})$. The particular functions used here are $\psi_{lm} = \hat{\psi}_{lm} \circ F_{K_{ij}}^{-1}$, where $\hat{\psi}_{lm} = \hat{\psi}_l(\hat{x}^1)\hat{\psi}_m(\hat{x}^2)$ and

$$\hat{\psi}_1(\hat{x}^1) = \begin{cases} \frac{2}{3}(\hat{x}^1 + 1), & -1 \le \hat{x}^1 \le \frac{1}{2}, \\ -2(\hat{x}^1 - 1), & \frac{1}{2} \le \hat{x}^1 \le 1, \end{cases}$$

$$\hat{\psi}_2(\hat{x}^1) = \begin{cases} 2(\hat{x}^1 + 1), & -1 \le \hat{x}^1 \le -\frac{1}{2}, \\ -\frac{2}{3}(\hat{x}^1 - 1), & -\frac{1}{2} \le \hat{x}^1 \le 1. \end{cases}$$

Clearly, since this approximation is a lower bound on the H^{-1} norm, we still have maintained reliability of the lower error estimate. The efficiency of this procedure will be investigated in the numerical experiments in the next section.

For the transmitted error we have the following bound.

Theorem 5.3 *For each cell K_{ij} there exists a constant $c_2(K_{ij})$ such that with $m = e^{-2\alpha \cdot \mathbf{x}}$,*

$$\frac{2}{c_2}\|me_{tr}\|_{L_2(K_{ij})}^2 + \|me_{tr}\|_{L_2(\partial_+ K_{ij})}^2 \le \|me\|_{L_2(\partial_- K_{ij})}^2 \tag{11}$$

Remark 6. If the inflow data $g \in P^1(\partial_- \Omega)$ then $e_{tr} = 0$ in $\bar{\Omega}$ and the error $e = e_c$.

Remark 7. When $g \notin P^1(\partial_- \Omega)$ then $\|e\|_{L_2(\partial_- \Omega)}$ can be estimated in terms of a computable seminorm of g.

Remark 8. The concept of local error introduced in this section is not restricted to the particular numerical method used in this paper. For other methods the cells K have to be defined appropriately, but there should be at least one node undetermined by the upstream data. The reasoning given above also extends to linear positively symmetric hyperbolic systems.

6 Numerical experiments

6.1 Example 1

For the first test case considered the convective velocity field is given by

$$\mathbf{a} = \left(\frac{(1+x^1)^{1+\beta}}{2\beta}, \frac{(1+x^2)^{1+\beta}}{2\beta}\right)^T,$$

which for $\beta > 0$ has both components strictly positive. With a forcing function given by

$$f = \left(\frac{(1 + \beta)((1 + x^1)^\beta + (1 + x^2)^\beta)}{2\beta} + 1 \right) e^{-(1+x^1)^{-\beta} - (1+x^2)^{-\beta}},$$

and $b = 0$, the solution of (1) with boundary conditions

$$u(x^1, x^2)|_{\partial_-\Omega} = \begin{cases} e^{-(1+x^2)^{-\beta} - 1}, & x^1 = 0 \\ e^{-(1+x^1)^{-\beta} - 1}, & x^2 = 0 \end{cases}$$

is

$$u(x^1, x^2) = e^{-(1+x^1)^{-\beta} - (1+x^2)^{-\beta}}.$$

Therefore the solution is smooth and we can increase the size of its gradient by increasing the value of β. For simplicity, the problem will be approximated on a uniform grid and in each cell we set

$$c_0(K_{ij}) = \frac{2\beta}{(1 + \beta)((1 + x_{i-1}^1)^\beta + (1 + x_{j-1}^2)^\beta)} \leq \frac{2}{\nabla \cdot \mathbf{a}}.$$

Fig. 1(a) shows a contour plot of the computed cell-vertex solution on a 32×32 cell grid which looks identical to the exact solution at plotting accuracy. Figs. 1(b) and (c) are of $\|e\|_{L_2(\Omega)}$ and $\bar{\epsilon}$, which have both been approximated using one-point Gaussian quadrature. From these figures it is clear that there is a good correlation between the error indicator and the actual error. The agreement is quantified in Table 1 which shows second-order convergence of the approximate L_2-norm of the error and the one-point quadrature approximation of $\bar{\epsilon}$. The third column of Table 1 shows an $O(h)$ convergence of the efficiency index $\bar{\theta} = \bar{\epsilon}/\|e\|_{L_2(\Omega)}$ of $\bar{\epsilon}$, to a value which is not much larger than unity. Therefore, for this problem the error appears to be dominated by locally generated cell errors and that the transmitted error in the interior of the domain is small by comparison. This is not entirely surprising since the inflow data is smooth.

The fourth column of Table 1 shows the performance of the lower bound estimate $\underline{\epsilon}$. We see that the error estimate is a true lower bound on the error. The efficiency index $\underline{\theta} = \|e\|_{L_2(\Omega)}/\underline{\epsilon}$ of $\underline{\epsilon}$ is given in the next column and shows that the error estimate converges at the same rate as the error. The degree of underestimation of the error is not excessive and could probably be improved by testing the residual against a larger set of functions. The choice of the linear functions in (10) may at first seem rather odd. However, numerical investigation with symmetric functions, both linear and quadratic, produced an error estimator which converged as $O(h^3)$ which is non-optimal. This can be partially explained by the fact that the residual is almost orthogonal to the test space of piecewise constant functions.

Table 2 and Fig. 2 show the corresponding results when $\beta = 10$. For the larger value of β we also observe good agreement between the distribution of error indicator and the true error and that both the upper and lower efficiency indices behave similarly to the case above.

To investigate the effect of quadrature on the approximation of integral norms, Table 3 and Fig. 3 show the results obtained when a 3×3 Gaussian quadrature rule is used for the calculation of $\|e\|_{L_2(\Omega)}$ and $\bar{\epsilon}$ when $\beta = 5$. What we find is that with the higher order quadrature on the L_2 norm the residual produces a poorer error indicator and a poorer error estimate than the one-point quadrature. In fact the error estimate converges like $O(h)$ whereas the true error converges as $O(h^2)$. Therefore the efficiency index behaves like $O(h^{-1})$ as $h \to 0$. Investigations using other high-order quadrature formulae give similar results. Therefore, as mentioned earlier, we find that $\|r\|_{L_2(\Omega)}$ is really $O(h)$. However, in the special case when we used one-point Gaussian quadrature we found that the error estimate converged at the same rate as the true error. Examination of the residual leads us to conjecture that the gradient approximation of the cell-vertex solution is superconvergent at the cell centroids. This is supported by the final columns of Tables 1 and 2, which show second-order convergence of the maximum norm of the gradient error at the quadrature points which is denoted here by $\||\nabla e|\|_{l_\infty(\Omega_q)}$. When higher order quadrature formulae are used, which sample the residual at points other than the cell centriod, we only observe first-order convergence of $\||\nabla e|\|_{l_\infty(\Omega_q)}$, as shown in Table 3.

6.2   Example 2

We now consider the performance of the error estimates and error indicators developed for a model non-smooth problem. For this example we have $\mathbf{a} = (2,1)^T$, $b = 0$, and $f = 0$. With the boundary conditions

$$u(x^1, x^2)|_{\partial - \Omega} = \begin{cases} 1, & x^1 = 0 \\ 0, & x^2 = 0, \end{cases}$$

the analytical solution is clearly

$$u(x^1, x^2) = \begin{cases} 1, & x^2 \geq \frac{1}{2}x^1, \\ 0, & \text{otherwise.} \end{cases}$$

Fig. 4(a) shows a contour plot of the cell-vertex solution obtained on a grid of 32×32 cells. As expected, the areas of highest nodal error are at the discontinuity of the exact solution but due to the dispersive error of the method there is a pollution of the solution in a region on one side of the discontinuity. We should note that no form of numerical dissipation has been added to the basic cell-vertex discretisation.

For this problem, which has $\nabla.\mathbf{a} = 0$ and $b = 0$, we need to introduce a positive function $m(\mathbf{x})$ which satisfies the stability requirements of Theorem 5.2. For this example in each cell $m(x, y)$ has been chosen locally to be

$$m(x^1, x^2) = \exp(-2[(x^1 - x^1_{i-1}) + (x^2 - x^2_{j-1})]), \tag{12}$$

which ensures that each cell is given an equal weight. Since we have simply chosen $\alpha = (1,1)$, we can take $c_1 = 1$ for each cell.

When the convective velocity field is constant, it is easy to show that

$$(r, p)_{K_{ij}} = 0, \qquad \forall p \in Q_0(K_{ij}).$$

If the problem is also homogeneous, then the residual is a bilinear function over each cell and the calculation of $\|mr\|_{L_2(K_{ij})}$ using one-point Gaussian quadrature gives a zero value. Therefore, Figs. 4(b) and (c), which are of $\|me\|_{L_2(\Omega)}$ and $\bar{e}$, have been obtained using a 3×3 Gaussian quadrature formula. The plot of the error estimator suggests that the error is largest at the point of introduction of the discontinuity into the domain and along a line which is at an angle of $\tan^{-1}(2)$. However, this is in disagreement with the plot of the error which is large and increasing along the line of discontinuity in the exact solution. The reason for this disparity is twofold. Firstly, the residual for this problem is simply $r = \mathbf{a}.\nabla u^h$, which is close to zero when either ∇u^h is close to zero or when $\mathbf{a}$ is close to being perpendicular to ∇u^h. By looking at the cell-vertex solution we find that ∇u^h is significant in regions other than immediately adjacent to the discontinuity in the exact solution. Moreover, ∇u^h is significant in directions other than perpendicular to $\mathbf{a}$. Therefore, although ∇u^h is largest at the discontinuity, it is also close to perpendicular to $\mathbf{a}$. This then explains the shape of the plot of the error estimator $\bar{\varepsilon}$. The second reason why the error indicator does not resemble the error is that $\bar{\varepsilon}$ is only an error indicator for e_c. Since $e|_{\partial_-\Omega}$ is large at the bottom left corner of the domain, we can expect that the transmitted error is significant in the interior of the cells along the line of the discontinuity. Therefore, this is an example of where the transmitted error is an important contribution to the total error.

Quantitative results on a sequence of subdivisions of the domain are given in Table 4. We observe that the error converges at an estimated rate of $\dot{0}.3$. From Table 4 we can also see that the $\bar{\varepsilon}$ behaves like $h^{-1/2}$, which results in the efficiency index growing rapidly as $h \to 0$. The reason for the divergence of $\bar{\varepsilon}$ can be explained quite simply by considering a linear interpolant of a one-dimensional step function with the discontinuity lying midway between the mesh points x_i and x_{i+1}. If the approximation,

$$u^h = \begin{cases} 0 & : \ 0 \le x \le x_i \\ (x - x_i)/h & : \ x_i \le x \le x_{i+1} \\ 1 & : \ x_{i+1} \le x \le 1 \end{cases},$$

was obtained from an advection problem $a(x)u'(x) = 0$, the residual function would be

$$r = \begin{cases} 0 & : \ 0 \le x \le x_i \\ a(x)/h & : \ x_i \le x \le x_{i+1} \\ 0 & : \ x_{i+1} \le x \le 1 \end{cases}.$$

Measuring the residual in the L_2-norm we have

$$\|r\|_{L_2(x_i,x_{i+1})} = \left\{ \frac{1}{h^2} \int_{x_i}^{x_{i+1}} a^2(x)\mathrm{d}x \right\}^{1/2},$$

which if $a(x)$ does not change sign in (x_i, x_{i+1}), behaves like $h^{-1/2}$. This is what happens with our model two-dimensional non-smooth problem. Therefore, the L_2-norm of the residual appears to be an inappropriate error estimator for problems which have the type of discontinuity found in this example.

Table 4 also shows the performance of the lower error estimate which shows an $h^{1/2}$ convergence rate. Therefore, since the error appears to be converging at a slower rate than $h^{1/2}$, we find that the efficiency index increases as $h \to 0$. However, the disparity between the convergence rate of $\underline{\varepsilon}$ and e is much smaller than that between $\bar{\varepsilon}$ and e.

7 Summary and conclusions

In this paper we have developed upper and lower error estimates for the cell-vertex finite volume approximation of scalar two-dimensional linear advection problems. A natural splitting of the error into a locally generated part and a propagated part enabled local bounds to be obtained. Numerical investigation of two model problems showed that

Table 1: Performance of error estimators for 2D linear advection with $\beta = 5$, where the integral norms have been approximated using one-point Gaussian quadrature.

N	$\|e\|_{L_2(\Omega)}$	$\bar{\epsilon}$	$\bar{\theta}$	$\underline{\epsilon}$	$\underline{\theta}$	$\|\|\nabla e\|\|_{l_\infty(\Omega_q)}$
4	1.42E-2	4.49E-2	3.13	9.80E-4	14.5	1.05E-1
8	3.33E-3	8.35E-3	2.51	2.21E-4	15.0	4.15E-2
16	8.26E-4	1.84E-3	2.23	5.22E-5	15.8	1.59E-2
32	2.06E-4	4.36E-4	2.10	1.26E-5	16.4	6.07E-3
64	5.15E-5	1.05E-4	2.04	3.07E-6	16.8	2.27E-3
128	1.29E-5	2.59E-5	2.02	7.58E-7	17.0	8.04E-4
256	3.22E-6	6.44E-6	2.00	1.88E-7	17.1	2.67E-4
512	8.05E-7	1.60E-6	1.99	4.69E-8	17.2	7.98E-5

the error estimates are reasonably tight for problems with smooth solutions. To obtain a tight upper bound on the local error, use was made of observed superconvergence properties of the gradient of the solution at cell centriods. For non-smooth problems, one has to ensure that the solution does not contain excessive dispersive errors otherwise residual based error estimates can produce misleading results.

The use of the finite element residual as an error indicator has already been applied to the compressible Euler equations [2]. With the aid of the mathematical results and numerical experience gained in the work reported here, it is hoped that such a process has been partially justified.

References

[1] K. Eriksson and C. Johnson. Adaptive streamline diffusion finite element methods for stationary convection-diffusion problems. *Math. Comp.* (60) pp 167-188 (1993).

[2] J. A. Mackenzie, D. F. Mayers and A. J. Mayfield. Error estimates and mesh adaption for a cell-vertex finite volume scheme. Technical Report NA92/10, Oxford University Computing Laboratory, 1992.

[3] K.W. Morton and E. Süli Finite Volume Methods and their Analysis. *IMA J. Numer. Anal.*, (11) pp 241–260 (1991).

[4] Th. Sonar. Strong and weak norm error indicators based on the finite element residual for compressible flow computations. DLR Gottingen Report 92/07, 1992.

[5] E. Süli. The Accuracy of Cell Vertex Finite Volume Methods on Quadrilateral Meshes. *Math. Comp.* (59) pp 359-382 (1992).

Table 2: Performance of error estimator for 2D linear advection with $\beta = 10$, where the integral norms have been approximated using one-point Gaussian quadrature.

N	$\|e\|_{L_2(\Omega)}$	$\bar{\epsilon}$	$\bar{\theta}$	$\underline{\epsilon}$	$\underline{\theta}$	$\|\|\|\nabla e\|\|\|_{l_\infty(\Omega_q)}$
4	4.31E-2	2.07E-1	4.81	3.04E-3	14.2	7.59E-1
8	8.57E-3	2.75E-2	3.21	5.50E-4	15.6	3.17E-1
16	2.04E-3	5.17E-3	2.54	1.19E-4	17.1	8.77E-2
32	5.08E-4	1.19E-3	2.24	2.79E-5	18.2	3.17E-2
64	1.27E-4	2.68E-4	2.11	6.70E-6	19.0	1.24E-2
128	3.17E-5	6.49E-5	2.05	1.63E-6	19.4	5.26E-3
256	7.93E-6	1.60E-5	2.02	4.03E-7	19:7	2.29E-3
512	1.98E-6	3.97E-6	2.00	1.00E-7	19.8	1.02E-3

Table 3: Performance of error estimator for 2D linear advection with $\beta = 5$, where the integral norms have been approximated using 3×3 Gaussian quadrature.

N	$\|e\|_{L_2(\Omega)}$	$\bar{\epsilon}$	$\bar{\theta}$	$\|\|\|\nabla e\|\|\|_{l_\infty(\Omega_q)}$
4	1.053E-02	6.080E-02	5.77	3.153E-01
8	2.173E-03	1.736E-02	7.99	2.071E-01
16	5.162E-04	7.253E-03	14.1	9.508E-02
32	1.276E-04	3.377E-03	26.5	4.724E-02
64	3.180E-05	1.636E-03	51.4	2.328E-02
128	7.945E-06	8.057E-04	101	1.190E-02
256	1.986E-06	3.999E-04	201	5.952E-03
512	4.964E-07	1.993E-04	401	2.977E-03

Table 4: Performance of residual error estimator for non-smooth 2D linear advection, where the integral norms have been approximated using 3×3 Gaussian quadrature.

N	$\|me\|_{L_2(\Omega)}$	$\bar{\epsilon}$	$\bar{\theta}$	$\underline{\epsilon}$	$\underline{\theta}$
4	1.55E-1	5.43E-1	3.50	2.86E-3	54.2
8	1.45E-1	8.61E-1	5.94	2.28E-3	63.6
16	1.29E-1	1.29E+0	10.0	1.59E-3	81.1
32	1.11E-1	1.89E+0	17.0	1.14E-3	97.4
64	9.17E-2	2.71E+0	29.6	8.09E-4	113
128	7.53E-2	3.87E+0	51.4	5.74E-4	131
256	6.12E-2	5.49E+0	89.7	4.06E-4	151
512	4.94E-2	7.78E+0	157	2.87E-4	172

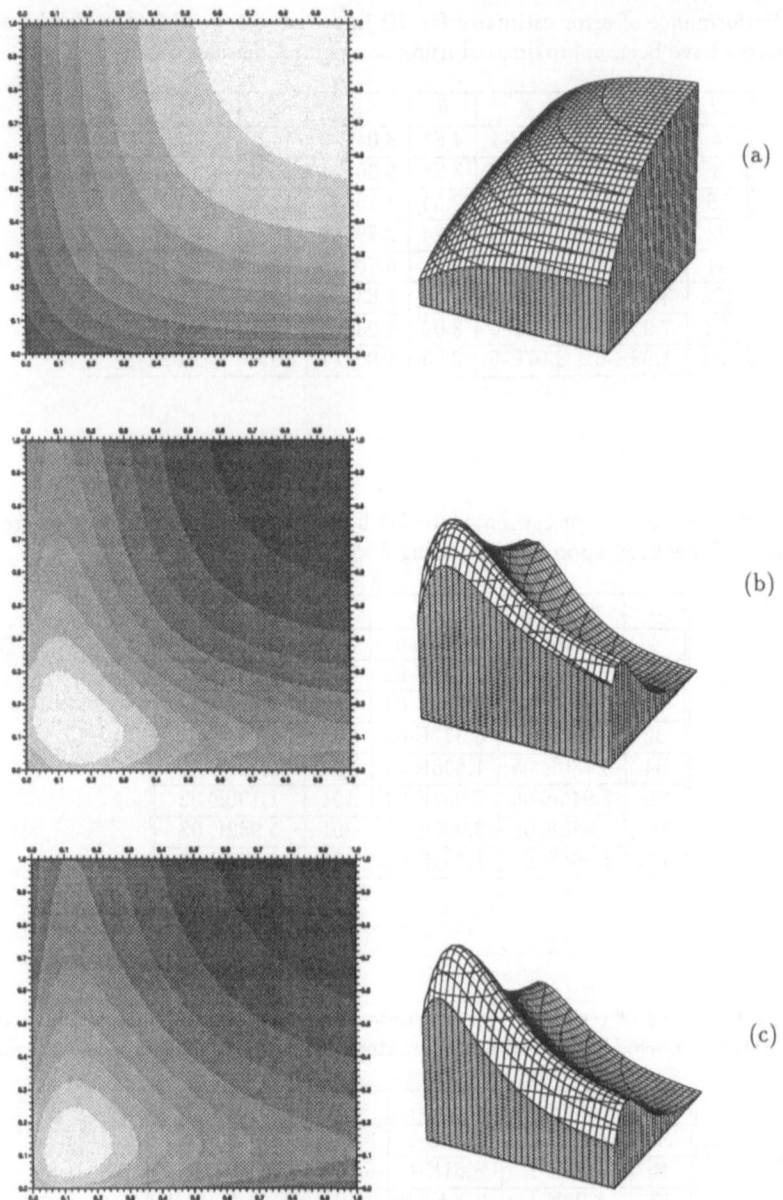

Figure 1: Contour plots of (a) approximate solution, (b) $\|e\|_{L_2(\Omega)}$, and (c) $\bar{\epsilon}$ where $\beta = 5$, where the integral norms have been approximated using one-point quadrature.

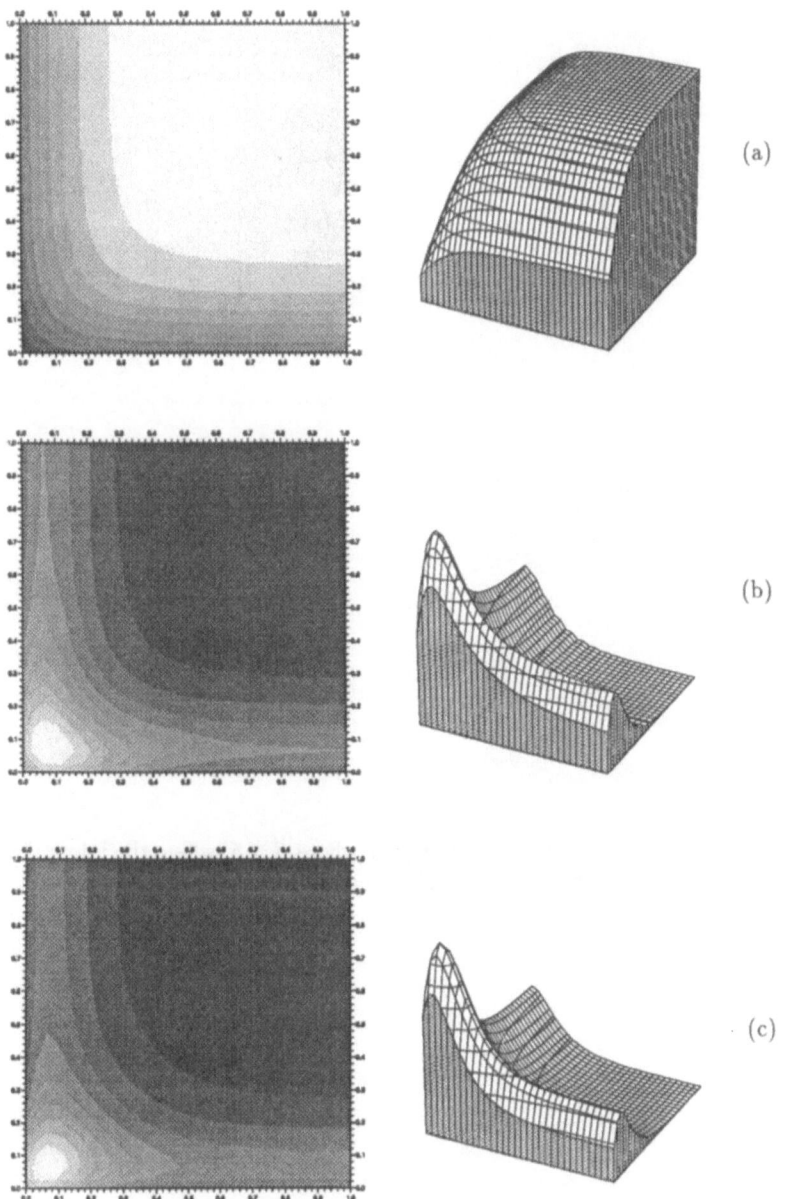

Figure 2: Contour plots of (a) approximate solution, (b) $\|e\|_{L_2(\Omega)}$, and (c) $\bar{\epsilon}$ where $\beta = 10$, where the integral norms have been approximated using one-point quadrature.

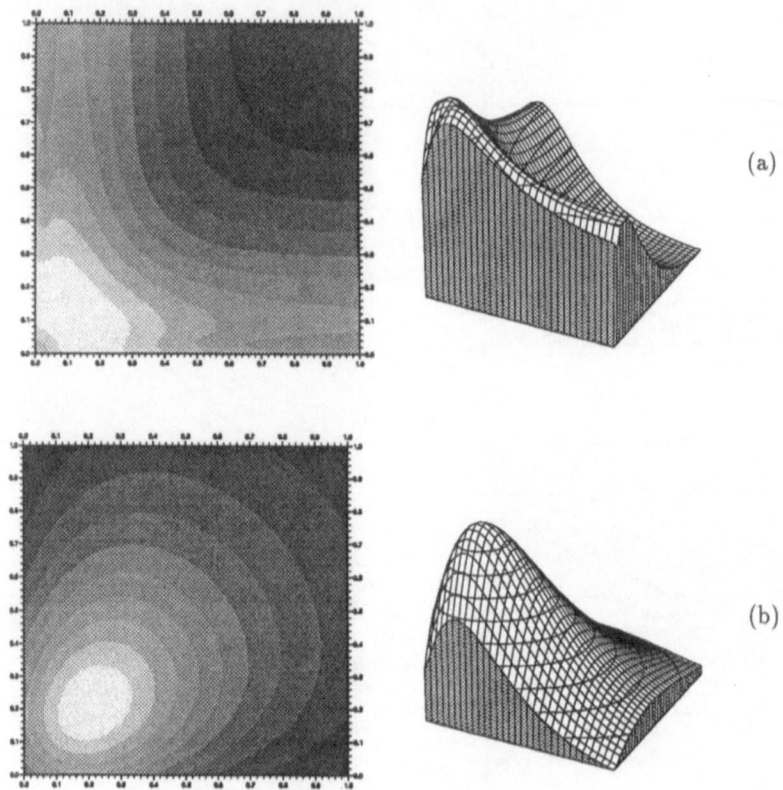

Figure 3: Contour plots of (a) $\|e\|_{L_2(\Omega)}$, and (b) $\bar{\epsilon}$ where $\beta = 5$, where the integral norms have been approximated using 3×3 Gaussian quadrature.

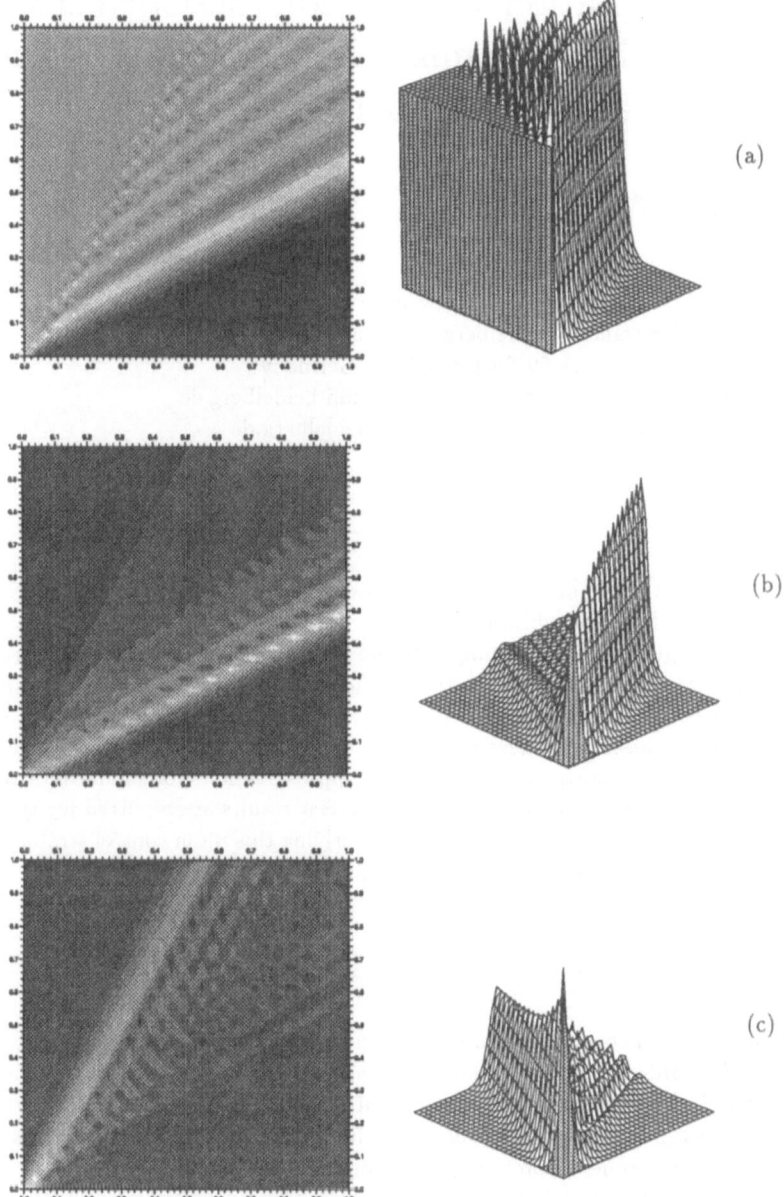

Figure 4: Contour plots of (a) approximate solution, (b) $\|me\|_{L_2(\Omega)}$, and (c) $\bar{\epsilon}$ for non-smooth advection, where the integral norms have been approximated using 3×3 Gaussian quadrature.

Mesh Adaptation via a Predictor–Corrector Strategy in the Streamline Diffusion Method for Nonstationary Hyperbolic Systems[*]

Rolf Rannacher   Guohui Zhou

Institut für Angewandte Mathematik
Universität Heidelberg, Im Neuenheimer Feld 293
D–69120 Heidelberg, Germany
E–Mail: rannacher@gaia.iwr.uni-heidelberg.de
zhou@gaia.iwr.uni-heidelberg.de

Summary

Streamline diffusion is a well-known damping strategy in the numerical approximation of transport–dominated transport–diffusion problems by the finite element method. It combines the effect of higher order upwinding with the flexibility of the variational approach. However, the reliable and accurate resolution of shock–like solution structures requires a refinement of the computational mesh in space and also in time. In this note we propose a strategy for such a mesh refinement within an implicit time stepping process which is based on a local predictor–corrector concept. This approach allows for significant savings with respect to storage and computing time. Some test results are reported for the 1–d Riemann shock tube problem. The mechanism underlying this mesh control strategy can be explained through a rigorous theoretical analysis.

1. Introduction

The streamline diffusion concept was introduced in the finite element method by Hughes et al., [5] and [6], in order to cope with the particular difficulties, like shocks and boundary layers, occurring in hyperbolic-type problems. Subsequently this methodology has been adapted to various types of practical problems, including for instance the Euler equations of gas dynamics, and a solid theoretical foundation has been laid in a series of research papers, see Johnson et al., [7], [8], [9], [11], [13], and the literature cited therein. Recently the streamline diffusion method has also been combined with strategies for a posteriori error estimation and mesh size control, see Johnson et al., [2] and [12]. This development has led to very flexible algorithms which work on largely unstructured meshes and are

---

[*]This work was supported by the Deutsche Forschungsgemeinschaft, SFB 123 and SFB 359 Universität Heidelberg

well adapted to the local features of the solution. However, the price to be paid for this flexibility is an unfavorable data structure and, in the case of an implicit scheme, relatively high computational costs in solving the algebraic systems. This effect becomes particularly evident in long time calculations when the time step is chosen relatively large compared with the minimum spatial scale. Here, one would like to keep the mesh as much uniform as possible and refine it only locally in the neighborhoods of the shocks. Such a strategy appears appropriate in cases when the shocks are well separated from each other by large areas where the solution is smooth. A typical example is the well-known "Riemann Shock Tube" problem. For such a situation a "Predictor-Corrector" process has been proposed in Zhou [20] for the local mesh adaptation near shocks within a fully implicit time stepping scheme. This article gives a brief account of the concept underlying this method and its practical realization, and presents some results on its theoretical justification.

The predictor–corrector method works as follows: Starting from a time level t_n, in the predictor step, one computes a first guess of the solution between t_n and t_{n+1}, on a space–time mesh with size h_{coarse}, which is fine enough to accurately capture the features of the solution in its "smooth" regions. From this "coarse grid" solution one detects the approximate position of the "shocks". Then, in the corrector step, the calculation is repeated in the neighborhoods of the shocks with a finer mesh sizes, $h_{\text{fine}} \ll h_{\text{coarse}}$, in space *and* time. The necessary boundary values for this local implicit time stepping are interpolated from the coarse grid solution. The theoretical analysis shows that these boundary values are of accuracy $O(h^{(12-d)/8}|\log h|)$ in d space dimensions, in a distance $O(h^{3/4}|\log h|)$ from the shocks. Hence, the neighborhoods for the re-computation of the shocks need to be only of width $O(h^{3/4}|\log h|)$. After having reached the time level $t_{n+1} = t_n + \Delta t_{\text{coarse}}$, the local solutions from the fine mesh computations are substituted for the corresponding parts of the coarse grid solution. Then, the whole process starts again from t_{n+1}. The described localization procedure has two major advantages. It requires significantly less work than the corresponding scheme on a globally refined mesh, and it preserves the favorable data structure related to a uniform mesh. Our theoretical analysis indicates that it also achieves the approximation accuracy of the uniformly refined mesh. These results are supported by preliminary numerical tests for the 1-D Euler equations (shock tube problem). Here, the predictor–corrector adaptive method saves up to 80% of the computational costs. The extension of this concept to more than one space dimension is possible.

2. The Streamline Diffusion Method

We consider a nonstationary convection–diffusion problem in one space dimension of the form

$$u_t + B(u)u_x - \varepsilon u_{xx} = f(x,t), \quad (x,t) \in Q \equiv (0,1) \times (0,T), \tag{2.1.a}$$
$$u(x,0) = u^0(x), \quad x \in \Omega \equiv (0,1), \tag{2.1.b}$$
$$u(0,t) = u_l(t), \quad u(1,t) = u_r(t), \quad t \in I \equiv (0,T), \tag{2.1.c}$$

where u is an M–vector, and $B(\cdot)$ is a real diagonalizable $M \times M$ matrix. The interest is here in the case of vanishing diffusion, $0 \le \varepsilon \ll 1$. We assume that for $\varepsilon = 0$ the boundary conditions are prescribed in such a way that the problem has a unique solution (cf. Smoller [17]).

For discretizing this problem we use the discontinuous (in time) Galerkin method (cf. Eriksson, el al., [3]). For a sequence of time levels $0 = t_0 < t_1 < \cdots < t_N = T$, let $I_n = [t_n, t_{n+1}]$ and

$$S_n = \bar{\Omega} \times I_n, \qquad L_n = \bar{\Omega} \times \{t_n\}.$$

Further, for $h > 0$, let T_h^n be quasi-uniform partitions of the space–time slab S_n into elements e (triangles or quadrilaterals) with element diameter $h_e \approx h$ (see Figure 1). The partition of S_n may be chosen independently of that of S_{n-1}, but it must satisfy the standard quasi-uniformity conditions for finite element meshes ("uniform shape and size" condition).

We set $\Delta t = \max_n \{t_n - t_{n-1}\}$, and Δx the maximal element width on the intersecting lines $\{L_n\}$, and $h = (\Delta t^2 + \Delta x^2)^{1/2}$. The nodal points may lie on the lines $\{t = t_n\}$ and $\{t = t_{n+1}\}$ or inside of S_n. However, the uniform shape condition requires that Δt is of the same order as Δx, that is, there exist two constants K_1 and K_2 such that $K_1 \Delta x \le \Delta t \le K_2 \Delta x$, for $h \to 0$.

On the slabs S_n, we introduce the finite element spaces

$$\mathbf{V}_{hj}^n = \{v \in \mathbf{C}(S_n) : v|_e \in \mathbf{P}_1(e), \forall e \in T_h^n\},$$

where $\mathbf{P}_1$ is the space of linear or (isoparametric) bilinear polynomials if e is a triangle or quadrilateral, respectively. Further, we set

$$\hat{\mathbf{V}}_{hj}^n = \left\{ v \in \mathbf{V}_{hj}^n : v(0, t) = \pi_h u_l^j, v(1, t) = \pi_h u_r^j \right\},$$
$$\mathring{\mathbf{V}}_{hj}^n = \left\{ v \in \mathbf{V}_{hj}^n : v(0, t) = 0, v(1, t) = 0, t \in I_n \right\}, \qquad \text{for } j = 1, 2, \cdots, M$$

and

$$\mathbf{V}_h^n = \prod_{j=1}^M \mathbf{V}_{hj}^n, \quad \hat{\mathbf{V}}_h^n = \prod_{j=1}^M \hat{\mathbf{V}}_{hj}^n, \quad \mathring{\mathbf{V}}_h^n = \prod_{j=1}^M \mathring{\mathbf{V}}_{hj}^n,$$

where $\pi_h : \mathbf{C}(S_n) \to \mathbf{V}_{hj}^n$ is the generic interpolation operator over the triangulations T_h^n. The finite element space $\mathbf{V}_{hj}^n$ is a subspace of $\mathbf{H}^1(S_n)$, $\hat{\mathbf{V}}_{hj}^n \subset \mathbf{V}_{hj}^n$ is the subset of functions with prescribed interpolated boundary values, and $\mathring{\mathbf{V}}_{hj}^n \subset \mathbf{V}_{hj}^n$ is the subset of functions with homogeneous boundary values.

Here we must take into account that the functions in $\hat{\mathbf{V}}_h^n$ have only to satisfy boundary conditions on $\partial\Omega \times I_n$; there are no constraints on the intersecting lines L_n and L_{n+1}. A function in $\mathbf{V}_h^n$ is continuous within each S_n, and in particular, continuous in x everywhere. It may be discontinuous across the intersecting lines L_n. Hence a function $\mathbf{W} \in \mathbf{V}_h^n$ always has two values on L_n denoted by $\mathbf{W}_+^n$ and $\mathbf{W}_-^n$, where

$$\mathbf{W}_+^n = \lim_{\substack{t \to t_n \\ t > t_n}} \mathbf{W}(\cdot, t), \qquad \mathbf{W}_-^n = \lim_{\substack{t \to t_n \\ t < t_n}} \mathbf{W}(\cdot, t). \tag{2.2}$$

For vector functions on S_n we define the inner products

$$(\mathbf{u}, \mathbf{v})_n = \sum_{j=1}^M \int_{S_n} u_j v_j \, dx \, dt, \qquad \langle \mathbf{u}, \mathbf{v} \rangle_n = \sum_{j=1}^M \int_{L_n} u_j v_j \, dx.$$

238

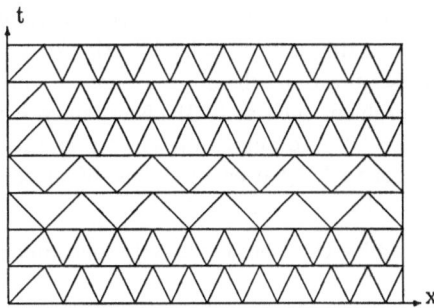

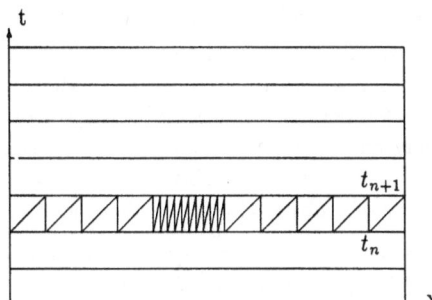

Figure 1: A triangulation of Q   Figure 2: A local spatial mesh refinement

Now the discrete approximation to problem (2.1) is successively defined as follows: Given $\mathbf{U}^n_-$, for some $n \in \{0, 1, \cdots, N-1\}$, find $\mathbf{U}^n \in \hat{V}^n_h$, such that

$$(\mathbf{U}^n_t + B(\mathbf{U}^n)\mathbf{U}^n_x, \mathbf{W}^n)_n + \varepsilon(\mathbf{U}^n_x, \mathbf{W}^n_x)_n \qquad (2.3)$$
$$+ \delta(\mathbf{U}^n_t + B(\mathbf{U}^n)\mathbf{U}^n_x, \mathbf{W}^n_t + B^T(\mathbf{U}^n)\mathbf{W}^n_x)_n + \langle \mathbf{U}^n_+, \mathbf{W}^n_+ \rangle_n$$
$$= (\mathbf{f}, \mathbf{W}^n + \delta(\mathbf{W}^n_t + B^T(\mathbf{U}^n)\mathbf{W}^n_x))_n + \langle \mathbf{U}^n_-, \mathbf{W}^n_+ \rangle_n \qquad \forall \mathbf{W}^n \in \mathring{V}^n_h,$$

where B^T is the adjoint of B, $\delta = Kh$, with K an appropriate constant, and $\mathbf{U}^0_- = \mathbf{u}^0$ the initial data.

The test functions in this Petrov–Galerkin scheme are $\mathbf{W}_t + B^T\mathbf{W}_x$ rather than $\mathbf{W}_t + B\mathbf{W}_x$. To justify this, we consider the simplest case where B is a constant matrix. Since problem (2.1) is of (essentially) hyperbolic type, the matrix B has M real eigenvalues and M linearly independent eigenvectors. Hence there exists an invertible real constant matrix D, such that $D^{-1}BD = \Lambda$, where $\Lambda = diag(\lambda_i)$. Set $\tilde{\mathbf{u}} = D^{-1}\mathbf{u}$. Multiplying the system (2.1.a) by the matrix D^{-1}, we get

$$\tilde{\mathbf{u}}_t + \Lambda\tilde{\mathbf{u}}_x - \varepsilon\tilde{\mathbf{u}}_{xx} = D^{-1}\mathbf{f}. \qquad (2.4)$$

The equations in this system are uncoupled, and hence the usual "scalar" streamline diffusion method can be applied to each single equation. The streamline-directional derivative in the i-th equation is $\tilde{u}^i_t + \lambda_i\tilde{u}^i_x$, and therefore the corresponding test function should be $\tilde{W}^i + \delta(\tilde{W}^i_t + \lambda_i\tilde{W}^i_x)$. Thus, for the diagonal system (2.4) we should take $\tilde{\mathbf{W}} + \delta(\tilde{\mathbf{W}}_t + \Lambda\tilde{\mathbf{W}}_x)$ as test functions, obtaining the scheme

$$(\tilde{\mathbf{U}}_t + \Lambda\tilde{\mathbf{U}}_x, \tilde{\mathbf{W}}) + \varepsilon(\tilde{\mathbf{U}}_x, \tilde{\mathbf{W}}_x) + \delta(\tilde{\mathbf{U}}_t + \Lambda\tilde{\mathbf{U}}_x, \tilde{\mathbf{W}}_t + \Lambda\tilde{\mathbf{W}}_x) + \langle \tilde{\mathbf{U}}_+, \tilde{\mathbf{W}}_+ \rangle_n$$
$$= (D^{-1}\mathbf{f}, \tilde{\mathbf{W}} + \delta(\tilde{\mathbf{W}}_t + \Lambda\tilde{\mathbf{W}}_x)) + \langle \tilde{\mathbf{U}}_-, \tilde{\mathbf{W}}_+ \rangle_n \qquad \forall \tilde{\mathbf{W}} \in \mathring{V}^n_h.$$

Here, the indices n are dropped. Now, replacing $\tilde{\mathbf{U}}$ by $D^{-1}\mathbf{U}$ and $\tilde{\mathbf{W}}$ by $D^{-T}\mathbf{W}$, we get

$$(\mathbf{U}_t + B\mathbf{U}_x, \mathbf{W}) + \varepsilon(\mathbf{U}_x, \mathbf{W}_x) + \delta(\mathbf{U}_t + B\mathbf{U}_x, \mathbf{W}_t + B^T\mathbf{W}_x) + \langle \mathbf{U}_+, \mathbf{W}_+ \rangle_n$$
$$= (\mathbf{f}, \mathbf{W} + \delta(\mathbf{W}_t + B^T\mathbf{W}_x)) + \langle \mathbf{U}_-, \mathbf{W}_+ \rangle_n \qquad \forall \mathbf{W} \in \mathring{V}^n_h,$$

which is the streamline diffusion scheme (2.3).

The system (2.3) is nonlinear in $\mathbf{U}^n$ if B depends on $\mathbf{u}$, which then requires a certain linearization. For this there are many possibilities, a simple fixed-point iteration or the more sophiscated Newton method. However, these iteration schemes may not be very efficient especially for a weekly positive definite problem such as (2.3). Now, since $\mathbf{U}^n$ is only an approximate solution, one may solve for $\mathbf{U}^n$ also approximately rather than exactly. A quite natural idea is to linearly extrapolate $\mathbf{U}^{n-1}$ over $\mathring{\mathbf{V}}_h^n$. This, of course, may cause problems in the neighborhood of the shocks.

Let (x,t) be an arbitrary point in $S_n (n \geq 1)$, and let $\mathbf{U}^{n-1}$ be the discrete solution on the slab S_{n-1}. We define $\hat{\mathbf{U}}^n$ as follows:

$$\hat{\mathbf{U}}^n(x,t) = \mathbf{U}^{n-1}(x,t_{n-1})\frac{t_n - t}{t_n - t_{n-1}} + \mathbf{U}^{n-1}(x,t_n)\frac{t - t_{n-1}}{t_n - t_{n-1}}.$$

For $n = 0$, we define $\hat{\mathbf{U}}^0$ as

$$\hat{\mathbf{U}}^0(x,t) = \mathbf{u}^0(x) + \mathbf{u}_t(x,0)t, \qquad 0 < t \leq t_1,$$

where $\mathbf{u}_t(x,0)$ is computed from (2.1.a) as

$$\mathbf{u}_t(x,0) = \mathbf{f}(x,0) - B(\mathbf{u}^0)\mathbf{u}_x^0 + \varepsilon\mathbf{u}_{xx}^0.$$

In both cases $\hat{\mathbf{U}}^n$ is a linear extrapolation of the discrete solution on the slab S_{n-1} over the slab S_n. It is clear that such a $\hat{\mathbf{U}}^n$ approximates $\mathbf{U}^n$ with an error of size $O(h^2)$.

Now substituting $B(\hat{\mathbf{U}}^n)$ for $B(\mathbf{U}^n)$ in (2.3) we get a linearized system for $\mathbf{U}^n$:

$$(\mathbf{U}_t^n + B(\hat{\mathbf{U}}^n)\mathbf{U}_x^n, \mathbf{W}^n)_n + \varepsilon(\mathbf{U}_x^n, \mathbf{W}_x^n)_n \tag{2.5}$$
$$+ \delta(\mathbf{U}_t^n + B(\hat{\mathbf{U}}^n)\mathbf{U}_x^n, \mathbf{W}_t^n + B^T(\hat{\mathbf{U}}^n)\mathbf{W}_x^n)_n + \langle\mathbf{U}_+^n, \mathbf{W}_+^n\rangle_n$$
$$= (\mathbf{f}, \mathbf{W}^n + \delta(\mathbf{W}_t^n + B^T(\hat{\mathbf{U}}^n)\mathbf{W}_x^n))_n + \langle\mathbf{U}_-^n, \mathbf{W}_+^n\rangle_n, \qquad \forall\, \mathbf{W}^n \in \mathring{\mathbf{V}}_h^n.$$

If the diffusion coefficient ε in (2.5) is zero, this could lead to some stability problems, as one loses the control over the gradients of the discrete solutions. To avoid this, we introduce a certain mount of higher order artificial viscosity. With an appropriate constant K_0 we define

$$\varepsilon_m = \max\{K_0 h^{3/2}, \varepsilon\}. \tag{2.6}$$

Replacing ε in (2.5) by ε_m, we get a modified scheme for hyperbolic systems: Given $\mathbf{U}_-^n$, for some $n \in \{0, 1, \cdots, N-1\}$, find $\mathbf{U}^n \in \hat{\mathbf{V}}_h^n$, such that

$$(\mathbf{U}_t^n + B(\hat{\mathbf{U}}^n)\mathbf{U}_x^n, \mathbf{W}^n)_n + \varepsilon_m(\mathbf{U}_x^n, \mathbf{W}_x^n)_n \tag{2.7}$$
$$+ \delta(\mathbf{U}_t^n + B(\hat{\mathbf{U}}^n)\mathbf{U}_x^n, \mathbf{W}_t^n + B^T(\hat{\mathbf{U}}^n)\mathbf{W}_x^n)_n + \langle\mathbf{U}_+^n, \mathbf{W}_+^n\rangle_n$$
$$= (\mathbf{f}, \mathbf{W}^n + \delta(\mathbf{W}_t^n + B^T(\hat{\mathbf{U}}^n)\mathbf{W}_x^n))_n + \langle\mathbf{U}_-^n, \mathbf{W}_+^n\rangle_n, \qquad \forall\, \mathbf{W}^n \in \mathring{\mathbf{V}}_h^n.$$

To justify the choice (2.6) we recall some results from the error analysis for the streamline diffusion method for linear scalar problems (cf. Nävert [14]). For piecewise linear approximation and sufficiently smooth solutions the error in the energy norm is of order $O(h^{3/2})$. Now, taking such an ε_m results in an extra perturbation term $(\varepsilon_m - \varepsilon)(\mathbf{u}_{xx}, \mathbf{W})$, which is also of order $O(h^{3/2})$. The local error analyses discussed below shows that this choice is optimal for a general mesh.

For the purely hyperbolic case ($\varepsilon = 0$) another modification of the streamline diffusion method was proposed in Hughes et al., [4], [5], [6], which consists of introducing a certain "shock-capturing" term. This is supposed to provide some "crosswind" damping along the shocks. An error analysis of this procedure was started in Johnson el al., [13], in connection with an entropy variable transformation. Usually the entropy transformation is very complicated and its evaluation requires excessive computing time on each time level. Since the shock-capturing term is only used to damp the possible overshooting near the shocks, we prefer to apply its main principle to the original variables.

We introduce the residual $\mathbf{E(U)}$ on each S_n,

$$\mathbf{E(U)} = \mathbf{U}_t + B(\mathbf{U})\mathbf{U}_x - \mathbf{f},$$

and correspondingly the matrix

$$C(\mathbf{U}) = \mathrm{diag}(c_i(\mathbf{U})), \quad c_i(\mathbf{U}) = \frac{|\mathbf{E}_i(\mathbf{U})|}{h + |\nabla \mathbf{U}|}, \tag{2.8}$$

with $|\nabla \mathbf{U}| = (|\mathbf{U}_t|^2 + |\mathbf{U}_x|^2)^{1/2}$. Using this notation, we define the following scheme: Given $\mathbf{U}_-^n$, for some $n \in \{0, 1, \cdots, N-1\}$, find $\mathbf{U}^n \in \mathring{V}_h^n$, such that

$$(\mathbf{U}_t^n + B(\hat{\mathbf{U}}^n)\mathbf{U}_x^n, \mathbf{W}^n)_n + ((\varepsilon_m I + \bar{\varepsilon} C(\hat{\mathbf{U}}^n))\mathbf{U}_x^n, \mathbf{W}_x^n)_n \tag{2.9}$$
$$+ \delta(\mathbf{U}_t^n + B(\hat{\mathbf{U}}^n)\mathbf{U}_x^n, \mathbf{W}_t^n + B^T(\hat{\mathbf{U}}^n)\mathbf{W}_x^n)_n + \langle \mathbf{U}_+^n, \mathbf{W}_+^n \rangle_n$$
$$= (\mathbf{f}, \mathbf{W}^n + \delta(\mathbf{W}_t^n + B^T(\hat{\mathbf{U}}^n)\mathbf{W}_x^n))_n + \langle \mathbf{U}_-^n, \mathbf{W}_+^n \rangle_n \qquad \forall \, \mathbf{W}^n \in \mathring{V}_h^n,$$

where I is the unit matrix.

The main difference between the schemes (2.9) and (2.7) is the presence of the matrix $C(\hat{\mathbf{U}}^n)$. Let us consider the pointwise behaviour of the matrix $C(\mathbf{U})$. From (2.8) we see that $C(\mathbf{u}) = 0$ for the exact solution $\mathbf{u}$. If the exact solution $\mathbf{u}$ is smooth in some region Q_0, then the discrete solution $\mathbf{U}$ should approximate it well, and $C(\mathbf{U})$ would be small in Q_0. Otherwise, if the exact solution $\mathbf{u}$ is rough at certain points, e.g., at shocks, then the discrete solution $\mathbf{U}$ cannot approximate it well, and the diagonal elements of $C(\mathbf{U})$ become large. This provides the desired local diffusion which damps out possible overshooting or undershooting in the discrete solution at shocks. Since the matrix $C(\mathbf{U})$ essentially acts at shocks, it is called a "shock–capturing matrix", and the scheme (2.9) is called a "shock–capturing scheme".

We note that there is also a parameter $\bar{\varepsilon}$ in the shock-capturing term which is taken of size $\bar{\varepsilon} = O(h)$. This is because the matrix $C(\hat{\mathbf{U}}^n)$ has also some order of h as discussed above. This shock–capturing concept avoids the extra computations for the entropy functions. Compared to the scheme (2.7), the scheme (2.9) requires only the additional calculation of the matrix $C(\hat{\mathbf{U}}^n)$.

3. A Test Case: The Shock Tube Problem

As a test case for the scheme described in the preceding section we choose the "Riemann shock tube" problem: In a long, thin cylindrical tube filled with gas two states differing in pressure and density are separated by a thin membrane. At the initial instant $t = 0$, the membrane is removed. The problem now is to determine the ensuing motion of the gas. Let (ρ, v, p, e) denote the density, velocity, pressure and total energy of the gas. Then, neglecting diffusive effects the motion can be described by the differential system

$$\frac{\partial}{\partial t} \begin{bmatrix} \rho \\ \rho v \\ e \end{bmatrix} + \frac{\partial}{\partial x} \begin{bmatrix} \rho v \\ \rho v^2 + p \\ v(e + p) \end{bmatrix} = 0, \tag{3.1}$$

and

$$e = \frac{1}{2}\rho v^2 + \frac{p}{\gamma - 1}, \tag{3.2}$$

with the initial conditions

$$\begin{cases} \rho_l = 1.0 \\ v_l = 0.0 \\ p_l = 1.0 \\ e_l = 2.5 \end{cases} \text{for } x < 0.5, \qquad \begin{cases} \rho_r = 0.125 \\ v_r = 0.0 \\ p_r = 0.1 \\ e_r = 0.25 \end{cases} \text{for } x > 0.5, \tag{3.3}$$

where γ is a positive constant, usually $\gamma = 1.4$.

Taking (ρ, v, p) as the independent variables and differentiating the equations (3.1) with respect to each single component we obtain

$$\mathbf{u}_t + B(\mathbf{u})\mathbf{u}_x = 0, \tag{3.4}$$

where

$$\mathbf{u} = \begin{bmatrix} \rho \\ v \\ p \end{bmatrix}, \qquad B(\mathbf{u}) = \begin{bmatrix} v & \rho & 0 \\ 0 & v & 1/\rho \\ 0 & \gamma p & v \end{bmatrix}. \tag{3.5}$$

The matrix B has 3 different eigenvalues

$$\lambda_1 = v + c, \quad \lambda_2 = v, \quad \lambda_3 = v - c,$$

with $c = \sqrt{\gamma p/\rho}$. Hence, the matrix B is diagonalizable and the problem (3.4) is of purely hyperbolic type.

For discretizing the shock tube problem, we use uniform triangulations in each space–time slab S_n, as shown in Figure 1, i.e., all nodal points lie on the horizontal lines $\{t = t_n\}$. This gives the system in each time step, $t_n \to t_{n+1}$, a structure similar to a one-dimensional problem. Numbering the nodes vertically leads to a band matrix, with minimal band width $m = 23$, which is inverted by Gaussian elimination.

In the test computation the parameters in the streamline diffusion method were chosen as $\delta = 0.4h$, $\varepsilon_m = 0.4h^{3/2}$ and $\bar{\varepsilon} = 0.4h$, where the factor 0.4 was experimentally determined. The quality of the computational results depends sensitively on the proper choice of this factor: Taking it too large results in an undesirable smearing of the shocks, taking it too small leads to overshooting or even instability. The results of a first test on an equidistant space–time mesh with $\Delta x = \Delta t = 1/200$ are shown in Figure 7. This computation took 460 seconds CPU–time (on an IBM 4381–P12). The discrete solution does not show any overshooting at the shocks, but the shock resolution is not very good.

242

The mesh size 1/200 is still too coarse for this problem. Accordingly, **Figure 8** shows the results of a second test on a uniformly refined mesh with $\Delta x = \Delta t = 1/1000$, with all the other parameters chosen as before. Now the shock resolution appears satisfactory, but the computation took more than 3 hours CPU-time, even for this simple one–dimensional problem. How to make this method more efficient by a proper local mesh refinement is the main theme of the following discussion.

4. Local Mesh Refinement

The first tests on the shock tube problem have shown that in a hyperbolic system the accurate computation of a shock solution requires a very good resolution. Since the use of a sufficiently fine *equidistant* mesh leads to unacceptably long computing time, even in one–dimensional problems, it is natural to think of a *local* mesh refinement. The initial value of problem (2.1) may have isolated discontinuities which propagate with time in the characteristic directions, i.e., the exact solution may be discontinuous across the characteristics. Using the theory of ordinary differential equations one infers that the characteristic curve are continuous, possibly with some wrinkles. Thus, the discontinuity lines of the exact solution are continuous curves in the (x,t)–plane, which are well separated from each other. In a hyperbolic system of M variables at any point there are M characteristics along which discontinuities can travel. Therefore, a local mesh refinement has to follow all these M directions simultaneously. Figure 5 shows the refinement region for a situation with two characteristics.

This characteristic property indicates that it may be sufficient to refine the mesh only locally where the solutions are rough. For this there are several possible procedures. A refinement only in space results in a mesh as shown in Figure 2 which has the advantage that the number of unknowns increases only linearly with Δx. However, a test computation shows that the approximate solutions with $\Delta t = 1/200$ and the smallest step in the x–direction $\Delta x_{min} = 1/1000$ are not satisfactory, though we have reduced the mesh size near the shocks by a factor of 5 in space. In nonstationary problems mesh refinement *only in space* is insufficient for an accurate shock resolution.

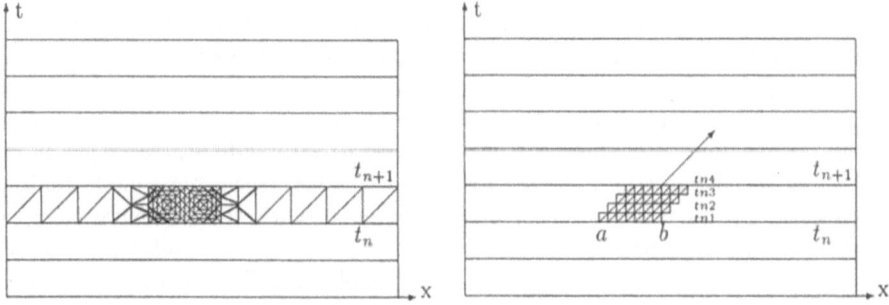

Figure 3: Another local refinement   Figure 4: Correction steps

For a local refinement of the mesh in space *and* time along the shocks one may think of a refinement strategy as shown in Figure 3. Such a refinement with slowly varying mesh size would be perfectly admissible within the streamline diffusion finite element method. However, due to the continuity requirement on the finite element functions on each space–time slab S_n, this would at least locally lead to a really two–dimensional system, i.e., to a matrix with a much bigger band width, and thereby significantly increase the computational costs. On the other hand, keeping the finite element functions discontinuous in time within the local refinement regions requires proper local boundary values for the implicit time stepping process. Thus, there is a conceptional problem in local space–time mesh refinement within an implicit scheme.

5. The Predictor–Corrector Scheme

Our solution to the dilemma of the missing boundary conditions for local space–time mesh refinement within an implicit time stepping scheme is a "Predictor–Corrector" concept. The idea is to preserve the band structure of the system matrix by consequently using finite elements discontinuous in time and computing boundary values for the refinement regions through a cheap predictor step. In this scheme each time step from t_n to t_{n+1} consists of two substeps. First, in the "Predictor Step", one computes a discrete solution on a uniform "coarse" mesh in the space–time slab S_n which is fine enough to capture the basic features of the solution, particularly the shock positions. From this coarse grid solution the approximate position of the shocks is detected and an appropriate neighborhood of each shock, say some strip $(a, b) \times I_n$, is selected for a subsequent mesh refinement. It is assumed that this coarse mesh solution is already sufficiently accurate in those region where the solution is smooth. Only in some neighborhood of the shocks the resolution may still be insufficient. In the subsequent "Corrector Step" the discrete solution is locally re-calculated within the strip $(a, b) \times I_n$ on a refined space–time mesh. The boundary values at $x = a$ and $x = b$ in this local implicit time stepping are taken from the coarse grid solution. This, of course, requires that the local boundaries are distant enough from the shocks in order not to feel their pollution effect too much. The quantification of this qualitative statement, which is the contents of our theoretical analysis of this predictor-corrector scheme, will be discussed below.

Now we describe the correction step in more detail. In each of the local refinement regions $(a, b) \times I_n$ the original problem (2.1.a) is considered with the initial value U^n_- from the previous slab and boundary values which are taken from the coarse grid solutions simply by restriction to the fine mesh. In order to adapt the mesh refinement even better to the shock structure the refinement interval (a, b) may be shifted accordingly to the movement of the shocks after each sub-time step. The local space–time mesh resulting through such a process is shown in Figure 4. Here we start with the computation of the local solution on the subslab $(a, b) \times (t_n, t_{n1})$, and then proceed from subslab to subslab until we reach the time level $\{t = t_{n+1}\}$. Finally the obtained local solutions are substituted for the coarse grid solutions at the time level $\{t = t_{n+1}\}$. We mention again that the advantages of the described procedure over the standard mesh refinement in space–time is mainly in its computational simplicity as it preserves the original band structure of the system matrix, corresponding to the spatially one–dimensional problem (2.1).

The predictor–corrector scheme requires the a posteriori determination of the shock position from the coarse grid solution and the choice of the subintervals (a, b) for the local mesh refinement. The first task is reached using a criteria based on the local behavior of the first and the second derivatives of the discrete solutions; details on this aspect will be given elsewhere. The strategy for estimating the necessary width of the subslabs $(a, b) \times I_n$ is based on a rigorous mathematical estimate of the error pollution caused by the shocks. In the streamline diffusion scheme (2.7) the artificial viscosity is of size $O(\varepsilon_m)$, and the shocks in the approximate solution are smeared over an interval of width $O(\varepsilon_m^{1/2} |\log \varepsilon_m|)$. Based on our theoretical considerations we take ε_m as in (2.6) and have for the shock smearing the width $O(h^{3/4} |\log h|)$; see the detailed discussion in the subsequent section. As a consequence the local mesh refinement has to be done in a strip along the shocks with a minimal width of size $O(h^{3/4} |\log h|)$.

The following results have been obtained using the described predictor–corrector strategy applied for solving the shock tube problem. They have to be compared with the results of the computations on uniformly refined meshes, c.f., Figures 7 and 8. Figure 9 shows the results of a computation with globally $\Delta t = \Delta x = 1/100$ in the prediction step, and locally $\Delta t = \Delta x = 1/500$ in the correction step. The width of the refinement regions was choosing as 0.1. It appears that the resulting shock resolution is significantly better than that on the uniform mesh with $\Delta t = \Delta x = 1/200$, shown in Figure 7, while the computational costs are nearly the same. To carry this test further, Figure 10 shows the results obtained with $\Delta t = \Delta x = 1/200$ in the prediction step, and with $\Delta t = \Delta x = 1/1000$ in the correction step. Now it appears that the shock resolution is nearly as good as that of the calculation on a uniform mesh with $\Delta t = \Delta x = 1/1000$, but takes only 20% of the CPU-time (on an IBM 4381-P12). These results demonstrate that the proposed mesh refinement strategy based on the predictor–corrector concept is functional and provides a significant gain in solution efficiency in the one–dimensional case. In higher dimensions we expect this gain to be even more significant at least in situations with a relatively simple shock structure. This will be the subject of further numerical tests.

6. Results from the Theoretical Error Analysis

Now, we present some results from Zhou [20] which provide a theoretical basis for the predictor–corrector scheme described above. Since this analysis is given for the linear case only, its implications for the nonlinear system (2.1) remain somewhat heuristic. The first result concerns the linear scalar problem in d space dimensions. Let (x_0, t_0) be any point in Q, such that the solution u is smooth in a dependence subdomain Q_0 of (x_0, t_0), as shown in Figure 6, which has width $O(h^{3/4} |\log h|)$ and extends backwards in the characteristic direction to the initial time $t = 0$. Then, on a quasi–uniform space–time mesh, there holds the local pointwise error estimate

$$|(u - U)(x_0, t_0)| \le C h_{loc}^{(12-d)/8} |\log h| \|u\|_{C^2(Q_0)} + C_\nu h^\nu \left(\|\varepsilon \Delta u\|_{L^1(Q \setminus Q_0)} + \|u\|_{L^\infty(Q)} \right),$$
$$(6.1)$$

where $\nu > 0$ is arbitrary, and h_{loc} denotes the "local" mesh size in the subdomain Q_0. In the purely hyperbolic case ($\varepsilon = 0$), the residual term $\|\varepsilon \Delta u\|_{L^1(Q \setminus Q_0)}$ does not occur in

(6.1). In the "essentially" hyperbolic case, $0 < \varepsilon \ll 1$, it appears reasonable to assume that $\varepsilon \Delta u \in \mathbf{L}^1(Q)$. The error estimate (6.1) can be extended to linear systems of the

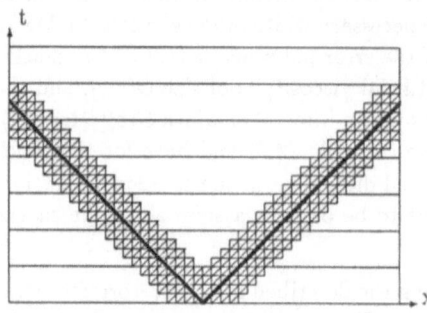

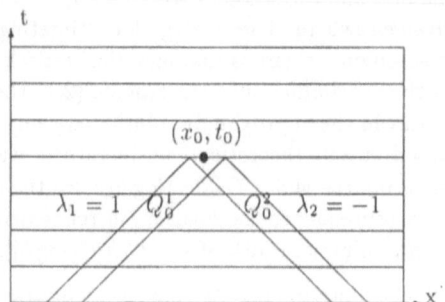

Figure 5: Following two shocks   Figure 6: Dependence subdomains

type (2.1) in one space dimension ($d = 1$),

$$|(\mathbf{u} - \mathbf{U})(x_0, t_0)| \leq C h_{loc}^{11/8} |\log h| \sum_{j=1}^{M} \|\tilde{u}_j\|_{C^2(Q_0^j)} \tag{6.2}$$

$$+ C_\nu h^\nu \left(\sum_{j=1}^{M} \|\varepsilon \tilde{u}_{jxx}\|_{L^1(Q \setminus Q_0^j)} + \|\tilde{u}\|_{L^\infty(Q)^M} \right).$$

Here, $\tilde{u} = D^{-1}\mathbf{u}$, and $Q_0^j \subset Q$ are the dependence regions of the point (x_0, t_0) in the M characteristic directions; Figure 6 shows this situation for the case $M = 2$. A more detailed discussion of this result together with a complete proof based on a careful $\mathbf{L}^1$-error analysis for discrete Green functions can be found in Zhou [20] and in Rannacher and Zhou [15]. The discontinuities in $\mathbf{u}$ may travel along all M characteristics intersecting at (x_0, t_0), but the transformed solution $\tilde{u}$ is piecewise smooth with well separated lines of discontinuity. This property allows one to distinguish "smooth" regions from "rough" ones for each single component of $\tilde{u}$. This may be illustrated through the following simple example: Suppose that each component $\tilde{u}_j$ has initially a jump at $x = 0.5$. Since the characteristic directions λ_j are usually mutually different, the solution $\tilde{u}$ contains M shocks travelling in M directions. More precisely, the component $\tilde{u}_j$ has a shock along the straight line $\{x - \lambda_j t = 0.5\}$, but is smooth outside this region. In contrast to this, under the variable transformation $\mathbf{u} = D\tilde{u}$ each component u_j has M shocks along the straight lines $\{x - \lambda_j t = 0.5\}$, $j = 1, ..., M$. Consequently, if we refine the mesh locally in the neighborhood of all M shocks, this is equivalent to refining the mesh near the shocks of $\tilde{u}$. This fact is of crucial importance for the convergence result. There, the discretization error at the point (x_0, t_0) depends upon the smoothness properties of the transformed solution $\tilde{u}$. If (x_0, t_0) happens to lie near a shock, one needs to refine the mesh only along this one shock rather than in all M characteristic directions. Note that it would not be sufficient if one refines the mesh only in a space–time neighborhood of (x_0, t_0) instead of the whole strip Q_0^j. Since the widths of the strips Q_0^j are only of size $O(h^{3/4}|\log h|)$, the refinement regions along the shocks in $\mathbf{u}$ also need to have only the widths $O(h^{3/4}|\log h|)$. At this distance from the shocks it is guaranteed that the discretization error is at most of the size $O(h^{11/8}|\log h|)$. This justifies taking the corresponding coarse grid values as boundary values for the local re-calculation in the $O(h^{3/4}|\log h|)$–neighborhoods of the shocks. The essence of this is that the error caused by the shocks, of course, propagates along the characteristics, but decays rather quickly in the crosswind direction. In fact, this decay appears to be even exponential, as was originally proved for the scalar *stationary* case in two dimensions by Johnson, Schatz and Wahlbin [10],

$$|(u - U)(x_0, t_0)| \leq Ch_{loc}^{5/4}|\log h|^{3/2}\|u\|_{C^2(Q_0)} + C_\nu h^\nu \left(\|\varepsilon\Delta u\|_{L^1(Q\setminus Q_0)} + \|u\|_{L^\infty(Q)}\right),$$
(6.3)

where $\nu > 0$ can be chosen arbitrarily large. The error estimates (6.1)–(6.3) are certainly not optimal in order. This is indicated by a result of Nävert [14], who proved related local error estimates in the $\mathbf{L}^2$–norm,

$$\|u - U\|_{Q_1} \leq Ch_{loc}^{3/2}\|u\|_{H^2(Q_1')} + O(h^\nu),$$
(6.4)

which are of the same order as the well-known error estimates for globally smooth solutions.

All the error estimates mentioned above have been obtained under the only assumption that the underlying meshes are quasi–uniform, i.e., the elements satisfy the uniform shape and size condition. If one imposes stronger restrictions on the meshes these results can be significantly improved. In a subsequent paper, Rannacher and Zhou [16], assuming a proper orientation and stretching of the meshes in the characteristic direction the stationary estimate (6.3) will be improved to the super–optimal order $O(h^{2-\frac{1}{8}}|\log h|)$. Further, there is another new result based on superconvergence property of finite elements on strongly uniform (but not necessarily oriented) meshes which contains a local error estimate of the type (6.3) with the almost optimal order $O(h^{3/2}|\log h|^{1/2})$; see Chen and Rannacher [1]. This result can even be carried further to a local asymptotic expansion of the error with respect to the several discretization parameters h, ε, δ, and ε_m, which provides the basis for local Richardson extrapolation or defect correction.

References

[1] Chen, H.; Rannacher, R.: Local error expansions and Richardson extrapolation for the streamline diffusion finite element method, 1993, to appear.

[2] Eriksson, E.; Johnson, C.: Adaptive finite element methods for parabolic problems IV: Nonlinear problems, to appear.

[3] Eriksson, E.; Johnson, C.; Thomée, V.: Time discretization of parabolic problems by the discontinuous Galerkin method, Math. Modelling and Numer. Anal., 19, 611-643(1985).

[4] Hughes, T.J.R; Mallet, M; Mizukami, A: A new finite element formulation for computational fluid dynamics: II. Beyond SUPG, Comput. Methods Appl. Mech. Engrg. 54, 341-355(1986).

[5] Hughes, T.J.R; Mallet, M: A new finite element formulation for computational fluid dynamics: III. The general streamline operator for multidimensional advective-diffusive systems, Comput. Methods Appl. Mech. Engrg. 58, 305-328(1986).

[6] Hughes, T.J.R; Mallet, M: A new finite element formulation for computational fluid dynamics: IV. A discontinuity–capturing operator for multidimensional advective-diffusive systems, Comput. Methods Appl. Mech. Engrg. 58, 329-336(1986).

[7] Johnson, C.: Finite element methods for convection-diffusion problems, in: Computing Methods in Engineering and Applied Sciences V, (R. Glowinski and J.L. Lions, eds.), North-Holland 1981.

[8] Johnson, C.; Nävert, U.: An analysis of some finite element methods for advection-diffusion, in: Analytical and Numerical Approaches to Asymptotic Problems in Analysis, (O. Axelsson; L.S. Frank, and A. van der Sluis, eds.), North-Holland 1981.

[9] Johnson, C.; Nävert, U.; Pitkäranta, J.: Finite element methods for linear hyperbolic problems, Comput. Methods Appl. Mech. Engrg. 45, 285-312(1984).

[10] Johnson, C.; Schatz, A.H.; Wahlbin, L.B.: Crosswind smear and pointwise errors in streamline diffusion finite element methods, Math. Comp., 49, 179, 25-38(1987).

[11] Johnson, C.; Szepessy, A.: On the convergence of a finite element method for a nonlinear hyperbolic conservation law, Math. Comp., 49, 180, 427-444(1987).

[12] Johnson, C.; Szepessy, A.: Adaptive finite element methods for conservation laws based on a posteriori error estimates, to appear.

[13] Johnson, C.; Szepessy, A.; Hansbo, P.: On the convergence of shockcapturing streamline diffusion finite element methods for hyperbolic conservation laws, Math. Comp., 54, 189, 107-129(1990).

[14] Nävert, U.: A finite element method for convection-diffusion problems, Thesis, Chalmers University of Technology, 1982.

[15] Rannacher, R.; Zhou, G.H.: Pointwise error estimate for nonstationary hyperbolic problems in the streamline diffusion method, 1992, to appear.

[16] Rannacher, R.; Zhou, G.H.: Mesh orientation and refinement in the streamline diffusion method, 1993, to appear.

[17] Smoller, J.: Shock Waves and Reaction–Diffusion Equations, Springer, Heidelberg, 1983.

[18] Walter, A.: Ein Finite–Elemente–Verfahren zur numerischen Lösung von Erhaltungsgleichungen, Thesis, Universität Saarbrücken, 1988.

[19] Walter, A.: Implementierung eines Finite-Elemente-Verfahrens für die Euler–Gleichungen der Gasdynamik, Preprint No. 500, SFB 123, Universität Heidelberg, 1989.

[20] Zhou, G.H.: An adaptive streamline diffusion finite element method for hyperbolic systems in gas dynamics, Thesis, Universität Heidelberg, 1992.

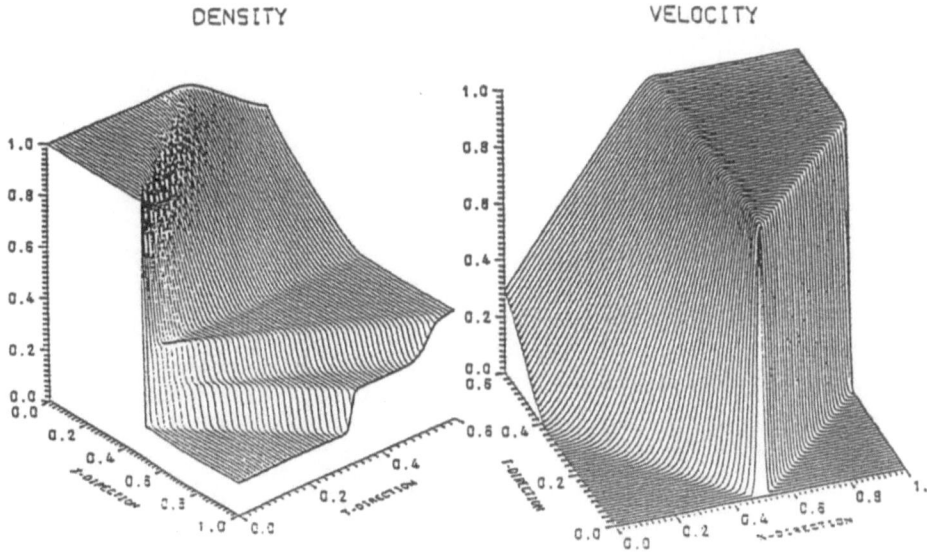

Figure 7: Results for an equi-distant mesh with $\Delta t = \Delta x = \frac{1}{200}$

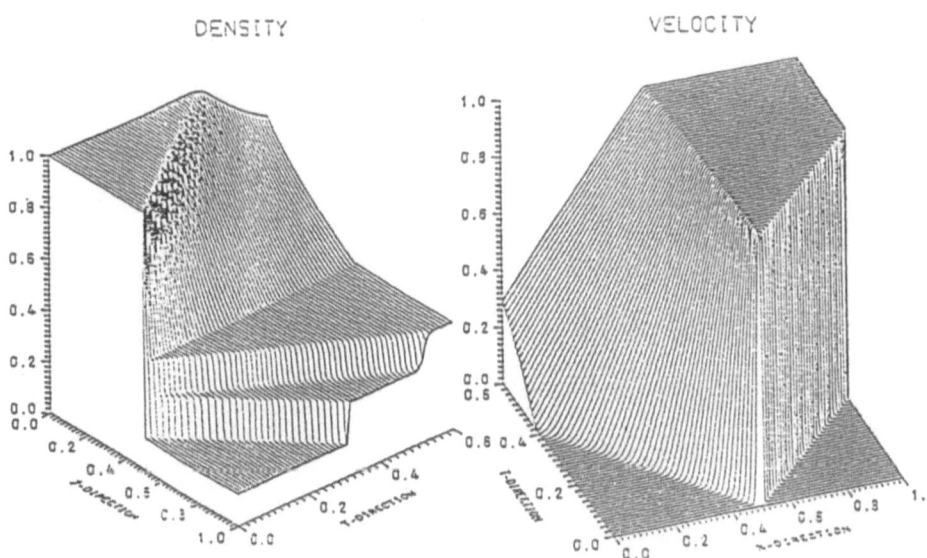

Figure 8: Results for an equi-distant mesh with $\Delta t = \Delta x = \frac{1}{1000}$

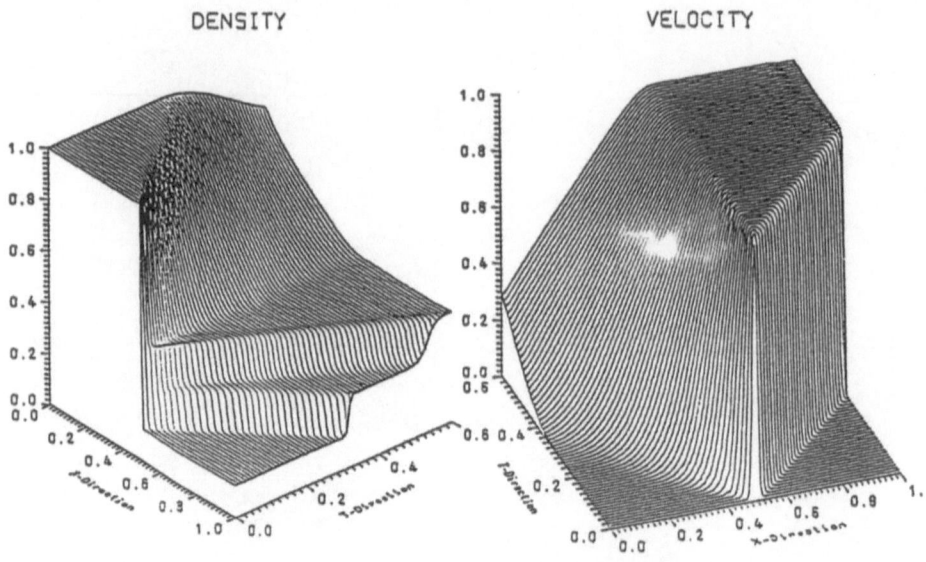

Figure 9: Results of the predictor-corrector scheme with $h_{\text{coarse}} = \frac{1}{100}, h_{\text{fine}} = \frac{1}{500}$

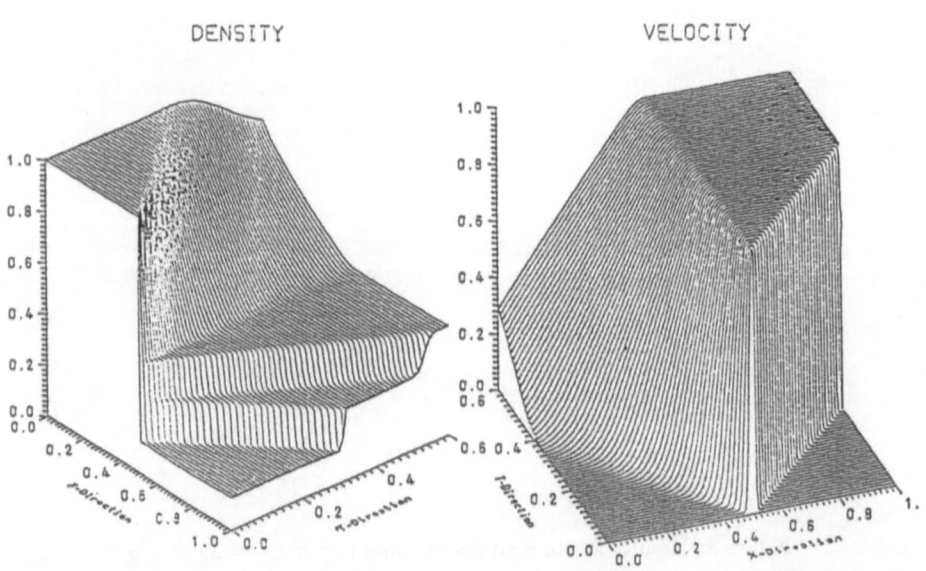

Figure 10: Results of the predictor-corrector scheme with $h_{\text{coarse}} = \frac{1}{200}, h_{\text{fine}} = \frac{1}{1000}$

On the V-Cycle of the
Fully Adaptive Multigrid Method

Ulrich Rüde
Institut für Informatik, Technische Universität
D-80290 München, Germany
e-mail: ruede@informatik.tu-muenchen.de

Summary

The Fully Adaptive Multigrid Method (FAMe) is a concept for the effective solution of elliptic problems including robust and efficient iterative solution, error estimation, and self-adaptive refinement. In this paper we introduce a variant of the FAMe similar in structure to a multigrid V-cycle and a multiplicative multilevel Schwarz method. This variant permits a convergence analysis showing that the FAMe provides optimal convergence rates when the classical methods do. The FAMe, however, will be more efficient in a local refinement context by exploiting the locality of the computations and will be more robust, because it naturally provides diagnostic information that can serve as rigorous error bounds.

Key words. Fully adaptive multigrid, multilevel iteration, error estimate, virtual global grids

AMS (MOS) subject classifications. 65F10, 65F50, 65N22, 65N50, 65N55.

1. Introduction

The *fully adaptive multigrid method* (FAMe) is a concept for the adaptive solution of elliptic partial differential equations combining *self-adaptivity, error estimation,* and *efficient iterative solution.* This technique is related to (product) iterative refinement (Widlund [16]), the fast adaptive composite grid method (FAC, McCormick [9]), hierarchical bases (Yserentant [17]), and adaptive multigrid algorithms (see e.g. Bai and Brandt [1], Bank [2]). However, it generalizes these techniques by giving up the strict separation of levels in the refinement and solution process and by providing an *adaptive solution strategy.*

We decide individually, whether an element is refined, using an integrated error indicator. The refinement process induces a multilevel structure and therefore a natural decomposition of the solution space. In contrast to the refinement algorithms of Bank [2], Rivara [11], or Leinen [6], we will not only extend the multilevel hierarchy by adding new subspaces, but can add elements on all existing levels, thus enlarging the old subspaces. This aids the construction of a well-balanced multilevel structure.

The flexibility in the mesh generation is reflected in the iterative solution process that is organized in two nested loops, where the inner one can be interpreted as an approximate subspace solver, the outer one as an iteration cycling between subspaces.

In Rüde [12] we have suggested to use *adaptive sequential relaxation* as approximate subspace solvers. In this paper, we introduce a new variant of adaptive relaxation based on Jacobi (or Richardson) relaxation, augmented with an *active set* strategy. The iteration is restricted to *active* unknowns. In each iteration we construct a new active set according to the dependencies between unknowns.

An outer loop cycles between subspaces. In addition to the usual transfer of numerical values between subspaces, we pass information to correctly initialize the active sets. In particular, if a subspace is visited more than once, it is likely that the solution has changed in a subdomain only, for example, where new refinement has been added.

This may be done for the full hierarchy of spaces, keeping all the computations completely *local* (see Rüde [14]). This includes relaxation within a subspace, as well as the transition between subspaces. Once the algorithm has collected enough knowledge about the structure of the problem, it will perform computations only in subdomains where the error is large and it will correctly keep track of the dependencies within a level and between the levels.

Monitoring the progression of the residual in all subspaces additionally provides an error estimate and a stopping criterion, see Section 2 and Rüde [13]. The resulting multilevel algorithm is fast and efficient for a wide range of problems. It is also robust, because sharp error estimates are enforced by the definition of the termination criterion.

Furthermore, with suitable data structures all the required processes can be performed with minimal overhead. The work remains proportional to the number of unknowns processed. This requires an efficient way to store sets of nodes. A suitable implementation is based on lists combined with flags for each element.

In the paper we present a description of the algorithm and discuss its theoretical foundation, based on results for hierarchical bases and the associated theory of multilevel preconditioners (cf. Yserentant [17]).

2. Multilevel algorithms

We study self-adjoint, scalar, elliptic equations in two dimensions that can be written as variational problems

$$\min_{u \in H(\Omega)} \left(A_\Omega(u, u) - 2(f, u)_\Omega \right). \tag{1}$$

For second order equations $H(\Omega) = H_g^1(\Omega)$ is the usual Sobolev space enforcing the essential boundary condition

$$u = g(x, y) \text{ on } \partial\Omega.$$

We assume that the problem is discretized by a nested sequence of finite element spaces

$$V_1 \subset V_2 \subset \cdots \subset H(\Omega) \tag{2}$$

induced by a nested sequence of triangulations. The corresponding discrete systems of dimension n_k in the nodal basis are denoted by

$$A_k x_k = f_k, \quad \text{for } k = 1, 2, 3, \cdots. \tag{3}$$

This multilevel structure induces two kinds of operations:

Relaxation: We will concentrate on the case of simultaneous relaxations, like

$$x_k \leftarrow x_k + \omega D_k^{-1}(f_k - A_k x_k), \tag{4}$$

where D_k is the diagonal of A_k and ω is an (under-)relaxation parameter.

Interpolation: Between any two subspaces $V_k \subset V_l$ an interpolation P_k^l is naturally defined by embedding, and its transpose (with respect to the discrete scalar product) defines an *projection* $P_l^k := (P_k^l)^T$.

The interpolations are related by

$$P_k^l = \prod_{\nu=k}^{l-1} P_\nu^{\nu+1}$$

and coarse mesh operators satisfy the *Galerkin* condition

$$A_k = P_l^k A_l P_k^l \quad \text{for } k < l. \tag{5}$$

For $V_l \subset V_K$ ($l \leq k$) we define the damped Jacobi V_l-*subspace correction* as

$$x_K' = x_K + \omega P_l^k D_l^{-1} P_K^l (f_K - A_K x_K), \tag{6}$$

or, written for the error $x_K - x_K^*$, where $x_K^* = A_K^{-1} f_K$:

$$x_K' - x_K^* = (I - \omega P_l^k D_l^{-1} P_K^l)(x_K - x_K^*).$$

Furthermore, we define the *current scaled residual* of x_K on level l ($l \leq K$)

$$\bar{r}_l = D_l^{-1} P_K^l (f_K - A_K x_K). \tag{7}$$

Successively applying subspace corrections for all embedded spaces $K, K-1, \cdots, 2, 1$, as shown in the pseudocode of Figure 1, leads to one of the simplest multigrid methods of V-cycle type.

This method is not competitive to well designed multigrid methods with several smoothings on each level. However, under suitable regularity assumptions it is already an asymptotically optimal method whose convergence rate is independent of the number of levels. Because of its simple structure we can describe the Simple-V-cycle by a plain product of operators.

$$\begin{aligned}
(u_K' - u_K^*) &= (I - \omega \bar{P}_1^T D_1^{-1} \bar{P}_1 A)(I - \omega \bar{P}_2^T D_2^{-1} \bar{P}_2 A) \\
&\quad \cdots \quad (I - \omega \bar{P}_K^T D_K^{-1} \bar{P}_K A)(u_K - u_K^*),
\end{aligned} \tag{8}$$

```
proc Simple-V-Cycle( $x_K$ )
    for $l = K$ to 1 step -1
            relax according to equation (6)
    end for
end proc
```

Figure 1: Simple multigrid V-cycle with 1 step of pre-smoothing only

where $u_K - u_K^*$ is the error before, and $u'_K - u_K^*$ is the error after an application of a Simple-V-cycle.

Convergence rates can now be bounded by estimating the norm of the operator product of equation (8). Corresponding V-cycle theories have e.g. been developed by Braess and Hackbusch [3], Maitre and Musy [7], Mandel, McCormick and Ruge [8], and Decker, Mandel, and Parter [5]. Recently, related statements have been proved in the theory of multilevel preconditioners that are free of regularity assumptions, see Yserentant [17], Bramble, Pasciak and Xu [4], Zhang [18], and Oswald [10]. The following Theorem is a direct consequence of these results.

Theorem 2.1 *Assume that the matrices A_k, $1 \leq k \leq K$, $A_K = A$ originate from a discretization of a linear, scalar, elliptic partial differential equation on a nested sequence of uniformly refined, shape regular triangulations T_k of a plane, polygonal domain with linear elements. Further let P_k^l be the discrete representations of the associated interpolation operators. Then there exist constants $0 < c_1 \leq c_2 < \infty$, such that*

$$c_1 \left\| \sum_{k=1}^{K} \bar{P}_k^T \bar{r}_k \right\|_2 \leq \|x_K - x_K^*\|_2 \leq c_2 \left\| \sum_{k=1}^{K} \bar{P}_k^T \bar{r}_k \right\|_2 , \tag{9}$$

where the current scaled residual $\bar{r}_k$ is defined in equation (7).

A proof can be found in Rüde [15].

Theorem 2.1 states that if the scaled residuals are small *on all levels*, then the error must be small. Equation (9) thus provides an error estimate based on the multilevel structure. By keeping track of the residuals on all levels we get sharp bounds on the L_2-norm of the error. The scaled residuals on the other hand can simply be interpreted as an Jacobi relaxation step on each level.

Most obviously equation (9) can be applied to bound the *algebraic error*, however, it is also useful to estimate the *discretization error*. If the solution x_k on some level k has been calculated, then the scaled residuals of the solution interpolated to the next finer level can be used as local error indicators. Based on equation (9) we get sharp error bounds by taking $k \longrightarrow \infty$ and by using regularity assumptions to find bounds for the contributions of levels $k > K$ not directly available.

It is important to realize that statements of the kind of equation (9) link together the essential components of elliptic solvers, that is *fast solution, error estimation,* and *local refinement.* The multilevel hierarchy is the key to devise good solutions to all these problems.

3. Virtual global grids

For this refinement technique we use globally refined grids covering the full domain. To accommodate adaptivity *ghost nodes* are introduced. These nodes need not be allocated in memory nor are there any computations performed with them. Ghost nodes do have associated numerical values defined by interpolation from the coarser grid, without needing to store or process them. This interpolation may have to be applied recursively, in the case that there are ghost nodes on the coarser grid, too.

If the value of a node is defined through interpolation, this can be interpreted that its *hierarchical transformation (hierarchical displacement)* vanishes. In the language of hierarchical bases, a ghost node is interpreted to have value = 0. The technique of ghost nodes is therefore analogous to sparse matrix techniques where a rectangular matrix is stored in a compact form, by using that many of its entries vanish.

The mechanism of multilevel algorithms can be reformulated so that nodes with vanishing hierarchical transform never appear in the real computations. A mechanism must be devised, however, to convert a ghost node to a *live node* if necessary to maintain the accuracy of the computation. This is accomplished by the *adaptive relaxation* algorithms as described in Rüde [14] and further below in Section 4.

With the ghost nodes we also relate the *discretization error* (caused by this node not being available for approximation) to the *algebraic error* (caused by this node having nonzero residual in the extended discrete system including the ghost nodes). Equation (9) quantifies this effect. Clearly, a ghost node must be converted to a live node, when the associated scaled residual becomes "large". Conversely, it should remain a ghost, as long as the residual is "small", because then it would be inefficient to allocate storage and perform computations for this node. "Large" and "small" must of course be interpreted in relation to the accuracy demands in the present stage of processing in the algorithm. The concept of performing computations only for subsets of nodal points is worked out in detail in the following sections.

4. Adaptive Jacobi relaxation

In this Section we will describe an adaptive variant of the Jacobi relaxation (6). For each level, we introduce the sets

$$S_k(\vartheta_k, x_k) = \{i | 1 \leq i \leq n_k, \; |\theta_i| > \vartheta_k\}, \tag{10}$$

where

$$\theta_i = \theta_i(x_k) = (\bar{r}_k)_i \tag{11}$$

is the current scaled residual of equation i on level k (see 7). As motivated in Section 2

255

```
proc AdaptiveJacobi( ϑ, x, S̃ )
    assert( S̃ ⊃ S(ϑ, x) )
    S = {i ∈ S̃|θᵢ(x) > ϑ}
    x ← x + ω∑ᵢ∈ₛ θᵢeᵢ
    S̃ ← Conn(S)
    assert( S̃ ⊃ S(ϑ, x) )
end proc
```

Figure 2: Simultaneous adaptive Jacobi relaxation

and Section 3, it may be appropriate to restrict a relaxation to the unknowns contained in $S_k(\vartheta_k, x_k)$, for an suitable tolerance ϑ_k, because it saves the full evaluation of the residual without affecting the convergence rate. Unfortunately, the set $S_k(\vartheta_k, x_k)$ is not known.

The following strategy allows to construct (and update) supersets of $S_k(\vartheta_k, x_k)$ with little overhead. The algorithm is described in Figure 2. This algorithm restricts the relaxation to equations that have large residuals, and where Theorem 2.1 indicates that large reductions in the error are possible. For sequential relaxations the reduction of the error can be calculated exactly, see Lemma 2.1 of Rüde [13]. The algorithm of Figure 2 needs a superset of $S_k(\vartheta_k, x_k)$ as its input, but will compute a new superset of S that may be used for further cycles.

The sparsity of the matrix is essential for this to work efficiently. For each unknown i, we define the set of *connections* by

$$\text{Conn}_k(i) = \{j|\, j \neq i, a_{j,i}^{(k)} \neq 0\}, \tag{12}$$

that is all nonzero entries in *column i* of the system matrix $A_k = (a_{i,j}^{(k)})$, excluding the diagonal. The sparsity structure of a matrix is expressed in the number of connections in each of its rows.

The algorithm of Figure 2 can be considered as a perturbation of the conventional Jacobi relaxation given in equation (6). This is expressed in the following lemma.

Lemma 4.2 *Assume routine AdaptiveJacobi is started with initial guess x_k, tolerance ϑ_k and an active set*

$$\tilde{S}_k \supset S(\vartheta_k, x_k) \tag{13}$$

and denote its result by x'_k. Then

$$x'_k = x_k + \omega D_k^{-1}(f_k - A_k x_k) + \mu_k, \tag{14}$$

where the perturbation is bounded by

$$\|\mu_k\|_\infty \leq \vartheta_k. \tag{15}$$

Furthermore at termination the new active set will again satisfy (13).

Proof. By the definition of $\theta_i(x_k)$ (see equation 11), we have

$$D_k^{-1}(f_k - A_k x_k) = \sum_{i=1}^{n_k} \theta_i e_i,$$

and correspondingly,

$$\mu_k = -\sum_{i \notin S} \theta_i e_i.$$

Therefore

$$\|\mu_k\|_\infty = \max_{i \notin S} |\theta_i| \leq \vartheta_k.$$

The set $\mathrm{Conn}(S)$ denotes all unknowns where the update of the unknown x can have effects, so that

$$\mathrm{Conn}_k(S(x_k, \vartheta_k)) \supset S(x'_{kr} \vartheta_k).$$

□

This lemma states that the AdaptiveJacobi routine for appropriately selected critical tolerance performs almost as good as regular Jacobi, however, it may be dramatically cheaper, if $|\tilde{S}_k| \ll n_k$. In this case a regular relaxation sweep is only effective where the residual is large. The computational work spent for equations with small residuals is mostly wasted.

5. Multilevel adaptive iteration

We will now describe how the adaptive Jacobi smoother can be used in the multilevel context. In routine Simple-V-Cycle of Figure 1, the standard Jacobi relaxation is replaced by the AdaptiveJacobi routine of Figure 2.

The key to exploiting the potential advantage of this algorithm is to initialize the characteristic sets appropriately for each level. Remember that an exact calculation of these sets would defeat the very purpose for their introduction. Fortunately, they can be determined by passing information between levels. In a nested iteration each level is visited several times. This is particularly true for our algorithm that has comparatively slow convergence so that several V-cycles must be applied on each level. Therefore it is important that information accumulated in the last visit to that level is re-used in the next visit.

When a level is relaxed, the adaptive Jacobi routine will construct a new active set for the next iteration. However, the structure of Simple-V-Cycle prescribes that before the same level is relaxed again, all other levels must be visited. The interpolation and restriction operators determine, how the levels influence each other.

These dependencies are local in nature. A change in an unknown on level k will influence only few unknowns on the neighboring levels $k-1$ and $k+1$. These dependencies must be traced, and used to properly initialize the active sets for each level. A detailed description of the necessary processes is given in Rüde [14]. We assume that this is done as described in Rüde [14] and concentrate on analyzing the numerical properties of the resulting method. In each V-cycle the perturbations described in Lemma 4.2 accumulate in the following way.

Theorem 5.3 *Let x be the input of routine* Simple-V-Cycle *(see Figure 1) for level K, where the relaxation is one sweep of* AdaptiveJacobi. *Let $\tilde{x}$ be its output. Assume that the sets satisfy $\check{S}_k \supset S(\vartheta_k, x_k)$ before the relaxation on level k for each k. Assuming*

$$\|I - \omega \bar{P}_k^T D_k^{-1} A\| \leq 1 \text{ for } k = 1, 2, \cdots, K, \tag{16}$$

then the perturbation caused by using AdaptiveJacobi *instead of regular Jacobi is bounded by*

$$\|\tilde{x} - x'\| \leq \sum_{k=1}^{K} \vartheta_k. \tag{17}$$

Proof. The result of the V-cycle with regular Jacobi smoothing is given in equation (8). The effect of AdaptiveJacobi has been described as a perturbation of regular Jacobi smoothing in Lemma 4.2 so that we can describe $\tilde{x}$ recursively.

$$\begin{aligned}
\tilde{x}^0 &= x, \\
\tilde{x}^k - x^* &= (I - \omega \bar{P}_k^T D_k^{-1} A)(\tilde{x}^{k-1} - x^*) + \mu_k, \quad \text{for } k = 1, 2, \cdots, K \\
\tilde{x} &= \tilde{x}_K.
\end{aligned}$$

This leads directly to the following recursive formula for the accumulated perturbations

$$\begin{aligned}
\tilde{\mu}_1 &= 0, \\
\tilde{\mu}_k &= (I - \omega \bar{P}_k^T D_k^{-1} A)\tilde{\mu}_{k-1} + \mu_k \quad \text{for } k = 1, 2, \cdots, K, \\
\tilde{x} - x' &= \tilde{\mu}_K.
\end{aligned}$$

With assumption (16) we have

$$\|\tilde{\mu}_K\| \leq \sum_{k=1}^{K} \|\mu_k\| \leq \sum_{k=1}^{K} \vartheta_k. \tag{18}$$

□

This theorem links the V-cycle with adaptive smoothing to a regular multigrid V-cycle. Therefore any convergence result for regular multigrid extends to the adaptive version. Though the analysis is far from being sharp it shows why and when the multilevel adaptive iteration is superior to a classical multigrid algorithm. This is always the case, when the characteristic features of the problem vary in the solution domain, so that certain subdomains need more attention than others. Typical examples are boundary and interior layers, jumps in the coefficients, singularities, and effects caused by adaptive mesh refinement.

As additional consequence of the adaptivity, the algorithm will be robust for a wide class of problems. The natural termination criterion for the cycling is that the active sets are empty on all levels, so that equation (9) directly yields an error bound.

The assumptions of Theorem 2.1 essentially say that we must have a suitable multilevel decomposition of the solution space (a *stable splitting*), in the sense that the projections are *approximately orthogonal*. It is important to realize that this, as for classical multilevel algorithms, is the core feature that makes the algorithm work. In extension of classical multigrid, the algorithm is robust even if the orthogonality is (mildly) violated. Strong violations can be diagnosed because the algorithm cannot reduce the residuals when the iteration in one subspace feeds into the residuals of another one. Though the algorithm will not be efficient if forced to find a solution by continued iterations, it stays robust and safe.

Acknowledgements. The idea to consider the FAMe as a perturbed multigrid algorithm has been developed in discussions of the author with P. Leinen and H. Yserentant in the fall of 1991.

References

[1] D. Bai and A. Brandt. Local mesh refinement multilevel techniques. *SIAM J. Sci. Stat. Comput.*, 8(2):109–134, March 1987.

[2] R. Bank. *PLTMG: A Software Package for Solving Elliptic Partial Differential Equations.* Frontiers in Applied Mathematics. SIAM, Philadelphia, 1990.

[3] D. Braess and W. Hackbusch. A new convergence proof for the multigrid method including the V-cycle. *SIAM J. Numer. Anal.*, 20:967–975, 1983.

[4] J. Bramble, J. Pasciak, and J. Xu. Parallel multilevel preconditioners. *Math. Comp.*, 31:333–390, 1990.

[5] N. Decker, J. Mandel, and S. Parter. On the role of regularity in multigrid methods. In S. McCormick, editor, *Multigrid Methods: Theory, Applications, Supercomputing: Proceedings of the Third Copper Mountain Conference on Multigrid Methods, April 5-10, 1987*, New York, 1988. Marcel Dekker.

[6] P. Leinen. Ein schneller adaptiver Löser für elliptische Randwertprobleme auf Seriell- und Parallelrechnern. Dissertation, Universität Dortmund, 1990.

[7] J.-F. Maitre and F. Musy. Multigrid methods: Convergence theory in a variational framework. *SIAM J. Numer. Anal.*, 21:657–671, 1984.

[8] J. Mandel, S. McCormick, and J. Ruge. An algebraic theory for multigrid methods for variational problems. *SIAM J. Numer. Anal.*, 25(1):91–110, February 1988.

[9] S. McCormick. *Multilevel Adaptive Methods for Partial Differential Equations*, volume 6 of *Frontiers in Applied Mathematics*. SIAM, Philadelphia, 1989.

[10] P. Oswald. On discrete norm estimates related to multilevel preconditioners in the finite element method. In *Proceedings of the International Conference on Constructive Theory of Functions, Varna, 1991*, pages 203–214, Sofia, 1992. Bulg. Acad. Sci.

[11] M. Rivara. Algorithms for refining triangular grids suitable for adaptive and multigrid techniques. *International Journal for Numerical Methods in Engineering*, 20:745–756, 1984.

[12] U. Rüde. Adaptive higher order multigrid methods. In W. Hackbusch and U. Trottenberg, editors, *Proceedings of the Third European Conference on Multigrid Methods, October 1-4, 1990*, pages 339–351, Basel, 1991. Birkhäuser. International Series of Numerical Mathematics, Vol. 98.

[13] U. Rüde. On the multilevel adaptive iterative method. In T. Manteuffel, editor, *Preliminary proceedings of the 2nd Copper Mountain Conference on Iterative Methods, April 9-14, 1992*. University of Colorado at Denver, 1992. Accepted for publication in SIAM J. Sci. Stat. Comput. and also available as TU-Bericht I-9216.

[14] U. Rüde. Fully adaptive multigrid methods. *SIAM J. Numer. Analysis*, 30(1):230–248, February 1993.

[15] U. Rüde. *Mathematical and computational techniques for multilevel adaptive methods*, volume 13 of *Frontiers in Applied Mathematics*. SIAM, Philadelphia, 1993.

[16] O. B. Widlund. Optimal iterative refinement methods. In T. F. Chan, R. Glowinsky, J. Périaux, and O. B. Widlund, editors, *Domain Decomposition Methods*, Philadelphia, 1989. SIAM.

[17] H. Yserentant. Two preconditioners based on the multi-level splitting of finite element spaces. *Numer. Math.*, 58:163–184, 1990.

[18] X. Zhang. Multilevel additive Schwarz methods. Tech. Report 582, New York University, Courant Institute, Department of Computer Science, 1991.

Wavelets and Frequency Decomposition Multilevel Methods

R. Schneider

FB Mathematik, THD

Schloßgartenstr.7, D-6100 Darmstadt

Summary

An additive decomposition of $L^2(\Omega)$ is defined through a suitable bases. The definition of such a basis and the transformation between the bases is performed by a cascade type algorithm using the prolongations of frequency decomposition multigrid methods. If this basis is an unconditional Schauder basis in $L^2(\Omega)$ it shares the main properties with wavelets. Particularly, it is, up to a renormalization, also an unconditional basis in a wide scale of Besov- and Triebel-Lizorkin spaces. Such a basis is appropriate for a good approximation of functions with local singularities. For Petrov Galerkin schemes for pseudodifferential equations or singular integral equations this basis gives rise to sparse representations.

1   Introduction

Recently, the development of *wavelets* have gained interest of several mathematicians and engineers. Main interest in applications is concentrated to the problem of image data compression and time frequency localization in signal processing. In view of various similarities to situations in the numerical solution of partial differential equations we like to review the problem of image data compresion.

Storing a picture pixelwise is often too expensive. In practice one want to store less datas, but without an essential loss of information. Appropriate tools for this purpose are mainly based on fast Fourier transform. Since Fourier transform serves an approximation by long waves $e^{i\omega t}$, this approach runs into difficulties when the function to be approximated has local discontinuities, e.g., it may be piecewise smooth. These problems are in spite of the high frequency components caused by these discontinuities. It seems to be an appropriate approach to approximate such a functions in terms of short waves in order to capture the local discontinuities. From the Fourier analytical point of view, this requires a suitable simultaneous localization in domain (or time) and frequency. One has to find a compromise which is limited in principle by the well known Heisenbergs uncertainty principle. There exist several attempts to tackle this problem, e.g. by windowed Fourier transform and other mainly Fourier based approaches. Recently *wavelets* becomes an alternate tool for this purpose. It seems to be particularly advantageous in situations where one deals with functions which are smooth up to a set of singularities of comparable small Hausdorff dimension.

In contrast to the Fourier based methods, wavelets are closely related to finite element methods, in particular to adaptive and multigrid methods. Here we want to emphazise that the present paper and our recent investigations are not mainly addressed to the purpose of an economical approximation of local singularities of the solutions by local refinements. We focuse our prior interest to an adaptive approximation of the singularities in the Schwartz kernel of nonlocal operators, e.g. integral operators, pseudodifferential operators etc. [7, 8, 9]. Usually, the discretization of nonlocal operators by means of local functions gives rise to full matrices. Typically this situation arises in boundary element methods and vortex methods, for example. This restricts the applicability of these methods seriously. Indeed, dealing with full matrices seems to be prohibitively expensive for large scale computations. For special kernels, e.g. Coulomb potential, a suitable approximation of their singularities by means of multipole expansion [31] and panel clustering [17] have been the first successful attempts to a sparse realization of the corresponding matrix vector multiplication. Beylkin, Coifman and Rohklin [1] found a generalization of these methods in terms of a wavelet Galerkin method. In their approach an explicit multipole expansion of the Schwartz kernel is not required. Based on these ideas, wavelet approximation methods, including a wide class of Petrov-Galerkin methods for periodic pseudodifferential equations, containing e.g. collocation methods, have been investigated in [7, 8].

From an independent point of view, [4] have described a method based on multigrid techniques which is intimately related to the wavelet method of [1]. While multigrid prolongations are mainly based on reconstruction by interpolation, [20] extended the method of [4] by applying ENO type reconstructions. For these approaches a rigorous analysis like in [8] is yet not available.

Unfortunately, all known constructions of compactly supported wavelets are based on multiresolution analysis in $I\!R^n$ or $[0, 1]$, and are restricted to uniform grids. The explicit orthogonalization used in [22] results in only in exponentially decaying wavelets and seems to be quite expensive.

Parallel to the development of wavelets, there has been recent progress in the development of iterative solvers for elliptic problem combining the strategies of multigrid methods with domain decomposition ideas. In [3], the difficulty of constructing a stable basis have been avoided by replacing the direct decomposition by a redundant decomposition. Recognizing the connection of these schemes to Besov norm characterizations [29, 10, 2], it has been shown that the BPX preconditioner applied to the stiffness matrix of strongly elliptic boundary value achieves uniformly bounded condition numbers.

According to the closed similarity between the V-cycle of the frequency decomposition multigrid methods and the pyramide algorithm, sometimes called fast wavelet transform [1], we applied the prolongations of frequency decomposition multigrid methods as a (nonuniform) mask or filter coefficients for the definition of wavelets or better of wavelet like functions. While a theoretical justification is in preparation, first numerical experiments have been performed. In the present paper we can only outline some basic principles of the construction of a stable multilevel decomposition, or a nonuniform multiresoluition analysis [22], together with a short discussion. Whereas a systematical treatment containing detailed proofs and representation of the numerical experiments are defered to forthcoming papers.

In contrast to classical multigrid methods, which is an ingenious combination of a multilevel approach with an iteration scheme, we seperate the multi level strategy from the iteration procedure and confine ourselves to the multilevel decomposition. A thouroughly study of suitable smoothers may be done seperately. Showing the intimate connection beween multigrid and wavlet methods we want to point out that the advantage of multilevel methods is not completely covered by iterative solution methods. Throughout this paper we reserve the expression multigrid method for iterative multilevel solvers whereas the expressions multilevel or multiscale schemes are used in a more general context.

2 Multilevel Decomposition and Wavelets

2.1 Function spaces

In the sequel we consider Ω to be a bounded domain or a compact manifold in $I\!\!R^n$.

Basically, we consider a nested sequence of function spaces

$$V^0 \subset V^1 \subset \cdots \subset V^j \subset V^{j+1}$$

respectively grids

$$\square^0 \subset \square^1 \subset \cdots \subset \square^j \subset \square^{j+1},$$

subordinated to a *regular* (cf. [5]) triangulation with mesh size $h_j \sim 2^{-j}$.

The starting point for any concrete discretization is the choice of a suitable basis. Conventionally, we suppose V^j to be generated by a local basis, i.e.,

$$V^j = \text{span } \{\varphi_k^j : k \in \square^j\} \ . \tag{2.1}$$

Where we have used the normalization $\|\varphi_k^j\|_0 = \|\varphi_k^j\|_{L^2(R^n)} = 1$. Necessarily it will be assumed that this functions are suffiently regular, $V^j \in H^s(I\!\!R^n)$, $s < \lambda$, where $H^s(\Omega)$ denote the usual Sobolev spaces. For practical reasons we suppose

$$\text{diam supp}\varphi_k^j < c \ 2^{-j} \ .$$

We call $\{\varphi_k^j\}$ a scaling function basis. Generally, this basis must not be an interpolatory basis. Well known examples are provided e.g. by B-splines.

One major assumption is the **Stability** (Riesz-basis)

$$C_1(\sum_{k \in \square^j} |u_k|^2)^{\frac{1}{2}} \le \| \sum_{k \in \square^j} u_k\varphi_k^j\|_{L^2(\Omega)} \le C_2(\sum_{k \in \square^j} |u_k|^2)^{\frac{1}{2}} \ . \tag{2.2}$$

All these requirements are traditional in finite element methods subordinated to quasiuniform triangulations and also in spline analyis. The following properties are well known for this examples and are supposed to be valid in the sequel.

The spaces V^j are assumed to satisfy the well known **approximation property** (Jackson type inequality) and the **inverse Property** (Bernstein type inequality).

2.2 Hierarchical bases

We consider a (nonoverlapping) additive (Schwarz) decomposition

$$V^j = V^{j-1} + W^{j-1} = \sum_{l=-1}^{j-1} W^l \ , \quad W^l \cap V^l = \{0\} \ .$$

For notational convience, we have set $W^{-1} := V^0$.

The spaces W^l are again defined through a suitable basis

$$W^l = span\{\psi_k^l : k \in \Delta^l := \Box^l \backslash \Box^{l-1}\} \ .$$

The basis $\{\psi_k^l\}_{-1 \le l < j, k \in \Delta^l}$ is called a (generalized) hierarchical basis. From the inclusions $V^l \subset V^{l+1}$, $W^l \subset V^{l+1}$ we infer the following representations

$$\varphi_K^{j-1} = \sum_{k \in \Box^j (\text{finite})} a_{K,k}^j \varphi_k^j \ , \quad K \in \Box^{j-1} \ ,$$

and by definition

$$\psi_K^{j-1}(x) = \sum_{k \in \Box^j, \text{finite}} b_{K,k}^j \varphi_k^j \ , \quad K \in \Delta^{j-1} = \Box^j \backslash \Box^{j-1} \ . \tag{2.3}$$

These relations are refered in wavelet theory as refinement or two scale relations, whereas in multigrid methods the above relations are used for restrictions and prolongations. For given $f_k^j = (f, \varphi_k^j)$, $k \in \Box^j$, we compute $w_k^{j-1} = (f, \psi_k^{j-1})$, $k \in \Delta^{j-1}$, and $f_k^{j-1} = (f, \varphi_k^{j-1})$, $k \in \Box^{j-1}$, applying the refinement relations

$$w_K^{j-1} = \sum_{k' \in \Box^j} \overline{b_{K,k'}^j} f_{k'}^j \ , \quad K \in \Delta^{j-1} = \Box^j \backslash \Box^{j-1} \ , \tag{2.4}$$

and

$$f_K^{j-1} = \sum_{k' \in \Box^j} \overline{a_{K,k'}^j} f_{k'}^j \ , \quad K \in \Box^{j-1} \ . \tag{2.5}$$

If the functions φ_k^{j-1} and ψ_k^{j-1} are local, the transformation $(f_k^j) \to (f_K^{j-1}, w_L^{j-1})$ is performed by a sparse matrix $L^j = (A^j, B^j)$. The invertibility and stability of this transformations is an important requirement for a suitable stable decomposition.

Definition 1 : A hierarchical basis satisfies the (uniform) *2-level stability* if

$$\| \sum_{k \in \Box^j} a_k^j \varphi_k^j + \sum_{k \in \Box^{j+1} \backslash \Box^j} b_k^j \psi_k^j \|_{L^2(\Omega}^2 \sim \sum_{k \in \Box^j} |a_k^j|^2 + \sum_{k \in \Box^{j+1} \backslash \Box^j} |b_k^j|^2 \ ,$$

and for computational reasons we suppose further the locality of ψ_k^j,

$$\text{diam supp } \psi_k^l < c2^{-l} \ .$$

Example 1 : As an example we consider continuous piecewise linear functions subordinated to a regular [5] triangulation. Each triangle has been refined by division into four regular triangles like in e.g. [32]. A canonical choice of the scaling functions basis is the nodal basis. This fixes the coefficients $a_{K,k}^j$. We note that each $K \in \Delta^j = \Box^{j+1} \backslash \Box^j$ lies on some edge connecting two vertices $k_l, k_r \in \Box^j$. We set

$$b_{K,k} = \begin{cases} -1 & \text{if } k = k_l, k_r \\ 2 & \text{if } k = K \end{cases}$$

Similar choices of $b_{K,k}^l$ are proposed in [18, 19].

2.3 Pyramide or Casectionade (or Tree) Algorithm

Once the uniform 2-level stabilty is guaranteed, we can continue the procedure 2.5, 2.4, by recursion. The arising algorithm is called *Pyramide or Cascade (or Tree) Algorithm* for

Decomposition

$$f_k^j \;\rightarrow\; f_k^{j-1} \;\rightarrow\; f_k^{j-2} \;\cdots\; \rightarrow\; f_k^0$$
$$\searrow \qquad\quad \searrow \qquad\quad \searrow$$
$$w_k^{j-1} \qquad w_k^{j-2} \;\cdots\; \qquad w_k^0 \;.$$

It requires totally $\mathcal{O}(2^{jn} = N)$ nontrivial operations if the corresponding bases are local. The adjoint procedure has the form of the

Reconstruction

$$w_k^0 \qquad w_k^1 \qquad \cdots \qquad w_k^{j-1}$$
$$\searrow \qquad\quad \searrow \qquad\quad \searrow$$
$$f_k^0 \;\rightarrow\; f_k^1 \;\rightarrow\; f_k^{j-1} \;\cdots\; \rightarrow\; f_k^j$$

The reconstruction scheme is also a fast transform if we apply the adjoint matrices $((A^j)^*, (B^j)^*)^T$. We should be aware that this transform is **not** the inverse transform.

It is quite important to notice that due to the recursive nature of the above transformation, the 2-level stability does **not** imply the stability of the complete transformation. It gives only a very crude bound c^j, where $c > 1$. Usually, this bound is not sufficient for our purpose. We need a stronger condition.

Definition 2: The hierarchical basis $\{\psi_k^l\}_{k\in\Delta^l, -1\le l<\infty}$ is called a (generalized) *wavelet basis* if it is a Riesz-basis in $L^2(\Omega)$.

Although 2-level stability is a severe requirement, for particular functions it is possible to construct the coefficients of B^l by hand applying e.g. discrete orthogonality. This approach is mainly used in [19] and applies e.g. to piecewise linear functions. For piecewise constant functions φ_k^j Harten [20] use ENO reconstructions for a systematical development. It seems to be very difficult to find similar constructions for function spaces W^j corresponding to higher regularity in case of nonuniform triangulations.

Unfortunately, two-level stability is not sufficient to define a wavelet basis. The classical hierarchical basis studied in [32] provides an immediate counterexample. This basis is not a Riesz basis in $L^2(\Omega)$, because the condition number of the complete transformation is bounded from below by

$$\text{cond}\; \Pi_{l=0}^j L^{(l)} \ge c\; 2^{jn/2} \;.$$

The classical hierarchical basis can be shown to be a Riesz basis in $H^s(\Omega)$ if $s > \frac{n}{2}$. Nevertheless, it satisfies the 2-level stability and the corresponding inverse transformation is of the form (C^l, D^l). This inverse was described in [32]. It seems to be remarkable that the appearing matrix $(D^l)^* = B^l$ is the same as used in example 1 for the definition of ψ_k^l. This yields together with the dicrete orthogonality the validity of the 2-level stability of the basis proposed in the present paper.

Proposition 2.1 *The basis defind by example 1 satisfy the 2-level stability.*

We want to emphasize that even for uniform grids the construction of a stable decomposition, i.e. a biorthogonal wavelet basis is by no means trivial. For this developments we refer to the constructions of prewavelets and biorthogonal wavelets [23, 6, 13], for example.

By the way we have developed all ingredients to transform the stiffness matrix relative to the scaling function basis into the corresponding stiffness matrix relative to the wavelet basis. This transformation requires the application sparse matrices $(L^{(l)})^*$. This kind of transformation was based on the discrete wavelet transform [26] and has applied in [1]. A corresponding algorithm has been developed independently in [32] and it is also an ingredient of the frequency decomposition multigrid algorithm in [18, 19].

Observation: For the numerical solution of operator equations we usually need only the decomposition (restriction) and its *transpose or adjoint* transform (prolongation) for the reconstruction algorithm. We do not require explicitly the exact reconstruction which is the inverse of the complete transformation. This is in contrast to problems in image data compression and signal analysis where one needs decomposition as well as exact reconstruction.

2.4   Stable multilevel decomposition

All previous constructions of wavelets have been based Fourier transform techniques. On non-uniform grids Fourier transform seems to be a prohibited tool.

The following theorem, establishing a necessary condition, indicates that the question of stability of a hierarchical basis is not quite trivial.

THEOREM 2.2 *If a hierarchical basis with 2-level stability is a wavelet basis then the function*

$$\sum_{l=-1}^{\infty} \sum_{k \in \Box^l \backslash \Box^{l-1}} \psi_k^l (\psi_k^l, 1) \ \in \ BMO \, .$$

Here BMO denotes the space of functions of bounded mean oscillation. $\|f\|_{BMO}^2 = \sup_\Omega |\Omega|^{-1} \int |f(x) - |\Omega|^{-1} \int f(y) dy| dx$.

Here we will only sketch some ideas and postpone a detailed treatment to a seperate issue. It is natural, in this context, to consider virtually all grids and ψ_k^l, $-1 \le l < \infty$. We remind that a collection of functions $\{\psi_k^l\}$ constitutes a *frame*, if there exist constants $c_1, c_2 > 0$ such that

$$c_1 \sum_{l,k} |(u, \psi_k^l)|^2 \le \|u\|_0^2 \le c_2 \sum_{l,k} |(u, \psi_k^l)|^2 \, .$$

Frames are frequently used in signal analysis. If a frame consist of linear independent functions it constitutes also a Riesz basis. The left frame bound is equivalent to the boundednes of the composition of frame operators $T = F^* F$. Here T is explicitly given by

$$T u = \sum_{l \ge -1} \sum_{k \in \Delta^l} (u, \psi_k^l) \psi_k^l \, .$$

A key observations is that T is a (nonlocal) operator with Calderón Zygmund kernel [28]. Roughly spoken this means that it is the kernel of a generalized singular integral operator . The T1 Theorem [11, 28] asserts that T is bounded in L^2 if and only if $T1, T^*1 \in BMO$ and T is weakly bounded.

Theorem 2.2 gives a necessary condition for a stable decomposition in $L^2(\Omega)$. Mathematically, it is amazing to see the connection between harmonic analysis, the theory of function spaces and singular integrals, as a powerful instrument for analyzing partial differential operators and the multigrid method as a powerful tool for efficient numerical solution of partial differential equations.

We remark that the BMO condition of Theorem 2.2 is obviously valid for functions ψ_k^l satisfying the **Moment condition** M_d :

$$\int \psi_k^l(x) x^l dx = 0 \ , \quad 0 \leq |l| \leq d \ .$$

In view of its locality, these function are atoms and the BMO condition can be rephrased by the following assertion. The deviation of functions ψ_k^l from the property to be atoms should be appropriately bounded. This bound can be described by a Carleson measure condition [27, 28] which seems to provide a practical criterion. On uniform grids a moment condition is necessary.

A stable decomposition gives rise to uniformly bounded projections.

Proposition: *In case of a wavelet basis there exists uniformly bounded projections (in L^2) P_j, such that $P_{l+1} - P_l : L^2(\mathbb{R}^n) \to W^l$ and $P_j = P_0 + \sum_{l=-1}^{j-1} P_{l+1} - P_l$.*

The following important result is a known consequence of a stable decomposition, e.g. [10].

THEOREM 2.3 *For any*

$$u = \sum_{0 \leq l < \infty} \sum_{k \in \Delta^l} w_k^l \psi_k^l + \sum_{k \in \square^0} s_k^0 \varphi_k^0 \ ,$$

where $0 \leq s < d'$, the following norms are equivalent

$$\|u\|_s^2 \ \sim \ \sum_{l=0}^{\infty} |(2^{sl} w_k^l)_k|_{l^2(\Delta^l)}^2 + |(s_k^0)_k|_{l^2(\square^0)}^2$$

$$\sim \ \sum_{l=0}^{\infty} 2^{2ls} \|(P^l - P^{l-1}) u^j\|_0^2 \ . \tag{2.6}$$

An analogous result is valid for *Besov norms* and for *Triebel Lizorkin norms*. The result can be rephrased by the following assertion. If a hierarchical basis is an unconditional Schauder basis in $L^2(\Omega)$ it is, up to a renormalization, also an unconditional Schauder basis in a wide scale of function spaces containing for example Sobolev spaces, Hoelder(-Zygmund) spaces, BMO and real Hardy spaces as well as functions of bounded variation. This is basic property of wavelets which is exploited in most applications of wavelets.

Let us observe that Theorem 2.3 illustrates that the spaces W^l are subordinated to dyadic frequency regions (frequency decomposition).

We summarize here several perspectives which are consequences of the previous results. We are aware that wavelets are not the only possibility to achieve this perspectives.

- *Preconditioning* : [29, 10]

- *Multigrid methods for negative order operators* : [8, 30]

- *Robust Multigridmethods* : The frequency decomposition multigrid developed by Hackbusch [18, 19] uses the spatial anisotropy of certain wavelet basis. We do not discuss this topic here because it is related to the more complicated characterization of anisotropic Besov norms.

- *Adaptive approximation of functions* : [13, 24, 25, 16] On the smooth parts of a function the wavelet coefficients decay rapidly. An asymptotically optimal choice of coefficients is to use those with the largest values $2^{2ls}|w_k^l|^2$, which is achieved by a single thresholding. This strategy was used in image compression. It is clear that high compression rates require higher order functions, which can be easily achieved on uniform grids. The connection to a posteriory error estimators and further developments will be defered to future projects.

- *Adaptive approximation of operators* : Adaptive aproximation via wavelet approximation of the Schwartz kernel $K_A(x, y)$ of a Calderon-Zygmund operator was proposed by *Beylkin, Coifman, Rohklin 1991* [1]. For the numerical solution of periodic pseudodifferential equations this method has been extended and analysed in [7, 8, 9]. [20] proposed an extension of this algorithm containing the method of [4] as well. The advantage of the multiscale approach relies in the fact that the stiffness matrix with respect to a wavelet basis is approximately sparse. This is in contrast to use of the stiffness matrix with respect to the scaling function basis which is never sparse unless the operators are local. It is shown in [8] that, for each fixed error bound or for certain given convergence rates, the wavelet algorithms require only $\mathcal{O}(N)$ nontrivial entries to achieve the desired accuracy.

- *Parabolic and spatially 1-dim hyperbolic linear operators* : The inverse of a stiffenss matrix A^j is not sparse, even the operator is local. Classical multigrid methods can be viewed as a compression of $(A^j)^{-1}$. Generally the compression may be applied to operators of the form $f(A^j)$, where f is some analytical function. For example, [15] applies the wavelet compression to the resolving operators e^{-At} of an evolution equation.

- *Inverse problems* : The decomposition in Theorem 2.3 is subordinated to different frequency regions and can be handled like the singular value decomposition in order to regularize ill posed problems. [14]

3   Numerical experiments

The following numerical experiments were performed jointly with Dipl. Math. B. Kleemann, IAAS Berlin. For first numerical experiments supporting the results of [8, 9], we consider the double layer potential equation for the Laplacian on a polyhedron. In order to apply the cascade scheme we use the coefficients from example 1 for a first attempt. We compute approximately the collocation matrix using piecewise linear trial functions on an almost hexagonal grid. The numerical integrations are performed by a (3 point-)quadrature rule. The computed matrix is used as a scaling function representation.

We transform this matrix into a corresponding wavelet representation applying the cascade schemes and discard all entries below a single threshold for first experiments. Solving the corresponding compressed equation gives an approximate solution, which is compared with the solution of the uncompressed scaling function representation. We exert few results from our first experiments relating the compression rate to the relative l^2-error caused by the compression. The compression rate is defined as the ratio between the number of nonzero entries and the total number of matrix coefficients,

The results are as follows:

Compression

number of collocation points	threshold	compression rate	relative l^2 error
98	$3.0 \ 10^{-4}$	2.1	$1.2 \ 10^{-3}$
386	$3.0 \ 10^{-5}$	3.2	$1.3 \ 10^{-4}$
386	$1.0 \ 10^{-4}$	4.9	$7.5 \ 10^{-4}$
1538	$1.0 \ 10^{-5}$	38.7	$3.0 \ 10^{-4}$
1538	$3.0 \ 10^{-5}$	14.0	$7.2 \ 10^{-4}$
6146	$1.0 \ 10^{-6}$	17.2	$3.1 \ 10^{-5}$
6146	$1.0 \ 10^{-5}$	46.3	$4.1 \ 10^{-4}$
6146	$1.0 \ 10^{-4}$	121.	$2.8 \ 10^{-3}$

We use a GMRES iteration scheme for the solution of the compressed and original matrices and compare the CPU times.

number of collocation points	threshold	condition number	time scaling f.r.	time wavelet r.
98	$3.0\ 10^{-4}$	18.5 (2.5)	0.05	0.08
386	$3.0\ 10^{-5}$	27.5 (3.1)	0.90	0.98
1538	$3.0\ 10^{-5}$	33.6 (5.2)	20.0	5.1
6146	$1.0\ 10^{-6}$	~ 36	435	62
6146	$1.0\ 10^{-5}$		435	28
6146	$1.0\ 10^{-4}$		435	10

I want to thank Dipl. Math. B. Kleemann, IAAS Berlin, for his help to present these results here.

References

[1] Beylkin, G., Coifman, R., Rokhlin, V.: The fast wavelet transform and numerical algorithms. *Comm. Pure and Appl. Math.*, 141–183 (1991).

[2] Bornemann, F.A., Yserentant, H.: A basic norm equivalence for the theory of multilevel methods , Preprint 92-1, ZIB, 1992 .

[3] Bramble, J.H., Pasciak, J.E., Xu , J.: Parallel multilevel preconditioners, *Math. Comp.* **55**, 1 – 22 (1990).

[4] A.Brandt, A., Lubrecht, A.A.:Multilevel matrix multiplication and fast solution of integral equations, *J. Comp. Phys.*, **90**, 348 – 370 (1991).

[5] Ciarlet, P.L.: *The Finite Element Method for Elliptic Problems.* North Holland, Amsterdam, New York, Oxford 1978.

[6] Cohen, A.; Daubechies, I., Feauveau, J.-C.: Biorthogonal bases of compactly supported wavelets, *Comm. Pure and Applied Math.*, **45**, 485-560 (1992).

[7] Dahmen, W., Prössdorf, S.,Schneider, R.: Wavelet approximation methods for pseudodifferential equations I: Stability and convergence, IAAS-Preprint No. 7, Berlin 1992, to appear in *Math. Zeitschrift.*

[8] Dahmen, W., Prössdorf, S.,Schneider, R.: Wavelet approximation methods for pseudodifferential Preprint No. 82, RWTH Aachen, 1993, to appear in *Advances in Computational Mathematics.*

[9] Dahmen, W., Prössdorf, S.,Schneider, R.: Multiscale methods for pseudodifferential equations, TH FB Mathematik Preprint No. 1566 , to appear in Schuhmaker & Webb (eds.) *Topics in the Theory and Applications of Wavelets* .

[10] Dahmen, W., Kunoth, A.: Multilevel preconditioning, *Numerische Mathematik*, **63** 315–344 (1992).

[11] David G., Journée, J.-L.: A boundedness criterion for generalized Calderón-Zygmund operators, *Ann. of Math.*, **120**, 371–397 (1984).

[12] Daubechies, I.: Orthonormal bases of compactly supported wavelets, *Comm. Pure and Appl. Math.* **41**, 909–996 (1988).

[13] Daubechies, I.: *Ten Lectures on Wavelets*, CBMS-NSF Regional Conference Series in Applied Mathematics **61**, 1992.

[14] Donoho, D.L.: Nonlinear solution of linear inverse problems by wavelet-vaguelette decomposition. Technical Report, Department of Statistics, Stanford University 1992.

[15] Enquist, B., Osher, S., Zhong, S.: Fast wavelet based algorithms for linear evolution equations, ICASE Report No. 92-14, 1992.

[16] Glowinski, R., Lawton, W.M., Ravachol, M., Tenenbaum, E.: Wavelet solution of linear and nonlinear elliptic, parabolic and hyperbolic problems in one space dimension, Preprint, 1989, Aware Inc., Cambridge, Mass.

[17] Hackbusch W., Nowak, Z.P.: On the fast matrix multiplication in the boundary element method by panel clustering, *Numer. Math.* **54**, 463–491 (1989).

[18] Hackbusch, W.: The frequency decomposition multigrid method. I. Aplication to anisotropic equations. *Numer. Math.* **56**, 229 – 245 (1989).

[19] Hackbusch, W.: The frequency decomposition multigrid method. II. Convergence analysis based on the additive Schwarz method *Numer. Math.* **63**, 433 – 453 (1992).

[20] Harten, A., Yad-Shalom, I.: Fast multiresolution algorithms for matrix-vector multiplication, ICASE Report No. 92-55, October 1992.

[21] Hörmander, L.: *The Analysis of Linear Partial Differential Operators*, vol. 3 *Grundlehren series*. Springer Verlag, Berlin, Heidelberg, New York, Tokio 1985.

[22] Jaffard, S.: Wavelet methods for fast resolution of elliptic problems, *SIAM J. Numer. Anal.*, **29**(1992), 965 – 986.

[23] Jia, R.Q., Micchelli, C.A.:Using the refinement equation for the construction of pre-wavelets, in: Curves and Surfaces (P. Laurent, A. Le Méhanté, L.L. Schumaker, eds.), Academic Press, New York, 209 – 246, 1991.

[24] Liandrat, J., Perrier, V., Tchamichian, P.: Numerical resolution of the regularized Burgers equation using the wavelet transform. in "Wavelets" Meyer et al. Springer Verlag 1992.

[25] Maday, Y., Perrier, V., Ravel, J.: Adaptativité dynamique sur bases d'ondelettes pour l'approximation d'equations aux dèrivées partielles. Preprint 1991.

[26] Mallat, S.: Multiresolution approximation and wavelet orthonormal bases of $L^2(I\!R)$, *Trans. Amer. Math. Soc.*, **315** (1989), 69–87.

[27] Meyer, Y.: *Ondelettes et Opérateurs 1 : Ondelettes.* Hermann, Paris, 1990.

[28] Meyer, Y.: *Ondelettes et Opérateurs 2 : Opérateur de Caldéron-Zygmund,* Hermann, Paris 1990.

[29] Oswald, P.: Stable splitting of Sobolev spaces and fast solution of variational problems, Forsch.-Erg., FSU Jena, Math/92/5, 1992.

[30] Rieder, A.: Semi-algebraic multi-level methods based on wavelet decompositions. manuscript 1992.

[31] Rokhlin, V.: Rapid solution of integral equations of classical potential theory. *J. Comp. Phys.* **60** (1985).

[32] Yserentant, H.: On the multilevel splitting of finite element spaces. *Numer. Math.* **49**, 379 – 412 (1986).

Addresses of the Editors of the Series "Notes on Numerical Fluid Mechanics"

Prof. Dr. Ernst Heinrich Hirschel (General Editor)
Herzog-Heinrich-Weg 6
D-85604 Zorneding
Federal Republic of Germany

Prof. Dr. Kozo Fujii
High-Speed Aerodynamics Div.
The ISAS
Yoshinodai 3-1-1, Sagamihara
Kanagawa 229
Japan

Prof. Dr. Bram van Leer
Department of Aerospace Engineering
The University of Michigan
Ann Arbor, MI 48109-2140
USA

Prof. Dr. Keith William Morton
Oxford University Computing Laboratory
Numerical Analysis Group
8-11 Keble Road
Oxford OX1 3QD
Great Britain

Prof. Dr. Maurizio Pandolfi
Dipartimento di Ingegneria Aeronautica e Spaziale
Politecnico di Torino
Corso Duca Degli Abruzzi, 24
I-10129 Torino
Italy

Prof. Dr. Arthur Rizzi
Royal Institute of Technology
Aeronautical Engineering
Dept. of Vehicle Engineering
S-10044 Stockholm
Sweden

Dr. Bernard Roux
Institut de Mécanique des Fluides
Laboratoire Associé au C.R.N.S. LA 03
1, Rue Honnorat
F-13003 Marseille
France

Brief Instruction for Authors

Manuscripts should have well over 100 pages. As they will be reproduced photomechanically they should be typed with utmost care on special stationary which will be supplied on request.
In print, the size will be reduced linearly to approximately 75 per cent. Figures and diagrams should be lettered accordingly so as to produce letters not smaller than 2 mm in print. The same is valid for handwritten formulae. Manuscripts (in English) or proposals should be sent to the general editor, Prof. Dr. E. H. Hirschel, Herzog-Heinrich-Weg 6, D-85604 Zorneding.